PSIP: Program and System Information Protocol

McGraw-Hill Video & Audio Books

PSIP: Program and System Information Protocol

Naming, Numbering, and Navigation for Digital Television

Mark K. Eyer

Jerry C. Whitaker, Editor

McGraw-Hill

New York   Chicago   San Francisco   Lisbon   London   Madrid
Mexico City   Milan   New Delhi   San Juan   Seoul
Singapore   Sydney   Toronto

The McGraw·Hill Companies

Cataloging-in-Publication Data is on file with the Library of Congress

1 2 3 4 5 6 7 8 9 0 DOC/DOC 0 8 7 6 5 4 3 2

ISBN 0-07-138999-7

The sponsoring editor for this book was Steve Chapman and the production supervisor was Sherri Souffrance. The book was set in Times New Roman and Helvetica by Technical Press, Morgan Hill, CA.

Printed and bound by RR Donnelley.

McGraw-Hill books are available at special quantity discounts to use as premiums and sales promotions, or for use in corporate training programs. For more information, please write to the Director of Special Sales, McGraw-Hill Professional, Two Penn Plaza, New York, NY 10121-2298. Or contact your local bookstore.

 This book is printed on recycled, acid-free paper containing a minimum of 50% recycled, de-inked fiber.

For *Ben*

Contents

Foreword

Bernard Lechner

This book is a major contribution to the understanding and application of the ATSC PSIP Standard. The author, Mark K. Eyer, was a principal architect of the PSIP Standard and is today, unquestionably, the world's leading expert on PSIP. I am delighted that he found the time to write this outstanding book.

PSIP uses the basic MPEG-2 Systems toolkit to provide a means for broadcasters to include information about their current and future programs as an integral part of the transmitted signal. Once collected by the television receiver, this information can be used to provide a rich user interface that may include an interactive on-screen Electronic Program Guide to facilitate navigating the channels.

The PSIP Standard was developed by the ATSC Specialist Group on Service Multiplex and Transport Systems Characteristics (T3/S8). I was privileged to be the Chairman of T3/S8 from its inception in January of 1994 until April of 2002 and was thus able to witness and guide the development of the PSIP Standard. The work on PSIP began in the latter part of 1996 and the finished Standard (A/65) was adopted by the ATSC a little over a year later on December 23, 1997. Mark Eyer participated in the development process from start to finish and made major contributions to the resulting standard. He continues today, now as Chairman of T3/S8, to work on improvements and extensions to the standard. He is also an active participant in related standards work of CEA and SCTE.

The book not only describes the syntax and semantics of the PSIP Tables and Descriptors but also includes an excellent tutorial on the relevant aspects of MPEG-2 Systems. The relationship between the required Program Specific Information (PSI) elements of MPEG-2 Systems and PSIP is described. In addition to everything you ever wanted to know about PSIP, from two-part channel numbers to Directed Channel Change, the author has included a wealth of information about related EIA/CEA and SCTE standards. Especially informative is the discussion of how PSIP relates to digital cable-ready television receivers and the current, and planned future, practices for System Information used on cable television systems, as documented in SCTE standards.

Mark Eyer has provided a wealth of examples to help the reader understand how PSIP works and how to implement it in the broadcast plant and the consumer digital television receiver. This very readable book is destined to become the definitive reference on PSIP.

Bernard J. Lechner
Princeton, New Jersey
June 2002

Acknowledgments

Just after the ATSC *Data Broadcasting* book was published by McGraw-Hill last year, one of the authors, Michael Dolan, called to suggest that PSIP would be a natural topic for another book in the DTV series and that I ought to take on that challenge. My thanks goes out to Mike for the initial idea and for putting me in touch with Steve Chapman, Executive Editor at McGraw-Hill Professional who has supported the project and all my various needs throughout the process. My thanks also to Henry Derovanessian and Mike Fidler at Sony for supporting my request to take on this assignment.

I would like to express my deep appreciation to everyone who helped review the manuscript: Art Allison, Richard Chernock, Michael Field, Adam Goldberg, Matthew Goldman, Edwin Heredia, Michael Isnardi, Steve Johnson, Jeff Krauss, Bernie Lechner, Don Moore, Gomer Thomas, and Joe Weber. My sincere thanks goes to Jerry Whitaker for his support of the project from the beginning, for introducing me to the mechanics of the authoring process (and answering endless questions about it), and for encouraging me to tackle the production and layout aspects of the job. Many thanks to Sharon Sears for her diligent work in copy editing and helping to convert the manuscript to camera-ready format.

One of the tasks in preparation of this book involved creation of actual example PSIP tables. The folks at Triveni Digital were kind enough to loan me one of their PSIP generator products, the GuideBuilder, to assist in that work. My thanks to Russell Wise, Brian Lee, and Luis Don for their help and support.

The standard that is the topic of this book, ATSC A/65, came into being as the result of the collaborative effort of engineers representing various industries, including broadcasters, consumer electronics manufacturers, and those involved with digital cable television. I would like to especially thank Bernard Lechner, under whose leadership in the ATSC Transport Specialists group the ATSC A/65 Standard was crafted. Bernie's expertise and guidance created the environment that allowed all those involved to do their best and most creative work. In addition to Bernie, those who played a significant role in the initial PSIP standard include Jack Chaney, Mehmet Ozkhan, Andy Teng, Edwin Heredia, Art Allison, Warner W. Johnston, and Matthew Goldman. Matthew's contribution has, and continues to be, to help us keep strict adherence to the philosophy and terminology established in the MPEG-2 *Systems* standard.

PSIP: Program and System Information Protocol

1

Introduction

If you are involved in any technical way with digital television sent either via terrestrial broadcast or cable means, chances are good you will need to deal in some way with the Program and System Information Protocol, or PSIP. This book was written to serve as an introduction to the general concepts embodied in the protocol, to explain how PSIP builds on the MPEG standards, and to describe the design philosophy the architects had in mind when the protocol was conceived. It offers a variety of helpful guidelines and insights to engineers involved in the design of consumer electronic and professional-grade products that support the PSIP protocol. It will also be helpful to broadcast station engineers and cable headend operations managers, or anyone who is involved with the creation and transmission of PSIP data.

In this introductory chapter, we start at the beginning by answering the most basic question, "what is PSIP?" Next we look at the reasons why the protocol was needed in the first place. We then discuss the conventions used in the book for table syntax and semantics, and then outline the structure of the book.

What is PSIP?

Simply put, the Program and System Information Protocol, or PSIP, is the part of the US Digital Television Standard that lets the digital television receiver know such things as the name of the channel and the name and description of current and future programs on that channel. In addition, PSIP is actually much more than that, as this book will show.

PSIP defines "system information" (sometimes called "service information" or just SI) for the Advanced Television Systems Committee (ATSC) standard developed in the United States. The ATSC A/65 PSIP Standard describes a method for delivery of program guide and system data tables carried in any compliant MPEG-2 transport multiplex.

The primary purpose of PSIP is to facilitate acquisition and navigation among the analog and digital services available to a particular receiver or set-top box, but it also serves as a support platform for applications such as data broadcasting. Delivery of PSIP data is essential for digital terrestrial broadcasts in North America, and cable operators have pledged to support it as well for the benefit of cable-ready digital televisions.

Why PSIP?

One might ask "why is PSIP necessary?" To answer that question it is helpful to look at the difference between analog and digital television signals. An analog television broadcast or cable signal includes at most one video component and one or two audio components. One analog signal represents one "channel" of programming, so that if a receiver acquires the signal, it has acquired that channel. If a user commands an analog-only television receiver to go to channel 4, the receiver looks for an analog signal in the 66-72 MHz band because "channel 4" is known to map to this portion of the spectrum.

Digital television, on the other hand, provides for the possibility that one broadcast or cable signal includes several television channels. Digital compression allows as many as a dozen or more standard-definition programs to be delivered within the same multiplex signal. Each program has a video component, one or more audio tracks, and may include accompanying data as well.

The FCC ruled that the RF spectrum currently in use for analog terrestrial broadcast must be relinquished for use by other digital services by the year 2006. In compensation, each broadcast licensee has been assigned a second 6-MHz channel for transmission of digital TV. The FCC's table of DTV channel allotments, defined in 47 CFR 73.622, was designed to minimize use of the spectrum at channels 60-69 and 2 through 13 as well. As originally conceived each 6-MHz channel would carry a single High Definition TV (HDTV) channel. Presumably, users would re-learn the new channel number for each of their favorite local broadcasters when the shift to digital occurred.

Early on, the flexibility of the MPEG-2 video compression standard was recognized. ATSC defined a set of possible compression formats including not only HDTV formats but standard definition (SD) formats as well. Clearly a broadcaster could choose to deliver a signal that included not just one channel but several. Typically, part of the broadcast day would be devoted to HD content. During that time, the other channels would have to go off the air because the HD channel would consume the full bandwidth of the channel. For other parts of the day, several channels of programming could be provided—all in the same broadcast multiplex.

It soon became clear that some guidance would have to be provided to the receiver (and hence to the viewer) to make sense of such a multi-channel signal. How would channel numbers work when the familiar RF-related channel number could be associated with more than one "TV channel"? Broadcasters realized their brand-name recognition (the channel number they have used for decades to identify their product) was in danger of being lost. They looked to the ATSC Standard to help with the problem.

Another factor contributing to the PSIP work was consumers' expectation that digital television would be a richer experience than the old style NTSC analog television it replaces. Consumers expect digital to be "better" than analog. The Digital Versatile Disk (DVD) for example includes interactive menus, bonus materials, multiple languages and camera angles, and 5.1-channel surround sound. Digital satellite broadcast receivers include interactive programming guides as well as hundreds of channels of programming to choose from. From the point of view of the technology, the digital television multiplex signal has more than enough spare bandwidth to deliver a rich variety of television-related data services, including the electronic program guide data we have come to expect.

We use the acronym SI for "system information" or (as it is called in Europe) "service information." The PSIP standard can deliver the following types of SI data in a digital cable or terrestrial broadcast signal:

- **Service identification**: PSIP data includes data that associates each television channel with a textual and numeric label. The textual label chosen by a broadcaster is often some form of their call letters.

- **Time of day**: PSIP data includes an indication of the current time of day so that receiving equipment can automatically synchronize its internal clocks. Information about Daylight Saving Time is included as well.

- **Program guide data**: program titles and descriptions for programs currently on the air are provided, and broadcasters have the option of providing titles and descriptions for programming up to sixteen days in advance of air time.

- **Caption service data**: programming may have accompanying closed caption services, and PSIP offers the broadcaster a way to identify and name the available services. Caption services present in future programs may be announced as well.

- **Content advisories**: the PSIP Standard defines a flexible method that allows a given program to be "rated" for various types of content. In the US, it provides support for the so-called "V-chip" standard mandated by the FCC. As an example, if violence is one of the rated parameters (as it is in the US standard), a user can set up a receiver to block programming containing a level of violence deemed to be objectionable.

- **Support for multiple languages**: all the textual data delivered in PSIP tables can be provided multi-lingually.

To ease the transition from analog to digital broadcasting, PSIP data in a digital broadcast multiplex can describe analog programming as well as programming delivered digitally. The PSIP data tables can be considered to be a platform upon which many new features may be built. Data broadcasting is one example. Another is the Directed Channel Change (DCC) mechanism that is now part of the ATSC A/65 standard itself (DCC is discussed in detail in Chapter 13)

Essential Resources

The latest version of the ATSC PSIP Standard, A/65, can be downloaded from the ATSC website, http://www.atsc.org/. This book describes the "B" version published in 2002. Other highly recommended resources include:

- The ATSC PSIP Recommended Practice: *Program and System Information Protocol Implementation Guidelines for Broadcasters*, available for download at the ATSC website, http://www.atsc.org/.

- EIA/CEA CEB-12, *PSIP Recommended Practice*, available from Global Engineering Documents, http://global.ihs.com/

Table Syntax and Semantics

In various places in this book, we describe the bits and bytes that make up the transmitted PSIP and MPEG table sections. In the simplest form, the syntax description looks like a series of named fields, each with a size given in bits. Sometimes, to emphasize that the field is some number of 8-bit bytes in size, the "number of bits" field may show (for example, for a 3-byte field) "3*8" instead of "24."

The syntactic descriptions can be somewhat more complex than a simple list of fields. The value of a field may determine whether or not certain fields or groups of fields are present in the transmitted data. Or, a certain field or group of fields may be present more than once, again depending upon the value of a certain field.

A thorough discussion of the table syntax conventions used in the MPEG and ATSC Standards may be found in Annex D of the ATSC PSIP Recommended Practice.

This book adheres to the MPEG-2 convention for representation of hexadecimal numbers and uses the "0x" prefix to denote hex numbers. For example, a number whose value is indicated as 0x10 is an 8-bit number whose decimal equivalent is 16. Binary numbers in this book are written as a series of 1's and 0's terminated with a lowercase "b." The number 0x10 is equivalent to 00010000b.

Structure of This Book

This book is both an introduction to the concepts embodied in the Program and System Information Protocol and a detailed handbook for its implementation. The first eight chapters provide the introduction, philosophy and foundation; the next three dive into the bits and bytes of the PSIP tables and descriptors. The remainder of the book deals with various PSIP-related topics including Directed Channel Change and various aspects of PSIP on cable including System Information delivered in the out-of-band channel, how Emergency Alerts work on cable, and the agreements and commitments that have been made by the cable community in the US with regard to carriage of PSIP data. The last chapter deals with considerations for implementers of equipment that either creates, processes, or receives PSIP data.

Here is an outline of the chapters and appendices in the book:

- Chapter 1 is this overview and introduction.

- Chapter 2, **Design Requirements**—describes the technical requirements set down by the PSIP system architects. A review of the requirements will help one understand why PSIP is the way it is, and the problems it was intended to solve.

- Chapter 3, **MPEG-2 Transport**—gives a tutorial introduction to the MPEG Standards that are the foundation for today's digital television systems, focusing particularly on the MPEG-2 *Systems* Standard. A thorough familiarity with MPEG-2 *Systems* is essential for an understanding of PSIP.

- Chapter 4, **Virtual Channels**—PSIP introduces the concept of "virtual" channels, and this chapter gives an introduction and overview of the topic.

- Chapter 5, **Two-Part Channel Numbers**—The concept and mental model of the two-part channel number is introduced in Chapter 5, and examples are given of how one can use them to navigate among digital channels.

- Chapter 6, **Source IDs**—This chapter discusses the Source ID concept, a mechanism used in PSIP tables to link channel maps to the Electronic Program Guide data.

- Chapter 7, **Program Content Advisories**—PSIP supports the delivery of "content advisory" data and this chapter discusses the conceptual framework for how this is done.

- Chapter 8, **Caption Services**—Digital television supports an "advanced" captioning method, and this chapter discusses how PSIP ties in with it.

- Chapter 9, **Data Representation**—Before diving into the detailed bits and bytes that make up the PSIP tables, we discuss the various types of data provided in the tables and how each is represented in the transmitted table formats.

- Chapter 10, **Main PSIP Tables**—This chapter goes into the detailed syntax and semantics of each of the main PSIP tables starting with an overview of the purpose and structure of each.

- Chapter 11, **PSIP Descriptors**—Here, each of the descriptor structures defined in the A/65 standard is discussed in detail, again starting with structural diagrams and then going into the specifics of the bits and bytes.

- Chapter 12, **Electronic Program Guide Example**—The previous chapter describes the structure and organization of the Event Information Table. Chapter 12 illustrates an example Electronic Program Guide to show the relationship between it and the on-screen presentation.

- Chapter 13, **Directed Channel Change**—Chapter 13 introduces the Directed Channel Change Concept, a PSIP feature that allows broadcasters to request suitably-equipped receivers to change channels under certain prescribed conditions. The detailed syntax and semantics of the tables and descriptors associated with Directed Channel Change are also given here.

- Chapter 14, **PSIP Expandability**—This chapter discusses an important aspect of PSIP protocol design: the philosophy that allows ATSC to change and expand the protocol in the future without disrupting the behavior of equipment built to an earlier version of the standard.

- Chapter 15, **Private Data**—It is possible to include privately-defined data in a terrestrial broadcast or cable transmission, and this chapter explains how it is done.

- Chapter 16, **PSIP and the Cable-Ready Receiver**—PSIP plays a role in the design of consumer electronic devices that are directly connected to cable, and this chapter gives an overview of the relevant systems issues.

- Chapter 17, **Service Information for Out-of-Band Cable**—Here we go into detail about how the cable out-of-band channel works to deliver navigational data, some of which is very much PSIP-like.

- Chapter 18, **Emergency Alert System**—Cable systems support an Emergency Alert System to comply with the FCC-mandated system of notification in case of national emergency. This chapter describes how EAS signaling is done on cable, and its relationship to the PSIP protocol.

- Chapter 19, **PSIP Agreement**—This chapter discusses an important aspect of PSIP on cable, an agreement made in February, 2000 between consumer electronics manufacturers and the cable industry.

- Chapter 20, **PSIP Implementation**—We wrap up with a discussion and review of implementation considerations for manufacturers and designers of consumer electronics, broadcast and cable system operators, and PSIP data suppliers.

- Appendix A, **Acronyms**—provides a list of acronyms used in this book.

- Appendix B, **References on the Web**—lists helpful Internet addresses relevant to PSIP and digital television standards.

- Appendix C, **PSIP Agreement**—The text of the February 2000 PSIP Agreement is reprinted here for reference.

- Appendix D, **History of the PSIP Standard**—This appendix gives a brief history of the development of the PSIP standard in the ATSC committees, and goes into some detail about the various changes and updates that have been made since it was first published by ATSC.

PSIP Design Requirements

The PSIP Standard was developed by the ATSC T3/S8 Transport Specialists group in 1997. Prior to assembling the new specification, a comprehensive set of design requirements was drafted. Committee members were aware of work being done on Service Information in Europe's Digital Video Broadcasting (DVB) consortium, and in fact some members participated in that effort. It was clear to those involved that the needs of the US system differed in significant ways from the needs fulfilled by the DVB approach. T3/S8 documented the following system requirements to address the naming, numbering, and navigation problem:

- The system had to support direct access to any channel, meaning that the navigational model needed to support the ability to access any analog or digital channel by direct entry of its channel number. Such an access method was deemed a necessary user feature. It is required to support devices that use remote control commands to tune a receiver in order to make an unattended recording. This design decision precluded an approach where navigation of a Graphical User Interface (moving a cursor on-screen and selecting a channel from a list of choices) was the only method offered.

- The approach had to support grouping of selected digital services with an existing analog service, or with digital services on other multiplexes. For a period of time, broadcasters will operate an analog channel in addition to a digital multiplex. The navigation model had to include a grouping concept to support channel surfing within a set of related analog and digital channels.

- The approach needed to support the preservation of channel branding. The system had to allow a broadcaster, when starting a new digital service, to associate the new programming with the channel label that had been used to establish identity in past years of advertising.

- The resulting standard had to be harmonized with cable standards. There was a clear recognition that cable set-tops and cable-ready receivers would also employ navigational and channel naming methods. Typically, cable headends carry local terrestrial broadcast channels. PSIP designers had to consider the needs of cable operators to make sure harmony with methods defined for terrestrial broadcast was maintained. A big hurdle early on in gaining support for the new ATSC standard was the fact that cable operators employed existing proprietary practice in digital set-top boxes owned by the Multiple-System Operators.

- The approach had to recognize and accommodate the flexibility of digital transport. In a terrestrial broadcast scenario, just as with a cable or satellite system, the use and contents of any given Transport Stream varies from time to time. The system design had to gracefully accommodate such changes.

- The implementation solution had to support printable program guides. Even though electronic means are now available, printed program guides will still exist. In addition, even though the "look and feel" for the receiver is not specified, whatever channel numbering and labeling method is used in the receiver must align with channel numbering and labeling found in printed guides. The printed guides must be able to clearly describe cable channels as well as programming received via terrestrial broadcast.

- The design needed to allow a broadcaster to "package" or "market" some services separately from others on the multiplex. For example, as a public service, the owner of a digital broadcasting license may offer a couple of spare megabits of Standard-Definition TV bandwidth to a college or community access channel, or to a government affairs (city politics) channel. It had to be possible for that operator to label that channel separately from the other services offered on that multiplex.

- The design had to support channel naming, which involves downloading to the receiver the textual name or call-sign associated with each program source. Processing in the receiver of this text would be optional.

- The approved approach could not preclude the development or adoption of new navigation paradigms. The approach must be flexible and extensible enough so that alternative navigation methods may be accommodated in the future. For example, the solution must also accommodate the presence of interactive services.

- The approach could not preclude or limit the development of data broadcasting services. While the initial design did not need to explicitly support data services, it had to support backward-compatible extensions for data. In the years since PSIP was completed, of course, data broadcasting services have been specified for ATSC in the ATSC A/90 Standard.

Terrestrial broadcast requirements

Certain requirements were specific to the terrestrial broadcast application:

- The system had to accommodate terrestrial broadcast translators. Broadcast translators are used in some communities where terrain such as a hill or mountain prevents the transmitter's main signal from reaching the full service area. In such cases, the MPEG-2 Transport Stream may be transmitted on more than one broadcast carrier frequency. The receiver must recognize multiple instances of one Transport Stream as alternate physical access points to the same services. The system needed to support a receiver to provide consistency of labeling (channel name and number) for each instance.

- An additional requirement related to translators was that it must not be necessary to alter the Transport Stream in any way prior to retransmission at the translated frequency. In other words, the PSIP data in the un-translated signal must be usable as-is in the translated one.

- The system needed to take into account movable terrestrial broadcast receiving antennas. A receiver could encounter one digital broadcast on frequency X at one point in time and a different Transport Stream on frequency X at another time due to atmospheric conditions or the re-orientation of a movable antenna. Designers were also aware that some people have TV receivers in their recreational vehicles for use while driving or camping.

Cable requirements

Other requirements were specific to the cable application:

- The cable system operator had to be able to label digital services independently of the RF channel number used to transmit them on cable. The requirement came from the cable system operator's desire to be able to change the physical locations of any service offering's carrier frequency without affecting the user's notion of the channel number.

- It had to be practical for the cable system operator to create Transport Streams by assembling services from various input streams received via any transmission media. Because effective data rates on cable can be nearly double that of terrestrial broadcast, services from two terrestrial multiplexes may be combined into a single cable Transport Stream. The approach had to allow the cable operator to perform such remultiplexing with a minimum of difficulty.

Later, another cable requirement was added: It must be possible to label a channel with a three-digit channel number, for consistency with common practice in digital cable systems where channels are numbered within the range 1 to 999.

Desirable features

Some features of the system were not strictly mandatory but conformance was clearly desirable:

- The approach should incorporate familiar paradigms for operation. The new services should fit easily within the mental model already used by the TV-watching public. This means that channel surfing and the familiar numbering of channels should work pretty much like they always have.

- The naming method should support naming of analog channels as well as digital. This would allow the most integrated look and feel for the receiver during the transition to full-digital transmission.

- The approach should use standards wherever possible.

- Finally, the approach should harmonize with emerging satellite and MMDS[*] standards. The chosen solution for broadcast should also recognize other transmission media.

The naming, numbering and navigation (N^3) problem

PSIP was developed to address three primary issues related to digital television, none of which was adequately addressed in the MPEG-2 family of standards:

- Channel names. Broadcasters wanted to label each of their digital channels with a user-friendly name, such as the call letters of the station. Channel names benefit the user as well as a facilitator for finding the channel of interest.

- Channel numbers. Broadcasters realized they wanted the flexibility to be able to assign channel numbers independently of the RF channel used to broadcast the digital signal. Such flexibility would allow any digital service to be associated with the same RF channel that broadcaster has used for years for the analog NTSC service. For the television-viewing public, such a numbering scheme would group all the analog and digital channels together on the TV dial.

  The fact that the digital signal carried in a single 6-MHz slice of RF spectrum could deliver more than one "television channel" created a second number-related problem: how should the "sub-channels" be numbered and referenced in electronic and printed program guides?

- Program guides. Broadcasters wished to be able to deliver data to support Electronic Program Guides (EPGs) so a digital receiver would be able to display a channel's current and future program schedule.

---

[*]   Multichannel Multipoint Distribution Service

These requirements came to be known as the "naming, numbering and navigating" or N-cubed problem. The term "navigation" in this context refers to the need to support a user's need to "navigate" or find his or her way around among the many digital service offerings. Part of navigation involves being able to go directly to a desired programming service given a well-known or advertised reference. Another part of navigation involves supporting the familiar "channel grazing" or "channel surfing" activity users often enjoy.

The two-part channel number

In analog broadcast or cable television, if a user selects channel "4," a receiver knows to tune the frequency associated with channel 4—the 66-72 MHz band. The correlation between the user's notion of a channel number and the 6-MHz frequency band carrying the RF modulated signal has long been established by well-known standards. For terrestrial broadcast, the FCC rules in 47 CFR 73.603 define the mapping. For cable, EIA-542-A[1] applies.

The situation has changed, however, with the advent of digital transmission. PSIP offers digital broadcasters the freedom to define channel numbers independently of the RF frequency used to carry the signal. PSIP's architecture is designed around the concept of "virtual channels." A virtual channel is called "virtual" because its definition is given by indirect reference through a data structure called a Virtual Channel Table (VCT). So when a viewer selects "channel 4," he or she is actually selecting the channel record associated with user channel number 4. The definition of the channel as given in the VCT includes its textual name, channel type (analog audio/video, digital a/v, audio only, data), and the channel number the user may use to access it.

The A/65 protocol introduces a new navigational concept, the "two-part" channel number. Broadcasters declared the need, as new digital services are introduced, to retain the brand-identity they currently have as a result of years of marketing and advertising. For broadcasters in the US, the first part of the two-part number (called the "major" channel number) is required to be the same as the EIA/FCC channel number already in use for the analog service. The second part of the number (called the "minor" channel number) identifies one service within the group of services defined by the major number. From the point of view of the user, where before there was just "channel 4," now there may also be channels 4.1, 4.2, 4.3, and so on.

Note that the delimiter, or punctuation, between the major and minor numbers is not specified in any current standard. Common practice may gravitate towards use of a hyphen instead of a decimal point, or even something else, but as of this writing a favorite has not yet been established. In this book hyphens and decimal points are used at different times in various examples.

Example Broadcast Scenario

Now let's look at a hypothetical case in which four analog NTSC broadcast stations are operating in a community. Each of the four is given a license to broadcast digital services on a new transmission frequency and for a period of time each operates both analog and digital channels. Finally, the FCC declares that analog broadcasting shall cease and the community moves to all-digital format. We assume that one of the broadcasters needs to operate frequency translators for both the analog and digital services.

In this example, we look at the channel numbers that appear in printed guides and advertising—the ones recognized and used by the public to access these television channels. Table 2.1 lists the four stations (plus translator) and indicates the assigned frequency spectrum.

TABLE 2.1 Example NTSC Analog Broadcast Stations

Station Call Letters	Frequency	RF Chan. No.	User Chan. No.
KXA	180-186 MHz	8	8
KXB	192-198 MHz	10	10
KXC	476-482 MHz	15	15
KXC (translator)	488-494 MHz	17	17
KXD	620-626 MHz	39	39

In printed guides and advertising KXA's programming is listed under channel 8, KXB's under channel 10, and KXC's under channel 15. The newspaper may indicate that channel 17 can be used in some areas to receive KXC programming. KXD's programming is identified with channel 39. The RF channel number used to transmit the signal is identical to the number used by a receiver to tune the signal, and it is the same number the station uses to identify itself in advertising and program guides. A small amount of confusion may exist for the consumer in the area targeted by the translated signal. They must tune to channel 17 to receive KXC, while the logo and advertising probably associate it primarily with channel number 15.

Now, each of these four broadcast licensees is given a license to broadcast digital television conforming to the ATSC Standard. Table 2.2 indicates the channel assignments granted by the FCC, one for each NTSC analog licensee. The table introduces the concept of Transport Stream ID, or TSID, for the digital Transport Stream. A TSID label is a 16-bit number assigned uniquely throughout the US, Canada and Mexico to each Transport Stream to help receivers identify and keep

TABLE 2.2 Assignment of Digital Broadcast Spectrum

Station Call Letters	NTSC RF Chan. No.	Frequency	DTV RF TSID	Chan. No.	Frequency	TSID
KXA	8	180-186 MHz	0x106C	55	716-722 MHz	0x106D
KXB	10	192-198 MHz	0x106E	25	536-542 MHz	0x106F
KXC	15	476-482 MHz	0x1070	30	566-572 MHz	0x1071
KXC (translator)	17	488-494 MHz	0x1070	32	578-584 MHz	0x1071
KXD	39	620-626 MHz	0x1072	40	626-632 MHz	0x1073

track of received signals. The NTSC analog broadcasts use an analogous label called the Transmission Signal ID, sometimes referred to as the "analog TSID."

How does KXA expect the public to find and recognize programming they broadcast on RF channel 55? Because they want to retain their brand identity, KXA wishes to have the digital services that are broadcast on RF channel 55 associated with "Channel 8" for purposes of user access, advertising, and printed guides.

Broadcasters, as mentioned, also recognized that the flexibility of the digital service multiplex afforded them the option to provide more than one channel of programming in the same multiplex. These two desires gave rise to the concept of the two-part channel number. Each digital programming service must be associated with both a major and a minor channel number. In the US, broadcasters are required to use the major number that corresponds to the RF channel number of their analog broadcasting license. Minor numbers can be chosen from within the range 1 to 99 for broadcast television or audio-only services. Minor numbers 100 to 999 may be used for services such as data broadcasting. Minor channel number zero is reserved for analog programming and cannot be used with a digital service.

In Table 2.3, we show how an example collection of analog and digital programming services available to a community could look. For purposes of illustration, the example involves several aspects of the virtual channel concept, not only to demonstrate its flexibility but also to alert designers of receiving equipment to some of the situations their products may encounter.

KXA has decided to broadcast two digital services in the signal broadcast on RF channel 55. The first is to be known to the public as channel 8-1 and the second as 8-2. The existing analog NTSC channel will continue to be known as channel 8, or on the digital TV as 8-0. KXA has spare digital bandwidth available during parts of the day where standard-definition programming is broadcast and has offered it to a local community college. The college broadcasts programming on a channel known to the public as KEDU1, channel 97-1.

KXB has created three digital channels, known as 10-1 through 10-3.

KXC has elected to broadcast High Definition television all day long so it has created just one HD channel called KXC-DT. Note that both the normal digital sig-

TABLE 2.3 Example Analog and Digital Channel Numbering

Channel Call Letters	NTSC Chan. No.	Transmitter Frequency	Analog TSID	DTV RF Chan. No.	Transmitter Frequency	Digital TSID	User Chan. No.
KXA	8	180-186 MHz	0x106C				8
KXA-DT1				55	716-722 MHz	0x106D	8-1
KXA-DT2				55	716-722 MHz	0x106D	8-2
KEDU1				55	716-722 MHz	0x106D	97-1
KXB	10	192-198 MHz	0x106E				10
KXB-DT1				25	536-542 MHz	0x106F	10-1
KXB-DT2				25	536-542 MHz	0x106F	10-2
KXB-DT3				25	536-542 MHz	0x106F	10-3
KXC	15	476-482 MHz	0x1070				15
KXC (xlat.)	17	488-494 MHz	0x1070				17
KXC-DT				30	566-572 MHz	0x1071	15-1
KXC-DT (xlat.)				32	578-584 MHz	0x1071	15-1
KXD	39	620-626 MHz	0x1074				39
KXD-DT1				40	620-626 MHz	0x1075	39-1
KXD-DT2				40	620-626 MHz	0x1075	39-2
SPAN				40	620-626 MHz	0x1075	84-1
KEDU2				40	620-626 MHz	0x1075	97-2
KEDU3				40	620-626 MHz	0x1075	97-3

nal broadcast on RF channel 30 and the translated digital signal broadcast on RF channel 32 are completely equivalent and share the same digital TSID. From the user's point of view, channel 15-1 can be tuned from anywhere in the region. Some receivers will find it on RF channel 30 while those in other parts of the county will discover it on RF channel 32.

Expected and proper receiver design hides from the user the fact that a signal is received via a translator channel. Whether received in a digital multiplex on RF channel 30, or on RF channel 32, the DTV is able to offer the user access to virtual channel 15-1. It may be that both multiplexes can be received on a DTV without re-orienting the antenna. In that case, if the user requests access to channel 15-1, tuning to the multiplex on either RF channel 30 or 32 will suffice. The receiver has already determined that these two Transport Streams are identical, having noted that both carry identical values for their Transport Stream IDs. In case the strength or quality of one signal is better than the other, it would be logical to prefer to look for the stronger one. If one carrier fades out, the receiver can attempt to find a stronger signal on the alternate frequency.

Again referring to Table 2.3, KXD has created two HD channels, known to the public as KXD-DT1 on channel 39-1 and KXD-DT2 on channel 39-2. The station has offered spare bandwidth in its digital multiplex to two unaffiliated programming services: the Satellite Public Action Network (SPAN) identified with call letters SPAN, and an educational channel associated with a local community college. SPAN can be received by a digital television receiver by selecting channel 84-1.

Our community college station has two more programming services it would like to make available to the community. KXD agrees to carry these two in its multiplex during some portion of the broadcast day. These two stations are known to the public as KEDU2 and KEDU3 and are associated with channels 97-2 and 97-3, respectively.

In this example, some of the KEDU services associated with major channel number 97 are carried in the KXA multiplex (TSID 0x106D) while others are carried in the KXD multiplex (TSID 0x1075). Receivers must not assume that all services associated with a given major channel numbers are associated with the same RF frequency or TSID.

It may be that at some future time SPAN or one or more of the KEDU services will be bumped off of KXD's multiplex and be picked up by another broadcaster. If that occurs, once digital receivers have learned their new locations, these stations will continue to be accessible as before: KSPAN on channel 84-1 and KEDU on channels 97-1, -2 and -3. Such a change in the location of a programming service is another aspect DTV designers must take into account when designing their products.

References

1. EIA/CEA-542-A, "Cable Television Channel Identification Plan," Electronic Industries Alliance and Consumer Electronics Association, 2001.

MPEG-2 Transport

Before we can describe how PSIP works and how the data it delivers is organized and delivered, we need to lay down a foundation. That foundation is MPEG-2, a family of international standards developed by the Moving Picture Experts Group (MPEG) and published by the International Standards Organization (ISO). The audio, video, and data that make up a standard digital television signal, whether as a terrestrial broadcast signal received by an indoor or rooftop antenna, as a signal received from satellite via a parabolic dish, or as a signal received and demodulated via cable, is organized and delivered as a sequence of data packets. In this chapter we provide a brief overview of the MPEG-2 family of standards, then focus our attention on the one that defines this packet multiplex, the MPEG-2 *Systems*[1] Standard, formally known as ISO/IEC 13818-1, entitled *Information Technology— Generic coding of moving pictures and associated audio information—Part 1: Systems*.

MPEG-2 *Systems* is comprehensive and covers a lot of territory that is not relevant to the transport of PSIP data. For example, it specifies the methods that allow the various component parts of a digital television programming service (audio, video, and data) to be presented in proper temporal relationship, and methods relevant to the management of buffers in the decoder. It also specifies the encapsulation and transport formats used for the audio, video, and data components themselves. For the present discussion, we do not detail these aspects of MPEG-2 *Systems*. The reader is referred to the standard itself, or to the excellent treatise on the MPEG-2 systems layer found in *Data Broadcasting*[2] by Chernock et al.

MPEG-2 Standards

In 1988, MPEG formed to consider digital storage of audio and video at bitrates not exceeding about 1.5 Mbps. The first standard, MPEG-1 *Systems*, was drafted in 1991 and issued as International Standard ISO/IEC 11172[3] in 1992. Today, MPEG-1 video, defined in ISO/IEC 11172-2[4] finds use in the video Compact Disk (video

CD) market, popular in Asia but rarely seen in the United States. Audio coding in MPEG-1 is defined in three flavors, each using a different coding algorithm: Layer I, Layer II, and Layer III. MPEG-1 Layer III audio coding was later adopted as an Internet de facto "standard" and is commonly known as MP-3. Thanks to the Internet and the availability of software decoders that run on personal computers, MP-3 has recently experienced an explosive grown in popularity.

Although MPEG-1 was developed mostly as a compressed storage standard, and not a streaming media standard, one of the considerations for MPEG-1's bit-rate was driven by the streaming bitrate limitations of CD-ROM drives at the time. It wasn't possible with MPEG-1 to achieve the quality level required for television broadcasting. MPEG began work on the MPEG-2 standard in 1991 to extend MPEG-1 to cover applications beyond the scope of the initial effort. These applications included real-time broadcasting of digital television signals for cable, direct-to-home satellite, and terrestrial (i.e. traditional) TV broadcasting.

MPEG-2 was designed to be backward compatible with MPEG-1, so that any MPEG-2 decoder would be able to decode MPEG-1 data[*]. Some of the features MPEG-2 provides that were not available in the original MPEG-1 specification include:

- higher data rates for audio and video coding.

- improvements in video and audio coding efficiency, including new coding modes specifically designed for interlaced video.

- video formats and quality suitable for studio applications.

- flexible support of multiple resolutions and picture aspect ratios, up to High Definition TV and beyond.

- support for multiple audio/video/data streams in the multiplex, each using its own time base (which, along with compression, makes possible multi-channel television).

- definition of a transport method that is optimized for use in error-prone transmission environments (i.e. over-the-air transmission as contrasted with playback from a CD-ROM or a computer's hard disk).

- conditional access support, to enable subscription and pay-per-view services.

- support for System/Service Information, to enable a rich user experience for channel tuning and service-offering navigation.

---

[*] To be technically precise, for an MPEG-1 stream to be compatible with an MPEG-2 decoder it must conform to an MPEG-1 profile called a "constrained parameters" stream. Certain MPEG-1 features (such as "D frames") are not supported in MPEG-2 and hence are not allowed in this profile.

- support for delivery in the MPEG-2 multiplex of data formats not defined within MPEG.

The MPEG-2 International Standard ISO/IEC 13818 consists of ten parts. Together they describe and specify the methods, bit-stream syntax and semantics necessary and sufficient for demultiplexing and synchronized decoding and presentation of video, audio, and/or data; the methods and tests for verifying bit stream conformance; and the syntax and semantics for control of digital video streams. Note that MPEG-2 does not specify requirements for the encoding process itself. Those wishing to build an encoder must derive specifications for the multiplexing and encoding functions from the specification of the System Target Decoder defined in the MPEG-2 Standard. For encoder designers there is plenty of room for creativity and innovation.

Most of our discussion focuses on MPEG-2 *Systems* since it provides the framework for delivery of PSIP data. Other important parts are ISO/IEC 13818-2[5], MPEG-2 *Video*, which defines the mechanisms and syntax of video compression, and ISO/IEC 13818-6[6], *Digital Storage Media Command and Control (DSM-CC)*, which defines several interactive television control protocols including the syntactic structures used extensively in data broadcasting applications. Note that MPEG-2 defines an audio compression standard in ISO/IEC 13818-3[7]. Audio in the US ATSC system, however, does not use the MPEG-2 standard but instead uses a coding technique developed by Dolby Laboratories called AC-3 and defined in ATSC Standard A/52[8] *Digital Audio Compression Standard (AC-3)*.

MPEG-2 Terminology

To understand discussions involving the MPEG-2 standards, it is helpful to be familiar with the definitions of some of the terms and elements involved. In some cases, for example the term "program," MPEG employs a definition that is inconsistent with common usage of the word. Here are some important definitions from MPEG-2 *Systems*:

- **Program**: in MPEG-2 terminology, a program is a collection of "program elements," where a program element may be a component part such as a video, audio, or data track. The idea of an MPEG-2 program is that it is a collection of elements that are related in some way to one another. Program elements are not required to include any defined time base, but those that do must share a common time base with the other elements that make up the program. Thus, program elements may be time-synchronized with one another for presentation. A typical MPEG-2 program consists of a video component and one or more audio tracks. Data program elements may be present as well. It is also possible to define audio-only or data-only MPEG-2 programs.

- **Elementary Stream (ES)**: refers to one of the coded video, coded audio, or other coded bit streams that make up a program. In the case of the MPEG-2 Transport Stream, each ES component is delivered in TS packets having a common Packet Identifier (PID) value. Elementary Stream components for audio and video are carried in a sequence of Packetized Elementary Stream (PES) packets.

- **Packetized Elementary Stream (PES)**: an intermediate packetization layer for video, audio, or data elementary stream information before segmentation and encapsulation into either Transport Stream or Program Stream packets. See PES packet.

- **PES packet**: a variable-length data structure used to carry Elementary Stream data. The PES packet consists of a header followed by a number of contiguous bytes from an elementary data stream. Time stamps for synchronization of decoding and presentation can be included in the header portion of the PES packet.

- **program element**: a component part of an MPEG program. Examples of program elements include Elementary Stream components such as video streams and audio tracks, and data elements that might be coded as sections of data tables or as PES packets. Whereas an Elementary Stream must be formatted as a sequence of PES packets, the term program element is more general. A program element could be formatted as PES packets or as table sections, or in some other proprietary non-standard format.

- **Program Specific Information (PSI)**: MPEG-2 *Systems* defines several data structures needed by decoders for demultiplexing the received MPEG-2 stream. These are the Program Association Table (PAT), Program Map Table (PMT), Conditional Access Table (CAT) and the Transport Stream Description Table (TSDT). We will look at the first three in detail.

- **Program Stream**: The first of two MPEG-2 systems-layer codings, the Program Stream is a data structure consisting of a header followed by one or more PES packets. Suitable for use in error-free environments, the Program Stream format has been used in digital video storage and packaged media applications such as DVDs.

- **Transport Stream (TS)**: The second of two MPEG-2 systems-layer codings, the Transport Stream is a sequence of fixed-length data packets, each consisting of a 4-byte header followed by a 184-byte payload. The Transport Stream format is suitable for use for all applications in which potential packet loss or corruption could occur; hence, it is used for the transmission of digital audio/video data (i.e., television signals) over satellite, cable, and terrestrial broadcast media.

The MPEG-2 *Systems* Standard

The MPEG-2 *Systems* Standard specifies the coding syntax necessary and sufficient to synchronize the decoding and presentation of video and audio information, while ensuring that data buffers in the decoders do not overflow or underflow. From this information, one can implement a system which combines synchronized digitally-compressed audio and video as well as other data into a single muliplexed stream for delivery across an interface. The resulting synchronized bit stream is designed to be suitable for storage of packaged media, for delivery across a network, or for application of coding layers preparatory to transmission across a broadcast, cable or satellite channel.

MPEG-2 *Systems* defines system coding in two forms, the Program Stream and the Transport Stream. As mentioned, the digital television systems we're interested in here utilize the MPEG-2 Transport Stream format exclusively.

The MPEG-2 Transport Stream

The Transport Stream is a sequence of 188-byte packets, each with a 4-byte header followed by 184 bytes of packet data. Figure 3.1 diagrams the structure of the MPEG-2 Transport Stream. Numbers without parenthesis are byte counts; parenthesized numbers indicate the size of the field in bits.

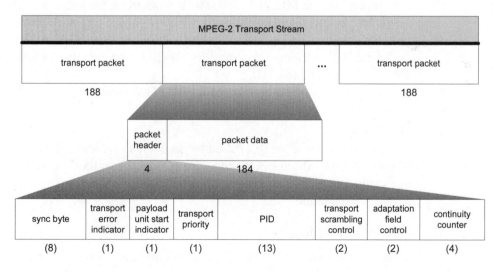

Figure 3.1 MPEG-2 Transport Stream Structure

Shown at the top of Figure 3.1 is the sequence of 188-byte packets. At the first level of decomposition, the packet can be viewed as a four-byte packet header followed by 184 bytes of packet data. Taking apart the 32 bits of the packet header we see these fields:

- an 8-bit sync byte used for packet synchronization.

- the transport_error_indicator flag that may be used to identify to the decoding circuits packets whose contents cannot be trusted.

- a flag called the payload_unit_start_indicator (PUSI) that tells whether the payload portion of the packet includes the starting position of certain data structures (details to follow).

- a transport_priority flag that can be used to identify certain TS packets as being higher priority than the others.

- a 13-bit Packet Identifier (PID) field used to label packets in the multiplex.

- two bits that indicate whether or not the payload of this packet is scrambled (and if so, whether an even or an odd key is used).

- two bits that control whether or not the packet data contains an adaptation field (see "Adaptation Field" on page 31).

- a four-bit continunity counter field that is helpful in detecting missing packets when processing a received Transport Stream.

In the discussions that follow, we focus primarily on aspects of the Transport Stream defined in MPEG-2 *Systems*, but where appropriate we will show how ATSC has extended and constrained MPEG-2 for the terrestrial broadcast application. For cable, extensions and constraints have been specified in standards coming out of the Digital Video Subcommittee (DVS) of the Society of Cable Telecommunications Engineers (SCTE).

Table 3.1 describes the syntax of the MPEG-2 Transport Stream Packet.

sync_byte

A fixed 8-bit value set to 0x47 used by decoders to achieve TS packet synchronization. If a decoder discovers the repeated occurrence of 0x47 occurring at a spacing of exactly 188 bytes in a bit stream, synchronization to the Transport Stream has probably been found.

transport_error_indicator

A flag that indicates the packet payload contains an uncorrectable error. The flag is always transmitted in the zero state, but it may be set by processes external to the MPEG-2 Transport Stream layer to indicate that the packet should be discarded. For example, a Reed-Solomon block code may be applied to the Transport Stream before transmission. If an uncorrectable error is discovered at the Reed-Solomon

TABLE 3.1 Syntax of the MPEG-2 Transport Stream Packet

Field Name	Size in Bits	Value	Description
sync_byte	8	0x47	First byte in the packet; decoders look for the sync_byte spaced 188 bytes apart to verify packet synchronization.
transport_error_indicator	1		Indicates, when set, that the packet contains uncorrectable errors and should be discarded. Conditionally set following detection of an error at the transmisison layer and used to signal decoders downstream.
payload_unit_start_indicator	1		For TS packets carrying PES packets, this flag indicates the packet payload is the start of a PES packet. For SI or PSI (table section) data, it indicates that the payload contains the start of a table section whose position is indicated by the pointer_field.
transport_priority	1		May be used to indicate the packet has a higher priority than other TS packets. Typically set to 0b.
PID	13		The Packet Identifier. Used to identify and associate TS packets with particular Elementary Streams or SI/PSI data in the multiplex.
transport_scrambling_control	2		When zero, indicates that the packet payload is not scrambled. Nonzero values are defined in conditional access standards to indicate the parity of the key used for descrambling.
adaptation_field_control	2		00b = MPEG-2 *Systems* reserved. 01b = No adaptation field present, only the packet payload. 10b = Adaptation field only, no payload. 11b = Adaptation field followed by payload.
continuity_counter	4		A 4-bit counter field used to help the decoder detect lost packets. See text below for a discussion.
if (adaptation_field_control==10b) || (adaptation_field_control==11b) {			
adaptation_field()	1 to 184 bytes		See below for an overview of the function of the adaptation field.
}			
if (payload_unit_start_indicator==1) && ("section-encapsulated payload") {			
pointer_field	8	0 to 182	Points to the first byte of a table section. It may be considered to be a byte offset from this position in the packet to that first byte. Value zero indicates the first byte of the packet payload is the first byte of a table section.
}			
if (adaptation_field_control==01b) || (adaptation_field_control==11b) {			
payload data bytes			Payload data, 1 to 184 bytes in length
}			

decoder, the transport_error_indicator flag may be set to ensure that the packet is disregarded.

payload_unit_start_indicator

This flag has a meaning that varies depending upon the type of payload. If the TS packet payload contains PES packet data, the payload_unit_start_indicator indicates whether or not the first payload byte commences with the first byte of a PES packet.

MPEG-2 *Systems* uses the term "section-encapsulated payload" to refer to SI/PSI data or data conforming to the MPEG-2 private_section() syntax. We describe MPEG-2 private sections later in this chapter. If the TS packet payload contains section-encapsulated payload, the payload_unit_start_indicator indicates whether or not the packet payload includes the first byte of a table section. If it does, the payload_unit_start_indicator is set to 1 and the first byte of the payload carries a pointer_field, discussed below.

transport_priority

When the transport_priority flag is set, the packet has a higher priority than packets having the same PID value which do not have the bit set to '1.' MPEG-2 *Systems* does not specify conditions or rules for setting transport_priority; ATSC makes no specifications regarding transport_priority and it is not typically used.

PID

The Packet ID, or PID, is a 13-bit number used to group packets in the Transport Stream. It is the label used by the demultiplexer in the decoder to collect all the parts of a given program element for decoding. A PID can be associated with TS packets carrying PES packet data, SI/PSI or table section data, or private data that may be any type including neither of the two. The process by which a decoder extracts TS packets with a given PID value is called PID filtering.

The choice of what PID values to use for transport of audio, video, or data is quite flexible, as the 13-bit number space covers 8,192 values. Certain PID values, however, are reserved for special uses or have been reserved by standards bodies for future assignment. Table 3.2 indicates allocations in current use.[*]

Note that standards bodies throughout the world have set aside small PID value ranges for use with System/Service Information. ATSC reserves PID values 0x1FF0 through 0x1FFE, DVB uses 0x0010 through 0x001F, and ARIB in Japan uses 0x0020 through 0x002F. In the US, there is general agreement that PID values

---

[*] In the range called "Available for general purpose use" the table reflects a proposed amendment to ATSC A/53B[14] that, as of this writing, has not passed ballot but is expected to pass in 2002.

TABLE 3.2 PID Value Allocations and Reserved Values

PID Value or Range	Description	Comment
0x0000	Program Association Table	PID 0x0000 carries the Program Association Table, and it is the only type of table that may be carried on this PID. See "Program Association Table (PAT)" on page 54.
0x0001	Conditional Access Table	PID 0x0001 carries the Conditional Access Table, and it is the only type of table that may be carried on this PID. See "Conditional Access Table (CAT)" on page 57.
0x0002	Transport Stream Description Table	PID 0x0002 carries the Transport Stream Description Table, and it is the only type of table that may be carried on this PID. See "Transport Stream Description Table (TSDT)" on page 60.
0x0003-0x000F	MPEG-2 *Systems* Reserved	MPEG has reserved this range for possible future standardization.
0x0010-0x001F	Reserved for DVB SI	PID values in this range are used in Transport Streams that carry DVB Service Information[9]. ATSC has recommended in A/58[10] against use of PID values in this range.
0x0020-0x002F	Reserved for ARIB SI	PID values in this range are used in Transport Streams that carry ARIB Service Information. Values in this range should not be used. An update to A/58[10] is expected to address this issue and state such a recommendation.
0x0030-0x004F	Reserved	Set aside for non-ATSC use.
0x0050-0x1FEF	Available for general purpose use	
0x1FF0-0x1FFA	Reserved	Reserved for possible future use by ATSC and/or SCTE.
0x1FFB	ATSC PSIP SI base_PID	PID 0x1FFB carries ATSC PSIP tables including the System Time Table (STT), Master Guide Table (MGT), the terrestrial and cable Virtual Channel Tables (VCT), and the Rating Region Table (RRT).
0x1FFC-0x1FFD	ATSC Reserved	Used by the now-obsolete ATSC A/55[11] and A/56[12] standards.
0x1FFE	DOCSIS	Used by OpenCable Data-Over-Cable Service Interface Specification (DOCSIS)[13] standard.
0x1FFF	MPEG-2 TS Null Packets	Reserved by MPEG-2 to identify Null packets.

reserved for SI in other standards (DVB, ARIB) are not to be used. The logic here is that a multi-standard receiver may look for tables carried on the well-known SI PIDs to determine which of the three standards an unknown Transport Stream conforms to. Such a receiver may be confused if something other than standard SI were to appear in TS packets where standard SI was expected.

Note also that there are currently no known plans for anyone to try to create a Transport Stream that simultaneously complies with more than one of the world's digital television standards. Due to differences in requirements in the construction of the Program Map Table such as specifications related to the inclusion of certain descriptors, such a multi-standard Transport Stream is currently problematical.

An additional consideration mentioned in MPEG-2 *Systems* is that the choice of PID values should be made so as to avoid emulation of the sync_byte value (0x47) in the TS packet header. However, typical transmission systems such as the QAM modulation used for cable and the 8-VSB modulation used for ATSC terrestrial broadcast replace the sync_byte with a code word derived from all of the bytes in the packet (in the case of cable, it is actually the bytes in the *previous* packet). Therefore, reliable packet synchronization can be achieved regardless of the choice of PID values.

transport_scrambling_control

These two bits indicate whether or not the payload portion of the transport packet is scrambled—in other words, whether conditional access security has been applied. Table 3.3 describes the meaning of the four possible bit combinations. MPEG-2 *Systems* does not define values other than 00b for transport_scrambling_control. The standards defining conditional access systems in the US have agreed on the meanings shown in the table. See for example the ATSC CA specification, A/70[15] and the ATSC A/53B[14] *Digital Television Standard*.

TABLE 3.3 Transport Scrambling Control

Bit Values	Description
00b	No scrambling of TS packet payload.
01b	Reserved.
10b	Transport packet scrambled with Even key.
11b	Transport packet scrambled with Odd key.

adaptation_field_control

We often speak of the 188-byte TS packet as consisting of a four-byte header and a 184-byte payload. That view is a bit over-simplified in that it does not acknowledge the existence of the adaptation field, a data structure that in effect allows the four-byte header to be extended some number of bytes into the payload portion of a packet (thus reducing the size of the payload for that packet). The adaptation_field_control bits indicate whether or not an adaptation field is present just following the 4-byte header. These bits are coded according to Table 3.4. Adaptation fields are discussed in detail below.

TABLE 3.4 Adaptation Field Control

Bit Values	Description
00b	Reserved for future use by ISO/IEC.
01b	No adaptation field, payload only.
10b	Adaptation field only, no payload.
11b	Adaptation field followed by payload.

continuity_counter

This 4-bit field is usually incremented with each TS packet using the same PID value. If the adaptation field value is 00b or 10b, the continuity_counter is not incremented. Decoders monitor the continuity_counter field to detect dropped packets. Packets received and filtered on a particular PID value are seen as continuous (no missing packets) if the continuity_counter in a particular TS packet differs by a positive value of one as compared with the continuity_counter on the previous TS packet on the same PID. Also, continuity is not considered broken when one of the non-incrementing conditions is met (adaptation field value 00b or 10b), or if a TS packet is exactly duplicated from the immediately preceding packet.

MPEG-2 *Systems* describes ways in which discontinuities of various sorts, including those related to the continuity_counter, can be signaled using the discontinuity_indicator flag in the transport packet adaptation field. As this method is not recognized in the ATSC standards, we won't describe it here.

packet data

The packet data portion of the TS packet is the data following the 4-byte header. These 184 bytes may be one of three things:

- payload data (Elementary Stream data, table sections, or other private data).

- a 184-byte adaptation field.

- an adaptation field of some length followed by payload data bytes.

The first packet payload byte is a pointer_field if the payload data conforms to the PSI table section structure, and the payload_unit_start_indicator field is 1b, indicating that somewhere in the payload is the first byte of a table section.

pointer_field

This is an 8-bit field that points to the first byte of a table section. It may be thought of as an index value if the payload bytes following the pointer_field are thought of as an array of bytes. Value zero of the pointer_field indicates that the byte immediately following is the start of a table section. Decoders must rely on the pointer_field mechanism to find table sections within TS packets, as there is no other reliable way to determine where one table section ends and another begins.

Transport Stream Null Packets

Any given Transport Stream is transmitted at a fixed rate based on the transmission system and physical channel. The 8-VSB transmission system utilized in ATSC terrestrial broadcast, for example, delivers a Transport Stream bitrate of 19.39 Mbps, and a cable signal modulated using 64-QAM provides a bitrate of 26.97 Mbps.

A Transport Stream is assembled from a number of audio, video, and data streams in addition to PSI and SI table sections. Although audio streams run at a fixed rate, typically video bitrates are bursty. Sometimes a number of different video streams are statistically multiplexed in the hopes that at times when one needs more bandwidth another may be needing less. It's nearly impossible to build a packet multiplexer in an MPEG-2 encoder that can exactly match the combined packet rate of all the streams making up the full (multiple-program multiplexed) Transport Stream at every instant of time. At certain moments, none of the input streams will have a packet ready for insertion into the Transport Stream, so at those times the packet multiplexer creates and inserts Null TS packets.

In some applications, Null packets are used as placeholders in the multiplex. It is possible to create a stream of Null packets at a fairly constant packet rate. This enables a piece of equipment downstream from the multiplexer to replace the Null packets with Elementary Stream data or section-encapsulated data.

MPEG-2 *Systems* defines some special rules with regard to Transport Stream Null packets, but their fundamental attribute is that they are identified with PID value 0x1FFF (the 13-bit PID value is all ones). Table 3.5 describes the construction of the MPEG-2 Transport Stream Null packet.

As shown in the table, some of the other attributes of Null packets include:

- payload_unit_start_indicator is set to zero.

- transport_scrambling_control is set to zero, indicating the packet payload is unscrambled.

- adaptation_field_control is set to 01b, indicating there is no adaptation field present.

The continuity_counter is undefined and may contain any value in the 0x0 to 0xF range. Often, the continuity_counter field is held at a fixed value (zero for example) for all Null packets.

The payload data bytes may be set to any value and are always discarded. In practice, these bytes may be set to random values to help maintain a statistical balance between ones and zeroes in the multiplex.

TABLE 3.5 The MPEG-2 Transport Stream Null Packet

Field Name	Size in Bits	Value	Description
sync_byte	8	0x47	TS packets always start with the sync_byte, value 0x47.
transport_error_indicator	1	0b	Indicates, when set, that the packet contains uncorrectable errors and should be discarded. At the point of transmission, set to 0b.
payload_unit_start_indicator	1	0b	Set to 0b for Null packets.
transport_priority	1		May be used to indicate the packet has a higher priority than other TS packets. Typically set to 0b.
PID	13	0x1FFF	The Packet Identifier. Value 0x1FFF indicates a Null packet.
transport_scrambling_control	2	00b	Set to zero for Null packets, indicating the payload is unscrambled.
adaptation_field_control	2	01b	Set to 01b indicating no adaptation field present, only the packet payload.
continuity_counter	4	0x0 to 0xF	Ignored for Null packets.
payload data bytes	8*184	anything	Payload data, ignored for Null packets.

Adaptation Field

Beginning at the front part of the TS packet payload may be a data structure called an adaptation field. Adaptation fields can carry a large number of various flags, but the most important piece of data is probably the Program Clock Reference (PCR). Decoders use the PCR to set the clock reference used for time synchronization of the Elementary Streams making up an MPEG program. Other flags and fields defned in the adaptation field include:

- a flag indicating that the TS packet contains some information to aid random access of the stream—for example for a video ES, a video sequence header.

- splice point data, indicating a point in the Transport Stream where a splice may be appropriate along with decoder timing parameters relevant to the splice point.

- a variable amount of optional private data.

Figure 3.2 diagrams the possible TS packet structures with regard to adaptation fields. In case a) the adaptation field control is 01b indicating no adaptation field is present. The packet payload is either PES packet data or SI/PSI data in which no byte in the packet payload is the first byte of a table section.

Case b) shows the situation where again there is no adaptation field, but the payload contains SI/PSI data and one of the bytes is the first byte of a table section.

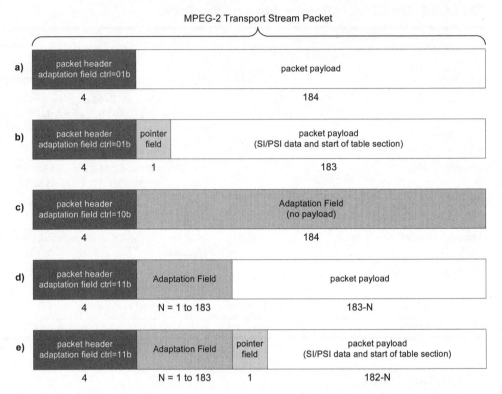

Figure 3.2 MPEG-2 Transport Stream Packet Structure with Adaptation Fields

In case c), there is no packet payload at all. The entire packet following the 4-byte header is an adaptation field.

Case d) shows the situation where the adaptation field control is 11b, indicating that an adaptation field follows but its length is less than 184 bytes. In this particular case, no pointer field is present because either the type of data carried in TS packets associated with this PID is PES packets or it is table-type data but no byte in the packet payload is the first byte of a table section.

Finally, case e) shows both an adaptation field and a pointer field. Here, the data is SI/PSI table sections and one of the packet payload bytes is the beginning of a table section. Again note that case e) cannot occur for any of the standard tables defined in A/65 PSIP because the Standard explicitly states that TS packets carrying these tables must have the adaptation_field_control bits set to 01b (no adaptation field present). Case e) could occur for some non-standard table sections, although it is unclear why an adaptation field would be useful for TS packets carrying table sections. It would be wise for decoder designers to accommodate the presence of adaptation fields in TS packets carrying section-encapsulated data, just to be safe.

MPEG-2 Tables

Fundamentally, there are three different types of data in the MPEG-2 Transport Stream. First, there are the Packetized Elementary Streams composed of a sequence of PES packets. This is the type of stream that carries compressed video or audio or certain types of streaming data, especially when that data must be accurately time-aligned with another program element such as video. Second, there is tabular data, used to deliver various types of standardized data structures in a robust manner. Tables are used to carry information necessary to demultiplex and present programs, as well as to give supplementary information about the multi-channel multiplex. The MPEG-2 *Systems* Standard itself defines the syntax and semantics of four table types, while standards built upon MPEG-2 (such as the ATSC PSIP Standard) define a half-dozen more. MPEG-2 *Systems* calls this type of tabular data "section-encapsulated data." We'll talk more about table sections shortly. Third, TS packets can carry data formatted in a proprietary, non-standard way that differs from both PES packets and the MPEG-2 table section formats.

MPEG-2 *Systems* defines a template for those extending the standard to use as a model for data tables, and the world's System Information standards all use it. By designing the structure of tables so they share a common basic structure, the mechanisms used for table section creation (in the encoder) and parsing/processing (in the decoder) are common for all types of tables. In general, MPEG uses the term "private" to designate something defined outside the scope of the MPEG-2 standards. The name of the template or data structure MPEG-2 defines for Private Tables is the private_section(), described next.

MPEG-2 Table Section Syntax

The MPEG-2 *Systems* Standard describes two different standard syntaxes for table sections. One, known as the short-form syntax, has only a few required syntactic elements. The other is known as the long-form syntax because it includes a longer common header structure designed to accommodate the need to segment large tables into smaller sections for later re-assembly. The long-form syntax also includes a versioning mechanism and a way to deliver sections of a segmented table that are to be considered part of the "next" table to become active. Finally, the long-form syntax includes a 32-bit cyclic-redundancy check (CRC) field so that received table sections can be confirmed (to a high probability) to be error-free before being used.

Figure 3.3 diagrams the short- and long-form MPEG-2 private_section() structure. The structures of both short and long versions have their first five fields in common, as shown. In the long form there are seven additional required fields. A

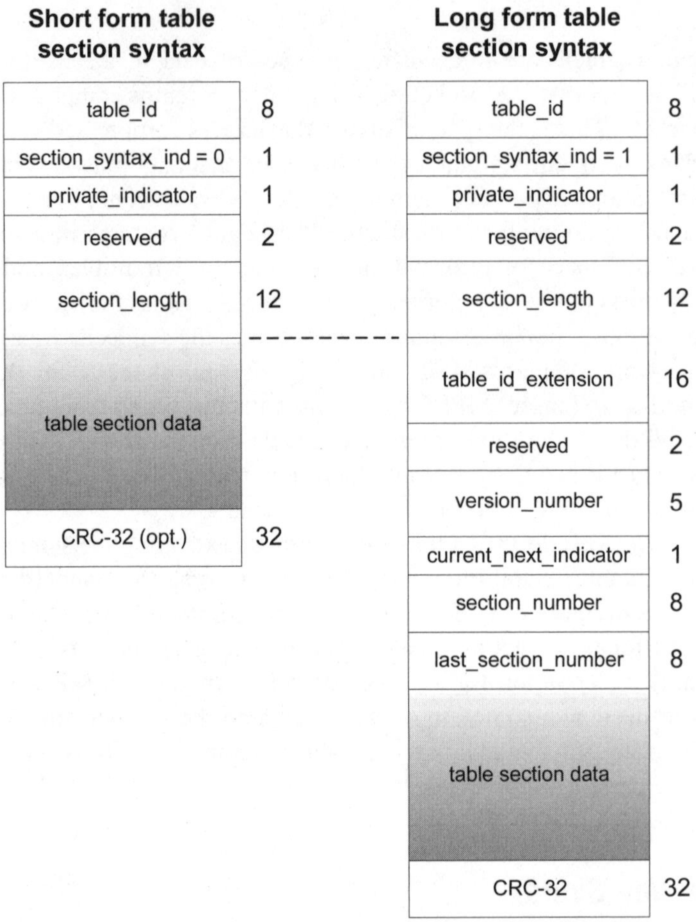

Short form table section syntax

table_id	8
section_syntax_ind = 0	1
private_indicator	1
reserved	2
section_length	12
table section data	
CRC-32 (opt.)	32

Long form table section syntax

table_id	8
section_syntax_ind = 1	1
private_indicator	1
reserved	2
section_length	12
table_id_extension	16
reserved	2
version_number	5
current_next_indicator	1
section_number	8
last_section_number	8
table section data	
CRC-32	32

Figure 3.3 Short- and Long-Form Table Section Syntax

device processing MPEG-2 Transport Packets can be designed to accept table section data in either form.

Table 3.6 describes the syntax of the MPEG-2 private table section. As can be seen, depending on the value of the section_syntax_indicator, private data either starts directly following the length field or following the fields used for version number control and segmentation/reassembly. Long-form tables always end with the 32-bit CRC defined in MPEG-2 *Systems*, while short-form sections may or may not have a CRC.

The fields common to both long and short forms are:

table_id

Each table section starts with an 8-bit field that identifies the type of table to follow. In processing the table_id field, a receiving device might first decide whether or not that type of table is supported. If tables of the indicated type are of interest, the receiver continues processing the table section. If not, the table section is skipped by jumping ahead the number of bytes given in the section_length field.

The table_id field establishes the syntax and semantics for the specific type of table, such that once the table ID is known the structure and definition of that table's fields and parameters are known.

Note: for the tables defined in the ATSC standards, an additional field is present in the header that has a direct bearing on the syntax and semantics of the table. This field is the protocol_version, and its function is to allow, in the future, a new table structure to be defined for the same value of table_id but with different syntax and semantics. Use of protocol_version can allow ATSC to reuse values of table_id at some future time in a way that does not break backward compatibility. At the current time, all the tables in PSIP are defined with protocol_version equal to zero. Receivers built to the current standards are expected to check protocol_version and discard the table section if a non-zero value is seen.

Within the SI standards, for both the ATSC Standard and the Digital Video Broadcasting (DVB) Standard developed in Europe, assignment of table ID values has been made such that all table ID value assignments are unique. A requirement for uniqueness of any code point arises from the need for receiving devices to unambiguously identify an object identified with that code point.

TABLE 3.6 Syntax of the MPEG-2 Private Section

Field Name	Number of Bits	Field Value	Description
private_section() {			Start of the private_section().
table_id	8		Identifies the type of table section and hence the structure of the private data included within.
section_syntax_indicator	1		When set, indicates the section is formatted in MPEG "long-form" syntax. When the flag is 0, data following section_length may be in any format.
private_indicator	1		A user-definable flag. MPEG-2 *Systems* states that the private_indicator will not be specified by MPEG.
reserved	2		Reserved by MPEG. Reserved bits are set to 1.
section_length	12		Specifies the length, in bytes, of data following the section length field itself to the end of this table section.

TABLE 3.6 Syntax of the MPEG-2 Private Section (continued)

Field Name	Number of Bits	Field Value	Description
if (section_syntax_indicator==0) {			Short-form section.
for (i=0; i<N; i++) {			In this case the value of N is equal to section_length.
private_data_byte	8		Data bytes comprising the short-form table section.
}			End of data bytes.
} *else {*			Long-form section.
table_id_extension	16		Extends the table ID. Used by different types of tables to differentiate instances of tables. See text.
reserved	2	11b	Reserved bits are set to 1.
version_number	5		Reflects the version of a table section, and is incremented by one (modulo 32) when anything in the table changes.
current_next_indicator	1		Indicates whether the table section is currently applicable (value 1) or is the next one to be applicable (value 0).
section_number	8		Indicates which part this section is when the table is segmented into sections.
last_section_number	8		Indicates the number of the last section (and hence the total number of parts) when the table is segmented into sections.
for (i=0; i<N; i++) {			Start of table section data. The value of N is given indirectly by section_length: it is section_length -9.
private_data_byte	8		Data bytes comprising the long-form table section.
}			End of data bytes.
CRC_32	32		A 32-bit CRC value that gives a zero result in the registers after processing the PAT section.
}			End of the long-form section syntax.
}			End of the private_section().

Sometimes, the context in which a table section appears must be used to help identify its contents. Even though the assignment of table ID values currently used for SI tables has been made unique world-wide, the tables defined by ATSC and those defined by DVB can always be distinguished from one another by context. The ATSC tables appear in TS packets with PID values associated with the ATSC standards, while the DVB tables are found in TS packets used by the DVB standards.

Standards makers have generally followed the path that world-wide uniqueness should be maintained for new assignments of table_id until such time that the number-space is exhausted. If no more new unique table ID values are available when a new table is invented, a standards body in one of the countries could choose to assign to it a table ID that has a different meaning in a different international standard. If this were to happen, receiving devices would need to identify the type of table by using both the table ID and the context in which it appears.

Table 3.7 lists table_id values defined or reserved by MPEG, ATSC, and other bodies.

TABLE 3.7 Table ID Values Defined by MPEG, ATSC, SCTE and Others

table_id	Name of Table	Where Defined	PID
0x00	Program Association Table (PAT)	MPEG-2 *Systems*	0x0000
0x01	Conditional Access Table (CAT)	MPEG-2 *Systems*	0x0001
0x02	Program Map Table (PMT)	MPEG-2 *Systems*	Any general use PID per Table 3.2.
0x03	Transport Stream Description Table (TSDT)	MPEG-2 *Systems*	0x0002
0x04-0x37	Reserved	MPEG-2 *Systems*	
0x38-0x39	Reserved	DSM-CC	
0x3A	DSM-CC Sections containing multiprotocol encapsulated data	DSM-CC	Any general use PID per Table 3.2.
0x3B	DSM-CC Sections containing User-to-Network Messages, except Download Data Message Sections	DSM-CC	Any general use PID per Table 3.2.
0x3C	DSM-CC Download Data Message Sections	DSM-CC	Any general use PID per Table 3.2.
0x3D	DSM-CC Sections containing stream descriptors	DSM-CC	Any general use PID per Table 3.2.
0x3E	DSM-CC Sections containing private data	DSM-CC	Any general use PID per Table 3.2.
0x3F	DSM-CC Addressable Sections	DSM-CC	Any general use PID per Table 3.2.
0x40-0x7F	Used in or reserved by DVB	DVB SI	
0x80-0x8F	Reserved for ATSC CA	ATSC A/70	Any general use PID per Table 3.2.
0x90-0xBF	User private		Any general use PID per Table 3.2.
0xC0-0xC1	Reserved by SCTE (in use in cable)		Any general use PID per Table 3.2.
0xC2-0xC5	Reserved (used in now-defunct A/56[12])		0x1FFC
0xC6	Reserved by SCTE (in use in cable)		
0xC7	Master Guide Table (MGT)	ATSC A/65	0x1FFB

TABLE 3.7 Table ID Values Defined by MPEG, ATSC, SCTE and Others (continued)

table_id	Name of Table	Where Defined	PID
0xC8	Terrestrial Virtual Channel Table (TVCT)	ATSC A/65	0x1FFB
0xC9	Cable Virtual Channel Table (CVCT)	ATSC A/65	0x1FFB
0xCA	Rating Region Table (RRT)	ATSC A/65	0x1FFB
0xCB	Event Information Table (EIT)	ATSC A/65	Any general use PID per Table 3.2.
0xCC	Extended Text Table (ETT)	ATSC A/65	Any general use PID per Table 3.2.
0xCD	System Time Table (STT)	ATSC A/65	0x1FFB
0xCE	Data Event Table (DET)	ATSC A/90	Any general use PID per Table 3.2.
0xCF	Data Service Table (DST)	ATSC A/90	Any general use PID per Table 3.2.
0xD0	Program Identifier Table (obsolete)	ATSC A/57	
0xD1	Network Resources Table (NRT)	ATSC A/90	Any general use PID per Table 3.2.
0xD2	Long Term Service Table (LTST)	ATSC A/90	Any general use PID per Table 3.2.
0xD3	Directed Channel Change Table (DCCT)	ATSC A/65	0x1FFB
0xD4	DCC Selection Code Table (DCCSCT)	ATSC A/65	0x1FFB
0xD5	Selection Information Table	EIA 775.2[16]	0x1FFB
0xD6-0xD7	ATSC/SCTE Reserved		
0xD8	Cable Emergency Alert	EIA-814, SCTE 18[17]	0x1FFB
0xD9-0xFD	Reserved for future ATSC/SCTE use		
0xFE	Splice Information Table	SMPTE 312M[27]	
0xFF	Reserved by MPEG-2 *Systems* for transport packet filler		

section_syntax_indicator

The section_syntax_indicator flag identifies the syntax as being short-form (value 0) or long-form (value 1). All the SI and PSI tables used in the ATSC Digital Television Standard use long-form tables.

private_indicator

MPEG-2 *Systems* includes a "private_indicator" flag in the section header syntax, but does not define its meaning or use, nor will it in the future. Users of the MPEG-2 standards such as ATSC, DVB, or ARIB are free to make use of this flag as they see fit. In all of the ATSC-defined tables, the flag is currently set to one and is considered a reserved field.

section_length

The 12-bit section length field gives the number of bytes remaining in the section immediately following the section_length field itself, all the way up to the end of the section. The value in this field cannot exceed 4093. The total number of bytes in the longest possible table section is therefore 4096.

Fields defined specifically for the long-form table section are listed below:

table_id_extension

We define an "instance" of a table as being an occurrence in the Transport Stream of a table section (or more than one if that instance is segmented for delivery) that carries a particular set of data. The table_id_extension is a 16-bit field that distinguishes one instance or occurrence of a table from another. The table_id_extension field, when used, is typically re-named to reflect the parameter used to make the distinction among various instances. We discuss the use of the table_id_extension field below.

By this definition, any given TS_program_map_section() is a table instance. We use this definition of "instance" because it applies to the SI tables based on MPEG-2 private section syntax.[*]

version_number

This is a 5-bit field providing the version of the instance of the table. The value of the version field by itself has no meaning. Equipment that generates tables conforming to MPEG-2 rules for private sections is required to increment the version field by one (modulo 32) whenever anything in the table instance changes. When a table instance is segmented for delivery, each section must indicate the version number of the full table instance. If it is necessary to change any piece of data in any section, even sections that have not changed must be retransmitted with the next-higher version number.

Receiving equipment may use the version_number field to avoid parsing tables that have already been processed and stored, thus saving processing overhead.

The version_number field is only applicable to a particular instance of a table. Each different table instance has a version_number that is independent of version

---

[*] We also have to disregard, for the moment, the fact that MPEG-2 *Systems* considers the PMT as one table composed of all TS_program_map_section() occurrences in the Transport Stream. For a discussion of this distinction please see "PMT sections" on page 57.

numbers for other occurrences of that type of table. For example, consider the following two tables:

1. table_id = m; table_id_extension = n; version = v

2. table_id = m; table_id_extension = n; version = v+1

The table instance is given by m and n so the second table is viewed as a new version of that same instance.

Here is another example:

1. table_id = m; table_id_extension = n; version = v

2. table_id = m; table_id_extension = p; version = v

In this case, because the table_id_extension fields differ the two tables are considered separate instances of a table with table_id value = m. This distinction holds even though the version number field for both is given as v.

The version_number field is discussed further in "Table version control" on page 47.

current_next_indicator

The current/next indicator is a flag that indicates whether the table instance is applicable for the current time (value 1) or will be the next version to apply (value 0). Use of this flag is discussed below (see "Table version control" on page 47).

section_number

This is an 8-bit number that indicates, for a segmented table, which part the current section is. Numbering starts at zero, so section_number = 0 indicates the first section of a segmented table or the only section of a non-segmented one.

last_section_number

The last_section_number field is an 8-bit integer that indicates which section_number is the highest or last. Again numbering starts at zero so a last_section_number value of zero indicates a non-sectioned table. If last_section_number is zero, the only valid value for section_number is also zero. "Part 1 of 1" is represented by section_number = last_section_number = 0.

Figure 3.4 shows six example table section headers, all representing sections with a common table ID value, T. Let's consider the case that all six are present in the same Transport Stream, appearing in TS packets with a common PID value. The figure illustrates the concept that inspection of each of the header fields is necessary for proper processing, and that each field distinguishes one section from another.

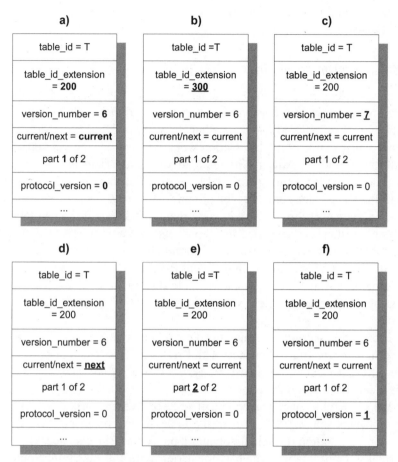

a)

table_id = T	
table_id_extension = **200**	
version_number = **6**	
current/next = **current**	
part **1** of 2	
protocol_version = **0**	
...	

b)

table_id =T
table_id_extension = **300**
version_number = 6
current/next = current
part 1 of 2
protocol_version = 0
...

c)

table_id = T
table_id_extension = 200
version_number = **7**
current/next = current
part 1 of 2
protocol_version = 0
...

d)

table_id = T
table_id_extension = 200
version_number = 6
current/next = **next**
part 1 of 2
protocol_version = 0
...

e)

table_id =T
table_id_extension = 200
version_number = 6
current/next = current
part **2** of 2
protocol_version = 0
...

f)

table_id = T
table_id_extension = 200
version_number = 6
current/next = current
part 1 of 2
protocol_version = **1**
...

Figure 3.4 Six Examples of Table Section Headers

The first section header, labeled a), is an instance given by table_id_extension value 200. It is version 6, defines the currently applicable data, and is part 1 of two. Each of the other five table sections differ in just one field from this reference.

The second header in the figure, labeled b), is recognized as a different instance of table ID T because the table_id_extension value is 300. The third header at c) is the same table instance as a) but a newer version, version 7.

The fourth header, d), is again the same instance as a) but the next table to be applicable. It cannot legally appear with a), i.e. in TS packets with the same PID value as a). MPEG-2 states that the version number must be incremented by one (modulo 32) above the version of the current table. If a receiving device were to see d) after already having processed a), it should throw d) away.

Header e) is simply part 2 of the table instance a). Finally, header f) must be considered an entirely different table, completely separate from table ID T, because the protocol_version is non-zero. If a receiving device does not recognize this form, it must discard the section.

Table instances and receiver processing

Receiving devices must distinguish among various instances of table sections in the Transport Stream in order to properly extract and make use of the data provided in each one. The table_id taken together with the table_id_extension field identify a particular instance of a table section when the sections are delivered in TS packets with common PID values. The table_id_extension field is always 16 bits to conform to the MPEG-2 long-form table syntax, but the meaning of the field may be redefined as needed for each type of table.

One Transport Stream may include many instances of a given table type. Often, multiple instances are carried in TS packets with common PID values. In this case, the table_id_extension is used to differentiate the instances. Whenever two tables are carried in TS packets with different PID values, we always consider the two to be separate instances even when the table ID and table_id_extensions are identical. That is, the scope of table instances is to the set of TS packets with the same PID value.

If one table section is received in TS packets with PID value A and another table section is received in TS packets with PID value B, one can infer that the two tables are different instances of that table type, even if *all* the values in the header fields are identical between the two.

Let's look at a concrete example. The PSIP Event Information Table is structured such that its source_id field is at the table_id_extension position within the table header. The Source ID field associates a particular EIT instance with an entry in the Virtual Channel Table, which is identified with the matching Source ID. Simply put, the Source ID in the EIT is what ties that event information to a particular virtual channel.

If a Transport Stream carries four channels of audio/video programming, four virtual channels may be defined. The Source ID values for each of the four are given in the VCT. Event information for current programming is carried in TS packets with a PID value identified in the Master Guide Table. The EIT tables describing the current three-hour time block are called EIT-0, and the tables describing the following time slot are called EIT-1.

Within the packets labeled with the PID value given for EIT-0, one can find EIT table sections describing each of the programming channels for the current three-hour time block. Likewise, all the programming channels' events for the subsequent three-hour time block can be found in EIT table sections labeled with the PID value given for EIT-1.

Context is used in processing SI tables, in that EITs with identical values in their table headers can appear in the Transport Stream, yet be recognized as separate EIT instances. The context is provided by the PID values associated with TS packets carrying the EIT sections. The Master Guide Table identifies one PID value as associated with EIT-0 and a different PID value as associated with EIT-1.

Figure 3.5 illustrates the concept. The tables at the top, a) and b), are carried in

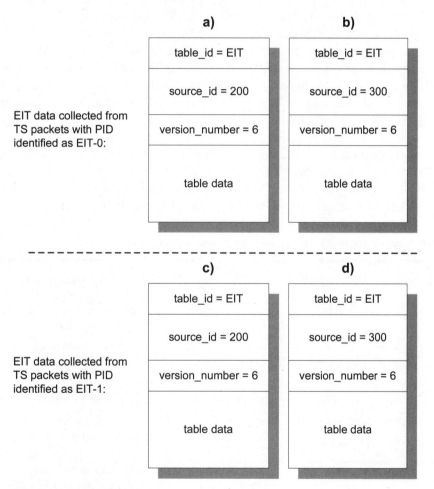

Figure 3.5 EIT Table Instances

TS packets with a PID value referenced in the MGT as EIT-0. The ones at the bottom, c) and d), are identified as EIT-1. Instance a) is differentiated from instance b) because the table_id_extensions are different. The Source ID for a) is value 200, while the Source ID for b) is 300.

Even though the header fields for a) and c) are identical, instance a) is differentiated from instance c) because the two were collected from TS packets having different PID values. In this case, a) came from the EIT-0 PID and c) came from EIT-1.

As we have seen, one of the functions of the table_id_extension field is to distinguish various table sections when they share a common table ID. As mentioned, in the PSIP tables and also within the PSI tables defined in MPEG-2 *Systems*, the 16 bits corresponding to the table_id_extension field are often re-defined according to the needs of each particular table definition. In every case, the table_id_extension field can be used to distinguish one instance from another.

Let's look at an example of how the table_id_extension field is used to distinguish instances for one of the table types defined in the MPEG-2 *Systems* Standard itself. The TS_program_map_section(), which delivers sections of the Program Map Table (PMT), defines the 16-bit table_id_extension field as program_number. This way, an instance of a TS_program_map_section() describing the program elements for program A can be distinguished from another instance describing ES components for program B. The two can be distinguished even when carried in TS packets with the same PID, and even when both are carried in the same TS packet.

Table 3.8 lists various MPEG-2- and PSIP-defined table types and describes the definition and use of the table_id_extension field for each type.

TABLE 3.8 Use of Table ID Extension for Some Table Types

Table ID	Table Type	Table ID Extension	Description
0x02	Program Map Table	program_number	Associates the PMTsection with the given MPEG-2 program number.
0xC7	Master Guide Table	0x0000	Not used. There is only one MGT.
0xC8, 0xC9	Virtual Channel Table	TSID	Associates the VCT with the TSID identified in the PAT.
0xCA	Rating Region Table	rating_region	Associates the RRT instance with a rating region.
0xCB	Event Information Table	source_id	Associates the EIT instance with a particular programming service, as identified by its Source ID.
0xCC	Extended Text Table	ETT_table_id_extension	Set to a value to maintain uniqueness when ETT table sections appear in TS packets with common PID values. The field carries no intrinsic meaning.
0xCD	System Time Table	0x0000	Not used, as there can be only one instance of the System Time table at any given time.

MPEG-2 table segmentation and reassembly

The designers of the MPEG-2 *Systems* standard envisioned the need, in some applications, to deliver very large tables in the Transport Stream. They could have achieved this objective simply by putting, say, a 32-bit field into the table header to indicate the table's total length, making it possible to deliver all the bytes in one very long sequence of packets. This approach would suffer from several problems:

- Receiving devices would not be able to begin to reliably process such a table until all bytes were received, which could be a long time.

- A complex method for detecting errors would have to be used, because even a 32-bit cyclic redundancy check probably would not be sufficient to detect missing or corrupt packets when so many might be involved.

- When one began to transmit a long table, it would not be possible to transmit any other table within TS packets of the same PID until the end of the long table was reached. If one wished to transmit a short table at a relatively frequent repetition rate, transmission of the long table would interrupt the transport of the smaller one.

- Buffering could be a problem for receiving devices in a system in which very long tables might appear in the Transport Stream.

The MPEG-2 architects addressed these issues as is commonly done in many communications protocols, by defining a general method for segmentation and subsequent re-assembly of long tables. Please note that the segmentation of the Program Map Table does not fit the model we're about to describe (please see "PMT sections" on page 57). The general method for segmentation and reassembly defined in the MPEG-2 *Systems* Standard involves the following aspects:

1. Tables may be divided into pieces called "sections."

2. Each section may be at most 4096 bytes in length (with exceptions as noted— some types of tables are limited to 1024 bytes).

3. Each table section conforms to a common syntax, which includes, as part of the table header, an indication as to the total number of sections in the full table, and which part the current section is. For example, examination of the header could show that this section is the third of five total sections, or the one and only section (when the full table fits into the size limit of one section).

4. A full table may be segmented into as many as 256 sections, making the largest full table that can be transmitted 256*4096 = 1MB.

Figure 3.6 depicts a long table on the left. Because the length exceeds 4096 bytes, it is segmented into two sections for placement into the Transport Stream, as shown in the middle of the figure. On the right, the original table is shown after re-assembly in the receiving device.

The figure illustrates a common situation with these types of tables in which different sections each carry one or more data blocks. In the example, the data blocks are called "table section data" and each is associated with a number in square brackets, similar to the way indexed records in a computer language might be depicted and referenced.

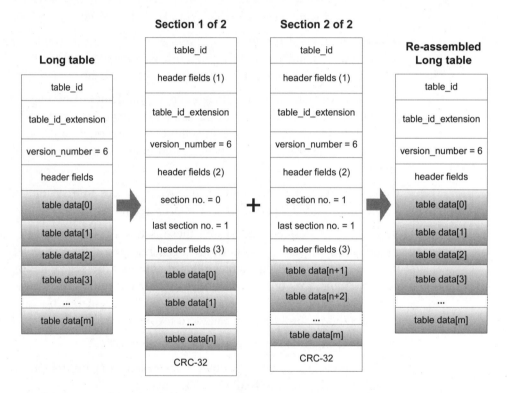

Figure 3.6 Example Segmentation and Re-assembly with "part M of N" Scheme

In the example, the large table includes some amount of header data common to the whole table and m variable-length data record blocks. Typically, for tables in which data records are small enough in size that at least several fit into each table section, the sections are organized so that each section provides some number of complete blocks. In other words, the table never ends in the middle of one of these blocks. One of the pieces of header data typically indicates the number of whole blocks to follow.

Segmentation involves creating two table sections, the first labeled "part 1 of 2" and the second labeled "part 2 of 2." The fields used to label the part number and total parts are actually the section_number (part number) and last_section_number (total number of parts). Numbering begins with value zero for the section_number and the last_section_number fields.

In the re-assembly process, all of the blocks are recovered and stored in the order corresponding to the order of the section numbering. Correspondingly, blocks recovered from the first section (labeled with section number zero) are placed at the top, with blocks arriving in each subsequent section stored next in sequence.

Note that table sections may arrive at a receiving device out of order. The MPEG-2 standard does not require sections to be transmitted in sequence, but it does recommend that practice. Clearly, any given table section may be found on reception to be corrupted due to noise; in that case the receiver must wait for the next repetition of the table in order to grab the missing section. It is the responsibility of the receiving device to restore the proper order by processing the section_number field.

The MPEG-2 sectioning method can also be used to deliver data that has no natural breakpoints. Let's say, for example, one wished to deliver a 100-kilobyte data block within sections of an MPEG-2 table. In this case, the syntax of the table could be defined such that each section delivers some number of bytes of the monolithic data block. Instead of a "for" loop indicating how many data records are contained in the section, the section could have a 12-bit length field to indicate how many bytes of the 100-kilobyte data block were present in that section.

An advantage to the method of sending an integral number of complete data records compared to sending pieces of a monolithic block is that receiving devices can process and derive useful information from any received section, even those received out of order.

The structure of PSIP tables follows the pattern established by MPEG *Systems*. For PSIP tables such as the terrestrial and cable Virtual Channel Tables and the Event Information Tables that can be segmented for delivery, the method shown in Figure 3.6 is used: the structure of the table sections is header data followed by a "for" loop containing one or more data record blocks. Note that several table types defined in PSIP are not allowed to be split into multiple sections, meaning that all the data comprising the full table must fit within a single section.

Table version control

At times, the contents of a transmitted table may change, and any changes to a table's content need to be communicated to receiving equipment. MPEG-2 *Systems* handles version control like many protocols do: each section of a table is labeled with a version number. Receiving equipment can check the version number against the version number of a previously saved section, and if the version number has changed the new table can be collected and used to replace the old one.

Let's say for example that a receiver has collected all three parts of a three-section table identified as version 6. If a table section is now seen for that same table instance, and the version number is found to be 7, the receiver knows that the one in memory is obsolete and should be replaced as soon as all the parts of the new table have been seen.

This simple method has the drawback that only one section of the version 7 table has been received at the time we discover that it replaces version 6. Should we use version 7 right away, or must we wait for all of its parts? Should we consider the version change to have occurred when the first new section was seen, or the last?

MPEG-2 *Systems* handles this problem with a simple one-bit flag called the "current/next indicator." When set, the flag indicates the section is currently applicable. When clear, the section should be considered to be part of the *next* table version—that is, one that will be valid at a time in the not-too-distant future. With this scheme it is possible to deliver all of the parts of the next table in advance. For the SI tables based on the MPEG-2 private section syntax, when one wishes to signal to all the receiving devices that it's time to use the new version, delivery of just one section indicating the version that was "next" is now "current" suffices. When a receiver notes that the version number of the current table has changed, it considers all sections labeled with that version number to be current.[*]

So it becomes possible for receiving devices to have, in memory, a pre-assembled table designated for future use ready to go at the moment it becomes current. Figure 3.7 describes the MPEG-2 table update method.

The example in the figure involves a table instance given by table_id value N and table_id_extension value 200. Initially, this instance is at version 6. Sections of this same table (table_id N and table_id_extension 200) identified as the "next" version (version 7) are then collected. These two parts are the tables shown as a) and b) in the figure. The two sections are reassembled into the version 7 table at c).

At this point in time, the receiver has a complete current table at version 6 and a complete next table at version 7. Now, if either part 1 or part 2 of the sectioned table is received, and version 7 is indicated as current, the receiver can respond by immediately replacing the old version 6 with the complete new version 7.

For many of the tables defined in the PSIP Standard it is not allowed to send a "next" version. Tables for which only the current version may be sent include the System Time Table, Master Guide Table, Rating Region Table, Event Information Table, and the Extended Text Table. In fact, the only tables for which delivery of a next version is permitted are the terrestrial and cable Virtual Channel Tables.

---

[*] Note that while the PMT can be in some sense be considered a "sectioned" table (with each TS_program_map_section() being one part), its versioning does not work as described here. Each PMT section is independent from the others as far as its version number is concerned. The special aspects of the PMT are discussed further on page 57 (see "PMT sections").

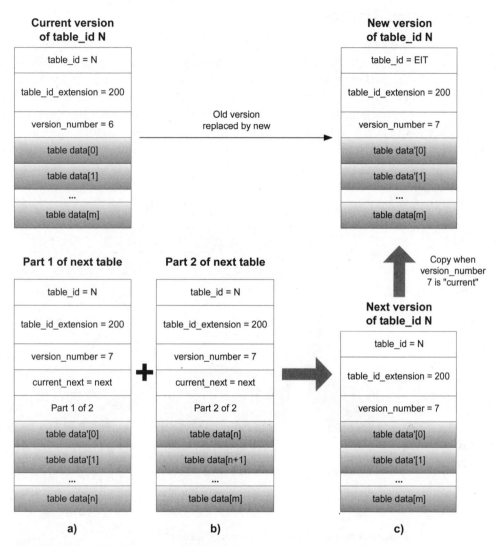

Figure 3.7 MPEG-2 Table Update Method

A note of caution: once a "next" table has been sent, there is no practical way to rescind or cancel it. One is forced to make it the current table, even if for a brief time. This assertion is derived from two facts: 1) each new version must increase the version number field by one; and 2) one cannot make a change to a table without increasing the version number. Since there is no way to send a correction or update to a "next" table, the general recommendation for PSIP (specifically for the VCT) is that delivery of next tables is discouraged.[*]

Table Section Transport

We've looked in detail at the structure of the private_section() syntax. Now we look at how private table sections are packetized into TS packets prior to insertion into the packet multiplex. We look at how a long table, for example the one at the left in Figure 3.8, is segmented into sections, and then how each section is packetized for delivery in the Transport Stream. We then look into the structure of the TS packets to see how the pointer mechanism works for table sections.

Figure 3.8 illustrates, at the top, a Table. As shown in the example, this Table is

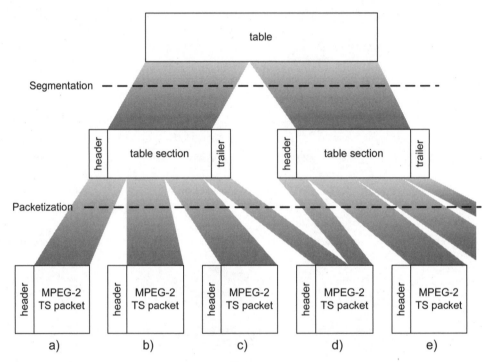

Figure 3.8 Table Segmentation and Packetization

segmented into three sections of differing lengths. This example has been constructed to illustrate principles of segmentation and packetization; it is not intended to represent anything typical with regard to table sizes or the sizes of the sections. Note that depending upon a table's size and the limitations on the size of table sec-

---

* MPEG-2 *Systems* defines a mechanism for signaling PSI table discontinuities using the discontinuity_indicator flag in the TS packet adapation field (see Sec. 2.4.3.5 of ISO/IEC 13818-1[1]) but this method is not specified for use in ATSC and SCTE standards, and hence not supported in encoders or decoders. For the SI tables, adaptation fields are disallowed.

tions for a table of its type, a table could be sent in one table section. Some types of tables *must* be sent in a single section. The maximum number of sections is limited to 256 by the 8-bit field sizes for section_number and last_section_number.

Each table section consists of a portion of the Table encapsulated within the header/trailer data structures we saw in the previous section. Table sections can be as large as 4096 bytes, although a smaller maximum length may be specified for some types of tables. Several of the MPEG-2 and PSIP tables, for example are limited to 1024 bytes in length.

Figure 3.9 shows an example of various possibilities for table section placement, beginnings and endings. The first table section, Section 1, includes the table section header, filling up the packet payload portion of the first TS packet, a). TS packet b) carries the next portion of the Section 1 and packet c) carries the end of Section 1 and all of the (very small) Section 2.

The figure also illustrates the pointer mechanism. Packet a), because it includes in its payload the first byte Section 1, has its Payload Unit Start Indicator (PUSI) flag set, and the first byte following the packet header (or following the adaptation field, if there is one) will be the pointer_field. In this case, the pointer_field is set to zero because the first byte of the table section is the first byte of the packet payload.

Packet b)'s payload is just the continuation of Section 1, so the PUSI flag is cleared and no pointer_field is present. Packet c) shows the case where the payload includes the start of a table section, Section 2, N bytes following the pointer_field; hence the value of the pointer_field is N. In packet c) we see a case where an entire table section is included within the packet payload. We also see a case where a new table section starts that is not pointed to directly by the pointer_field. When the TS packet is processed, the starting point of Section 3 should be derived by looking at the byte just following the end of Section 2. If the byte is not 0xFF, it is interpreted as the table_id field of a new table section. Otherwise, it indicates that the remainder of the packet payload is stuffing bytes that are to be discarded.

Packet d) shows the case where a table section ends somewhere within the packet payload, and a new table section does not directly follow. Important note: in the case where stuffing bytes are used in this way, since the packet payload does not include the start of a new section, the PUSI is cleared and no pointer_field is sent.

Transport Stream Packet Stuffing Bytes

A table section may not end at the last byte of a TS packet (in fact, chances of that are very good). It is not necessary for a new table section to directly follow the last byte of the CRC of the previous table section. Instead, the packet can be padded out with stuffing bytes. A table ID value of 0xFF is reserved for this purpose.

On the packet multiplexer side, as table sections are packetized for insertion in to the transport multiplex, it may occur that as TS packets are being filled no table

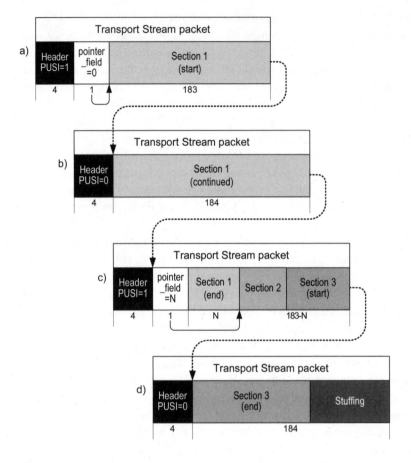

Figure 3.9 Example of Pointer Mechanism for Section Encapsulated Data

section is ready for insertion. The multiplexer may decide to fill out the rest of a packet payload with stuffing bytes rather than delay insertion of the packet containing the end of the last available table section.

If a table section that had started in a previous packet ends within a subsequent packet payload and no new table section directly follows, the Payload Unit Start Indicator is *not* set for that TS packet (because the payload does not contain the start of a new table section). This scenario is the one illustrated in Figure 3.9 d) above.

However, if stuffing bytes are used the PUSI flag may still be set (and hence the pointer_field will be present). Figure 3.10 below shows a couple of situations where a packet payload includes both the start of a new message as well as stuffing bytes.

The examples show one whole section in the packet payload; there could just as well be more than one.

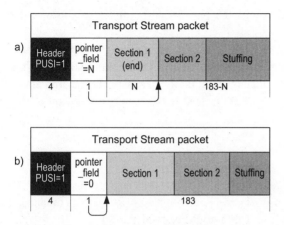

Figure 3.10 Examples of Pointer Mechanism with Stuffing Bytes

On the receiving side, as TS packets are parsed, the location of the CRC-32 field (and hence the end of the table section) is determined by processing the section_length field. If the packet payload has additional bytes, the byte following the CRC-32 field is checked. If that byte is anything other than 0xFF, it is considered to be the table ID of a new table section and subsequent bytes are interpreted as the header portion of this new section. If on the other hand the byte is 0xFF, packet payload stuffing is indicated and all remaining bytes in the TS packet payload are discarded. Processing begins again when a TS packet with the PUSI bit is set, at the byte indicated by the pointer_field.

Program Specific Information (PSI)

As we have noted, the MPEG-2 architects wished to define a packet multiplex in which more than one audio/video/data program could be present. To support multiple-program Transport Streams, they added another layer to the systems design whose function it was to establish the necessary groupings and associations among the component parts of each service. Even for single-program Transport Streams, a decoder needs to be able to quickly find the PID values—and other information necessary for audio, video, and data demultiplexing and presentation. To fulfill these needs, MPEG defined a set of tables called Program Specific Information (PSI) and used the table section format to carry the PSI within the transport multiplex.

The most fundamental of the PSI tables is the Program Association Table (PAT), which in essence provides pointers in the form of PID and program_number values to one or more sections of the Program Map Table (PMT). Each PMT section lists the program elements, including elementary streams that make up the program, and the PID values associated with TS packets carrying those audio, video, and/or data program elements.

Other tables defined in MPEG-2 *Systems* include the Conditional Access Table (CAT), used to indicate PID values of the elementary streams that are used for delivery of conditional access data in the Transport Stream, and the Transport Stream Description Table (TSDT), used for descriptors relevant to the entire Transport Stream.

Program Association Table (PAT)

PID value 0x0000 is reserved for TS packets carrying sections of the Program Association Table or PAT. There is at most one PAT per Transport Stream. As a large Program Association Table could exceed the table section size limit of 4096 bytes, the PAT can be sectioned for delivery.

As mentioned, the PAT provides pointers, in the form of PID and program_number values, to one or more Program Map Table sections also carried in the Transport Stream. In some systems the PAT also identifies the PID value associated with TS packets carrying the Network Information Table (NIT). MPEG-2 *Systems* does not specify the syntax and semantics of the NIT. Furthermore, only some of the SI standards in practice worldwide use the NIT pointer mechanism. Most place the base SI tables within TS packets with well-known PID values. For example, the ATSC system uses PID value 0x1FFB, while DVB SI uses PID values in the 0x0010 through 0x001F range, depending upon the type of table.

One might wonder why the program_number value is included with the PID as the pointer to an instance of a Program Map Table section when the PID by itself should suffice. The program_number is needed because MPEG-2 *Systems* specifies that two different PMT sections (describing different programs) can share the same PID value, so the PID by itself could be ambiguous. In this situation, the program_number is used to distinguish between the two. Decoder implementations must process and check the program_number field in the TS_program_map_section() to ensure they have found the one of interest.

Another important function of the PAT is that it provides the identification for the Transport Stream itself in the context of a "network" (MPEG-2 does not define the term). Transport Stream IDs are discussed in detail below.

Figure 3.11 describes the basic structure of the Program Association Table section.

As shown, the table section conforms to the standard long-form structure, with the transport_stream_id field appearing in the position of the table_id_extension. Note that the section length is limited to 1024 bytes.

The body of the table (the portion following the standard header fields) consists of one or more program data blocks. The number of program data blocks is deter-

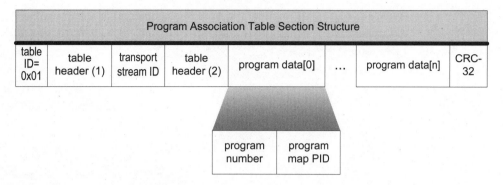

Figure 3.11 Structure of the Program Association Table Section

mined indirectly by the table's section_length field. One can process the PAT section until the section_length indicates that only four bytes are left. Those must be the CRC.

Each program data block consists of a program_number and a program_map_PID field. The program_number field associates this program data with a specific section of the PMT, which is carried in TS packets identified with the PID value given in program_map_PID.

Program Map Table (PMT)

The Program Map Table (PMT) perhaps ranks a close second for the most important type of table defined by MPEG-2 *Systems*. Each PMT section defines a programming service in terms of the component parts making up that service, and gives the types of each stream along with the Transport Stream PID values used to transport them in the packet multiplex. The PMT section syntax provides powerful flexibility in that it can include one or more descriptors pertinent to the program as a whole or to specific program elements comprising the service. Both MPEG-2 *Systems* and the ATSC Digital Television System Standards have defined several descriptors for carriage in the PMT section. We discuss these here and in later chapters in this book.

In accordance with MPEG-2 definitions and restrictions, any given program definition must be able to be described by a single PMT section having a maximum size of 1024 total bytes. In MPEG-2 *Systems*, the name of the data structure for the PMT section is TS_program_map_section().

Figure 3.12 describes the basic structure of the PMT section.

The PMT section conforms to the standard long-form table syntax. In the position of the table_id_extension field, the PMT section has the MPEG-2 program_number. Other examples of the use of the table_id_extension were listed in Table 3.8.

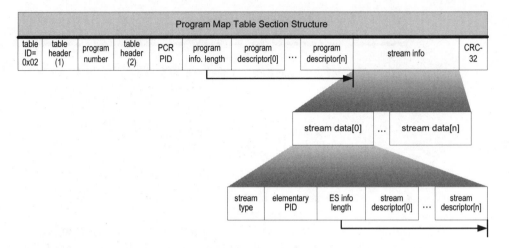

Figure 3.12 Structure of the Program Map Table Section

Referring to Figure 3.12, the PMT section can be considered to be structured in three parts. First, it identifies the PID value that contains the Program Clock Reference for this program. The next part consists of the program_info_length field followed by zero or more descriptors. These descriptors are sometimes called the "outer" descriptor loop, though more precisely they are called the program information descriptors. Descriptors placed in this position in the PMT section pertain to the program as a whole. Descriptors pertinent only to specific program elements are placed into the so-called "inner" loop—the descriptors following the ES_info_length field.

The third part of the PMT section is composed of one or more blocks of stream data, as it is shown in Figure 3.12. Stream data includes a field called stream_type which identifies the "type" of the Elementary Stream, a field identifying the PID value of the TS packets that carry this stream, and zero or more descriptors pertinent to the particular program element (these are the "inner loop" descriptors).

As mentioned, MPEG allowed for several different PMT sections to be placed into TS packets with the same PID value. In Annex C of A/53B[14], the ATSC terrestrial broadcast standard specifies several constraints on the MPEG-2 Transport Stream coding. Among others, these two rules are listed:

1. Only one definition (corresponding to only one TS_program_map_section) shall be allowed in TS packets of the same PID value. That is, there shall be a different program_map_PID value for each program described in the Transport Stream.

2. For terrestrial broadcast applications, the PMT section shall be the only PSI table contained in TS packets of a program_map_PID.

For the US cable application, SCTE 54 *Digital Video Service Multiplex and Transport System Standard for Cable Television*[18] specifies constraints on the use of the PMT for cable applications. SCTE 54[18] does not limit the construction of the Transport Stream to just one PMT section per PID. Furthermore, it states: "Private table sections in addition to Program Map Tables may be present in TS packets of the PMT_PID."

PMT sections

As mentioned, the Program Map Table is a "sectioned" table, but it doesn't fit the general pattern where the section_number and last_section_number fields in the header provide "part M of N" tags for each part. The MPEG-2 systems architects recognized that each PMT section needed to be able to be transported in TS packets identified with different PID values. They also decided to restrict the definition of any given program to just one TS_program_map_section(). Therefore, the section_number and last_section_number fields in the TS_program_map_section() are constrained to be zero.

Figure 3.13 compares the sectioning method applicable to PSIP tables with the sectioning used by the PMT. The example shows a PMT consisting of three TS_program_map_section() instances, each delivered in TS packets using a different value of PID. The full PMT for this Transport Stream is considered to be all three TS_program_map_section() instances taken together.

In the example, a PSIP table is sectioned for delivery into part 1 of 3, 2 of 3, and 3 of 3. Each section is delivered in TS packets identified with the same PID value (M). The three PSIP table sections taken together are considered the full PSIP table.

Another related difference between the PMT sections and sections of PSIP tables is that the version_number field among all the parts of a sectioned PSIP table has a common (coordinated) value, where the version_number field of each TS_program_map_section() is independent from the others. This difference again has to do with the practical needs of re-multiplexing equipment.

Conditional Access Table (CAT)

Not all television services can be received for free, of course. When a digital service is delivered in a scrambled form it must be possible for a receiving device to locate in the Transport Stream the information needed to 1) know if it is authorized to descramble (and if not, why not); and 2) actually descramble the ES components (audio, video, data) themselves. We now discuss the Conditional Access Table, which addresses the first of these requirements.

A brief review of the basic principles of conditional access may be helpful here. First, what exactly is "conditional access" anyway? It is the concept that not all devices capable of receiving the digital signal may be allowed to decode it. Or, it

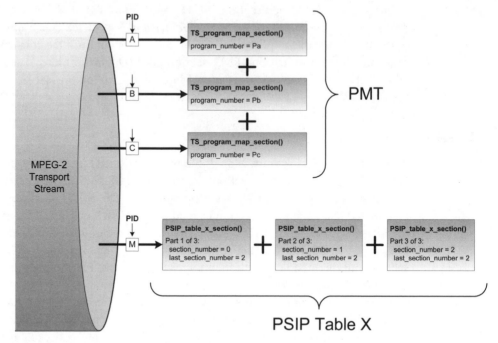

Figure 3.13 PMT Sectioning vs. General PSIP Table Sectioning

may be that viewing is possible if the user agrees to allow the signal provider (a cable or satellite system operator, for example) to charge his or her account a certain agreed-to amount. Certain channels can be designated "premium channels" and can be made available only to users who have agreed to pay the extra monthly charges for them.

Two distinct mechanisms regulate which receiving devices are allowed to decode and which are not. These two mechanisms may be viewed as *access rights* and *access requirements*. A decoder is given the rights to access certain programming based on the subscription and billing arrangements that have been made with the system operator/program provider.

In some cable systems, for example, channels are arranged in groupings called "tiers." One group, called the "basic tier" is available to those who pay a nominal monthly fee, while another group, called the "premium tier" is available to those who pay a higher charge. A pay-per-view (PPV) system can augment tier-based access control so that a program that is part of a programming tier not authorized for descrambling by subscription may be available if the user agrees to a one-time charge for viewing.

Access rights are granted to an individual decoder based on that subscriber's choice of programming packages. They come in the form of an individually-addressed message called an Entitlement Management Message or EMM. An

EMM delivers the cryptographic keys needed in the descrambling process and establishes exactly which tier or tiers of programming that particular descrambler is authorized to access. Another aspect of access rights relates to the pay-per-view system: a given decoder may or may not be granted access to PPV programming. On the other hand, if it is granted access, it may be limited to certain prescribed spending or purchase-count limits.

To have any possibility of decoding a scrambled program, a decoder must be equipped with the conditional access system used by that program provider. Digital cable-ready retail devices for sale in the United States, for example, may be equipped to support a Point of Deployment (POD) module. When a consumer arranges for cable service, they can ask the cable company to send a POD module for use with that device.

The term "point of deployment" is appropriate because the conditional access system embodied in the POD module is placed into service when and where the user connects the retail device into the cable system. This is in contrast to traditional cable systems in which the conditional access system is an inseparable part of the set-top box a cable operator has purchased from an equipment vendor. With the POD module concept, a consumer can buy a digital TV, bring it home and plug in a POD module supplied by cable operator A and enjoy premium programming. If that consumer moves to a different neighborhood or city, they can return the old POD module and replace it with a module supplied by cable operator B.

When a decoder equipped with conditional access support and access rights provided by an EMM encounters a program of interest, it must check that program's access requirements. Access requirements represent those requirements that must be met before descrambling/decoding is permitted. Examples of access requirements can include things like "must be a member of the basic programming tier" or "available on pay-per-view for a cost of $1.95." A program's access requirements, plus the cryptographic keys required for descrambling that program, are delivered in a message called an Entitlement Control Message or ECM.

These simple examples of EMMs and ECMs are intended to be illustrative only. Within the context of any proprietary CA system, much flexibility is possible and a wide variety of conditional access-related features can be implemented.

Nearly all conditional access systems are based on proprietary techniques, meaning that a technology license is needed in order to legally sell devices. It is pretty much a prerequisite that one must employ technology or algorithms protected under intellectual property laws when constructing a CA system because pirates will always try to build devices that circumvent the system. If an unlicensed (pirate) device uses a patented technique or algorithm (which, if all goes well, they will be forced to do) that device is breaking intellectual property laws. The product can then be taken off the market and legal action brought against the manufacturer.

With that background, what is the Conditional Access Table? It is the means by which a decoder can find the PID values associated with TS packets that carry Entitlement Management Messages which are relevant to the CA system or systems it supports. Structurally, as can be seen in Figure 3.14, the CAT is simply a container for descriptors related to the CA function. Just how many descriptors are present is given indirectly by section_length.

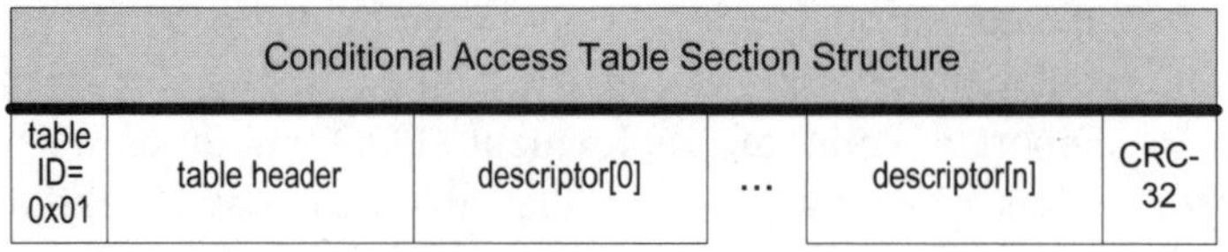

Conditional Access Table Section Structure					
table ID= 0x01	table header	descriptor[0]	...	descriptor[n]	CRC-32

Figure 3.14 Structure of the Conditional Access Table Section

Transport Stream Description Table (TSDT)

In 1996, a couple of years after the original MPEG-2 *Systems* was completed, the MPEG committee created Amendment 3 to the Standard and created the Transport Stream Description Table or TSDT. While the descriptor mechanism in place was already very powerful and provided a great deal of flexibility to the MPEG-2 architecture, up until the advent of the TSDT it wasn't possible to include a descriptor in one place in the Transport Stream that applied to all the programs in the multiplex. So the TSDT simply functions as a container structure in which descriptors can be placed that apply to the entire Transport Stream.

We won't dive into the details of the TS_description_section here because it is currently not mentioned in the ATSC Standard or by SCTE in the US. Structurally, it is exactly like the CAT. Its table_id value is 0x03, and its length is limited to a total of 1024 bytes. It may be sectioned if needed. PID 0x0002 is reserved for the exclusive use of the TSDT.

Relationship of PAT and PMT

Let's bring together the concepts of the Program Association Table and the Program Map Table with an illustration. Figure 3.15 shows a Transport Stream carrying two MPEG-2 programs identified as program_number 1 and program_number 2.

In the illustration, the full Transport Stream is depicted as a large pipe. Emanating from that pipe are streams composed of TS packets with common PID values. Specific PID values are labeled in the small rectangular boxes. Each of the two MPEG-2 programs in the example is also depicted as a pipe to show that a program is a grouping of related streams. In the example, the first program is composed of three Elementary Stream components: an MPEG-2 video stream, an English audio

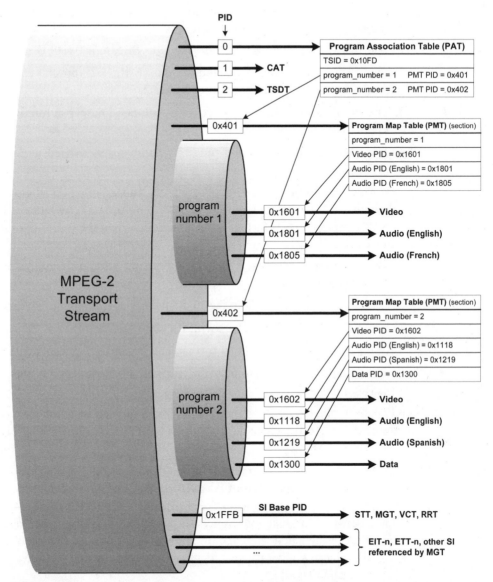

Figure 3.15 Transport Stream Showing the Relationship Between the PAT and PMT

track and a French audio track. The second program has four program elements: the video stream, English and Spanish audio tracks, and a data component.

Now let's look at how the Program Association Table lends the needed organization to what would otherwise be an unmanageable mess of TS packets. As shown, the PAT is carried in TS packets with PID 0x0000. For a decoder wanting to make sense of the packet multiplex, this is the place to start.

Collection of the table section from PID 0x0000 packets yields the Program Association Table shown in the figure. Inspection of the PAT shows several things:

1. The Transport Stream ID for this Transport Stream is 0x10FD.

2. This Transport Stream includes two MPEG-2 programs, identified as program_number 1 and program_number 2.

3. The PMT section describing program_number 1 is carried in TS packets with PID 0x0401.

4. The PMT section describing program_number 2 is carried in TS packets with PID 0x0402.

If we're interested in the first program, we can set our PID filter on 0x0401 and collect the PMT section identified as describing program_number 1. Inspection of this PMT section yields further information regarding the structure of this programming service:

1. The program is composed of three Elementary Stream components: one video and two audio tracks.

2. One of the audio tracks is in the English language, the other is French.

3. Video is carried in TS packets with PID 0x1601.

4. English audio, in Dolby AC-3 format, is carried in TS packets with PID 0x1801.

5. French audio, also in Dolby AC-3 format, is carried in TS packets with PID 0x1805.

6. The Program Clock Reference can be found in the adaptation field of TS packets with PID 0x1601 (which happens, as is typical, to be the video ES component of this program).

We're now ready to instruct our video decoder to begin collecting PES packets from TS packets labeled with PID 0x1601. We set up our clock recovery circuit to look for PCR timestamps in packets with this same PID value. We can make a choice between English and French audio either by asking the user directly or by using a pre-established language preference. If we choose French, we program the audio decoder to collect packets identified with PID value 0x1805 from the multiplex and begin decoding.

Figure 3.15 also indicates that PID 0x0001 carries the Conditional Access Table and PID 0x0002 carries the Transport Stream Description Table. PSIP tables are carried on the SI base_PID, 0x1FFB. Other PSIP tables such as Event Information and Extended Text Tables may be present as well.

PAT Syntax and Semantics

Table 3.9 describes the syntax and semantics of the Program Association Table section. Its formal name in MPEG-2 *Systems* is program_association_section().

TABLE 3.9 Syntax of the MPEG-2 Program Association Table Section

Field Name	Number of Bits	Field Value	Description
program_association_section() {			Start of the program_association_section().
table_id	8	0x01	Identifies the table section as a section of the Program Association Table.
section_syntax_indicator	1	1b	Indicates the section is formatted in MPEG "long-form" syntax.
'0'	1	0b	MPEG-2 *Systems* defines this bit as 0b for the program_association_section().
reserved	2	11b	Reserved bits are set to 1.
section_length	12	<=1021	An unsigned integer that specifies the length, in bytes, of data following the section_length field itself to the end of this table section. For the program_association_section(), section length cannot exceed 1021 bytes.
transport_stream_id	16		TSID value for the Transport Stream.
reserved	2	11b	Reserved bits are set to 1.
version_number	5		The version number reflects the version of a table section, and is incremented by one (modulo 32) when anything in the table changes.
current_next_indicator	1		Indicates, whether the table section is currently applicable (value 1) or is the next one to be applicable (value 0).
section_number	8		Indicates which part this section is when a PAT is segmented into sections.
last_section_number	8		Indicates the number of the last section is when a PAT is segmented into sections.
for (i=0; i<N; i++) {			Start of the program data loop. The value of N is given indirectly via the section_length field. Each iteration of the "for" loop is four bytes in length, so the value of N is equal to (section_length - 9) divided by 4.
program_number	16		Specifies the number to which program_map_PID is applicable. Value zero means that the PID specified applies to the Network Information Table (not used in ATSC/SCTE).
reserved	3	111b	Reserved bits are set to 1.
if (program_number == 0){			

TABLE 3.9 Syntax of the MPEG-2 Program Association Table Section (continued)

Field Name	Number of Bits	Field Value	Description
network_PID	13		The PID value associated with TS packets carrying the Network Information Table (not used in the ATSC standard).
} else {			
program_map_PID	13		The PID value associated with TS packets used to carry sections of the Program Map Table for the program given in program_number.
}			
}			End of the program data loop.
CRC_32	32		A 32-bit CRC value that gives a zero result in the registers after processing the PAT section.
}			End of the program_association_section().

section_length

Note that MPEG-2 *Systems* limits the maximum size of any program_association_section() to 1024 bytes. Therefore the section_length field cannot exceed 1021 (0x3FD).

transport_stream_id

The Program Association Table gives the Transport Stream ID (TSID) associated with the Transport Stream carrying the table. Transport Stream ID is a 16-bit identifier for the Transport Stream that is specified to be unique throughout the network. It is illegal for the Program Association Table to describe anything other than the current Transport Stream. Only one PAT can appear in any given Transport Stream, and that PAT gives the value of that Transport Stream's TSID. It is essential, however, to keep track of the value of the TSID whenever PSI data extracted from a given Transport Stream is stored for later reference and use.

TSID values are unique throughout a "network." MPEG does not provide a definition of the term "network," but in practical terms it means the set of Transport Streams that can be received by any one receiver. Therefore, all of the Transport Streams coming in to a digital cable set-top box must each have unique TSID values. For terrestrial broadcasting, since there are many receivers each with a potentially different set of signals that can be received and since service areas overlap, TSID values have to be unique on a regional basis. North America is one example of a region, Taiwan is another.

As of this writing, Maximum Service Television (MSTV), a national association of local television stations (visit http://www.mstv.org) maintains the current list of

TSID values for digital broadcasting in the US. A current link to this table can be found by navigating the ATSC website, http://www.atsc.org. If management of the table were to change, the ATSC link would likely be updated accordingly.

program_number

Every program in the Transport Stream must have a unique MPEG-2 program_number. In some systems program_number may be used as a user channel number (such usage may be occurring in some DVB systems in Europe), but in the US, program_number is not intended for consumer use. Its primary purpose is as the link between a program identified in the PAT and an instance of a PMT section describing the program elements making up that program.

network_PID

MPEG-2 architects envisioned that a Transport Stream might carry information about services and about the network itself, and indeed it does. In the spirit of flexibility they established in the PAT a pointer mechanism referring to the PID value of TS packets that would carry a table they called the Network Information Table or NIT. MPEG-2 *Systems* does not specify syntax or semantics for the NIT, but the Standard does indicate it would carry parameters related to the physical network such as carrier frequencies and (for satellite delivery) transponder numbers.

In none of the three dominant international digital television standards, DVB, ATSC, and ARIB, is the network_PID pointer used. Instead, SI standards have chosen to use well-known PID values to carry the tables giving physical network parameters. DVB, for example, uses PID 0x0010. ATSC uses PID 0x1FFB. The advantage of hard-coded values for SI PIDs is that acquisition time is decreased. Instead of having to look first for the PAT in PID 0 to find the value of the network PID for the Transport Stream, receivers can begin looking immediately for TS packets matching the well-known PID.

program_map_PID

This PID value gives the Packet Identifier for TS packets carrying the PMT section for this program. In some systems PMT sections describing different programs may be carried in TS packets with the same PID value. The program_number must always be used to identify the instance of the PMT section, even when the value of program_map_PID is unique in the PAT.

PAT repetition rate

Acquisition of the Program Association Table may be necessary for proper decoding of audio and video, therefore a fast repetition rate is recommended.

MPEG-2 *Systems* does not directly specify minimum or maximum repetition rates for the program_association_section(). It does specify that the average bitrate for "System data" shall not exceed 80,000 bps. System data is defined as combined data taken from TS packets with PID values 0x0000 (the PAT), 0x0001 (the CAT) and the PID value for the selected program (a PMT section).

The ATSC A/53B[14] *Digital Television Standard* and the SCTE cable transport specification, SCTE 54[18], both specify that section zero of the program_association_section() must be repeated every 100 milliseconds. The entire PAT, therefore, must repeat at a rate of ten repetitions per second or higher. The upper limit on the repetition rate is set by the MPEG-2 80 kbps bit-rate limit for System data.

Example PAT

Presented below in Table 3.10 is the bit-stream representation of an example PAT table section. This PAT is carried in a Transport Stream whose TSID value is 0x0E21. This PAT references three MPEG programs: the first is identified as program_number 0x1001 and the PMT section describing it can be found in TS packets with PID value 0x0060. The second program is program_number 0x1002, and its PMT section can be found in TS packets with PID value 0x0070. The third and last program is program_number 0x1003 and its PMT section may be found in TS packets with PID value 0x0080. Note that this example PAT does not indicate a value for network_PID (the ATSC Standard states no requirement for the PAT to identify a network_PID value).

TABLE 3.10 Program Association Table Example Bit-stream Representation

Field Name	Value	Description
table_id	0x00	Identifies the table section as being a PAT.
section_syntax_indicator	1b	The PAT uses the MPEG "long-form" syntax.
private_indicator	0b	Set to 0 in the PAT.
reserved	11b	Reserved bits are set to 1.
section_length	0x015	Length of rest of this section is 21 bytes.
transport_stream_id	0x0E21	TSID.
reserved	11b	Reserved bits are set to 1.
version_number	01101b	Version number is 13.
current_next_indicator	1b	Section is current PAT.
section_number	0x00	First section.
last_section_number	0x00	Last is first (only one section).

TABLE 3.10 Program Association Table Example Bit-stream Representation (continued)

Field Name	Value	Description
program_number	0x1001	First program is program_number 0x1001.
reserved	111b	Reserved bits are set to 1.
program_map_PID	0x0060	PMT section PID is 0x0060.
program_number	0x1002	Second program is program_number 0x1002.
reserved	111b	Reserved bits are set to 1.
program_map_PID	0x0070	PMT section PID is 0x0070.
program_number	0x1003	Third program is program_number 0x1003.
reserved	111b	Reserved bits are set to 1.
program_map_PID	0x0080	PMT section PID is 0x0080.
CRC_32	0xE088DC30	MPEG-2 CRC (example).

PMT Section Syntax and Semantics

Table 3.11 describes the table section syntax for the PMT section.

TABLE 3.11 Syntax of the MPEG-2 Program Map Table Section

Field Name	Number of Bits	Field Value	Description
TS_program_map_section() {			Start of the TS_program_map_section().
table_id	8	0x02	Identifies the table section as the Program Map Table.
section_syntax_indicator	1	1b	Indicates the section is formatted in MPEG "long-form" syntax.
'0'	1	0b	MPEG-2 *Systems* defines this bit as 0b for the TS_program_map_section().
reserved	2	11b	Reserved bits are set to 1.
section_length	12	<=1021	An unsigned integer that specifies the length, in bytes, of data following the section length field itself to the end of this table section. The maximum length of the full TS_program_map_section() is 1024 bytes.
program_number	16		The program_number ties this instance of the PMT with the corresponding MPEG-2 program listed in the PAT.
reserved	2	11b	Reserved bits are set to 1.
version_number	5		The version number reflects the version of a table section, and is incremented by one (modulo 32) when anything in the table changes.

TABLE 3.11 Syntax of the MPEG-2 Program Map Table Section (continued)

Field Name	Number of Bits	Field Value	Description
current_next_indicator	1		Indicates whether the table section is currently applicable (value 1) or is the next one to be applicable (value 0).
section_number	8	0x00	Must be set to zero for the PMT, as it is not allowed for an instance of the PMT to exceed a total of 1024 bytes in length.
last_section_number	8	0x00	Must be set to zero.
reserved	3	111b	Reserved bits are set to 1.
PCR_PID	13		Identifies the PID of TS packets which carry the Program Clock Reference values for this program.
reserved	4	1111b	Reserved bits are set to 1.
program_info_length	12	<1024	Gives the number of bytes taken up by the program information descriptors directly following the field.
for (i=0; i<N; i++) {			Start of the program information descriptors, the so-called "outer" loop. The value of N is determined indirectly by program_info_length. One can process successive descriptors until the total length of those processed equals program_info_length. Of course if program_info_length equals zero, no descriptors are present.
descriptor()			A descriptor, of type-length-data format, giving some information pertinent to the program as a whole.
}			End of the program information descriptors.
for (i=0; i<M; i++) {			Start of the program element "for" loop. The value of M is determined indirectly by section_length. Receiving devices are expected to determine they have reached the end of this "for" loop by keeping track of the position in the section. When the position reaches the CRC field, processing is complete.
stream_type	8		Indicates the type of ES being described in this iteration of the program element "for" loop. Receiving devices use stream_type to determine what kind of decoder is needed to process this stream.
reserved	3	111b	Reserved bits are set to 1.
elementary_PID	13		This is the PID value of the TS packets carrying the program element being described in this iteration of the "for" loop.
reserved	4	1111b	Reserved bits are set to 1.
ES_info_length	12	<1024	Indicates the total length of the ES info descriptors loop to follow. The two most-significant bits must be set to zero.

TABLE 3.11 Syntax of the MPEG-2 Program Map Table Section (continued)

Field Name	Number of Bits	Field Value	Description
for (i=0; i<P; i++) {			Start of the ES info descriptors, the so-called "inner" loop. The value of P is determined indirectly by ES_info_length. Receivers are expected to process descriptors until the total length of those processed equals ES_info_length. If ES_info_length equals zero, no descriptors are present.
descriptor()			A descriptor, of type-length-data format, giving some information pertinent to this program element.
}			End of the ES info descriptors.
}			End of the program element "for" loop.
CRC_32	32		A 32-bit CRC value that gives a zero result in the registers after processing the PMT section.
}			End of the TS_program_map_section().

section_length

The TS_program_map_section() can be at most a total of 1024 bytes in length, therefore the section_length field is limited to a maximum value of 1021 (0x3FD).

program_number

Each occurrence of a PMT section within a Transport Stream is known to be unique because each is identified with a unique value for program_number. Values for program_number are scoped to the Transport Stream, so that separate Transport Streams in a network may reuse program_number values. The combination of TSID and program_number uniquely identifies a programming service (MPEG-2 program) within a network, and it is this pair of coordinates that is used by the Virtual Channel Table as the reference mechanism.

A program_number, as mentioned, is not a number expected to be displayed on a consumer receiver. In any case, because program_number is a 16-bit field, it is quite possible that the values used are high numbers and therefore user-unfriendly.

PCR_PID

Each PMT section identifies the PID value for those TS packets that carry adaptation fields with samples of the Program Clock Reference (PCR). Decoders load the value of the PCR_PID into the clock recovery circuit, so those packets can be processed to extract clock reference samples and thereby establish and maintain synchronization with the time base used for decoding and presentation timing.

program_info_length

This field indicates the number of bytes of descriptors to follow. If no program-level descriptors are present, program_info_length is set to zero.

Program information descriptors

These descriptors pertain to the entire MPEG-2 program and not to just an individual program element. Each descriptor is in a standard format, which is an 8-bit descriptor_tag followed by an 8-bit length field, followed by zero or more bytes of data. Several descriptors are currently defined for use within the PMT section in the outer loop. Table 3.12 lists them.

TABLE 3.12 Common Program-Level Descriptors in the PMT

Descriptor Tag	Descriptor Name	Where Defined	Description
0x05	Registration	MPEG-2 *Systems*	In general, used to identify an entity supplying private data. As of this writing, for ATSC-compliant Transport Streams, the PMT section is required to include an MPEG-2 Registration Descriptor (MRD) containing the format_identifier value 0x47413934 ("GA94" in ASCII). A proposed revision to A/53B now making its way through committee would make the GA94 MRD optional. For cable Transport Streams, SCTE specifies the MRD may carry a format_identifier value of 0x53435445 ("SCTE" in ASCII).
0x09	Conditional Access	MPEG-2 *Systems*	The presence of the MPEG-2 Conditional Access (CA) Descriptor in the PMT section indicates that the program is scrambled—that is, subject to conditional access. The Conditional Access Descriptor indicates the PID values for TS packets where Entitlement Control Messages (ECMs) may be found.
0x0A	ISO 639 Language	MPEG-2 *Systems*	Identifies the language pertinent to the program as a whole. Typically, languages tags are placed at the ES level, for example to identify the language of an audio track.
0x80	Stuffing	A/65 PSIP	No function other than to make a placeholder of a given number of bytes. Stuffing bytes are disregarded by decoders.
0x87	Content Advisory	A/65 PSIP	Provides program content advisory information pertinent to the program.
0x88	ATSC Conditional Access	A/70	Indicates the program is subject to conditional access in accordance with ATSC A/70[15]. When it appears in the PMT, the ATSC CA Descriptor provides pointers to Entitlement Control Messages (ECMs) relevant to gaining access to the program via a conditional access system.
0xAA	Redistribution Control	A/65 PSIP	Used in conjunction with expected legislation defining content owners rights with regard to re-distribution of content.

stream_type

Identifies the type of stream being described in this iteration of the program element "for" loop. MPEG-2 *Systems* defines nearly two dozen standard stream types. Table 3.13 lists those stream types plus types standardized in ATSC and/or SCTE.

TABLE 3.13 Stream Type Codes

Value	Type	Notes
0x00	ITU-T | ISO/IEC Reserved	MPEG-reserved.
0x01	ISO/IEC 11172-2 Video	MPEG-1 Video[4].
0x02	ISO/IEC 13818-2 Video or ISO/IEC 11172-1 constrained parameter video stream	MPEG-2 Video[5].
0x03	ISO/IEC 11172-3 Audio	MPEG-1 Audio[19].
0x04	ISO/IEC 13818-3 Audio	MPEG-2 Audio[7].
0x05	ISO/IEC 13818-1 private sections	The long- or short-form private section syntax defined in MPEG-2 *Systems*.
0x06	ISO/IEC 13818-1 PES packets containing private data	A/90 uses this stream type to carry PES packets containing stream-synchronized data.
0x07	ISO/IEC 13522 MHEG	
0x08	ISO/IEC 13818-1 DSM-CC[6], Annex A	
0x09	ITU-T Rec. H.222.1	
0x0A	ISO/IEC 13818-6 DSM-CC[6] Type A	DSM-CC Multi-protocol Encapsulation (not the type used in the ATSC A/90 *Data Broadcast Standard*[23]).
0x0B	ISO/IEC 13818-6 DSM-CC[6] Type B	Used by the ATSC A/90 Download Protocol for asynchronous data.
0x0C	ISO/IEC 13818-6 DSM-CC[6] Type C	DSM-CC Stream Descriptors.
0x0D	ISO/IEC 13818-6 DSM-CC[6] Type D	DSM-CC Sections (any sections, including private data). Used by ATSC A/90 for use in the DSMCC_addressable_section.
0x0E	ISO/IEC 13818-1 auxiliary	A stream type that is MPEG-2 *Systems*-defined data but is not audio, video, PSI, or DSM-CC. Examples of stream_type 0x0E data include some data structures associated with the MPEG-2 Program Stream (not used in ATSC standards).
0x0F	ISO/IEC 13818-7 Audio with ADTS transport syntax	MPEG-2 *Systems*.
0x10	MPEG-4 Audio Elementary Stream	MPEG-2 PES-based.

TABLE 3.13 Stream Type Codes (continued)

Value	Type	Notes
0x11	MPEG-4 Visual Elementary Stream	MPEG-2 PES-based.
0x12	MPEG-4 SL-packetized Stream or FlexMux Stream in MPEG-2 PES	Defined in MPEG-4 *Systems*, ISO/IEC 14496.
0x13	MPEG-4 SL-packetized Stream or FlexMux Stream in MPEG-4	Defined in MPEG-4 *Systems*, ISO/IEC 14496.
0x14	ISO/IEC 13818-6 DSM-CC Synch. Download Protocol	Used in ATSC A/90. Sections containing non-streaming synchronized data.
0x15-0x7F	ISO/IEC 13818-1 Reserved	MPEG-2 *Systems*.
0x80	Same as ISO/IEC 13818-2 Video	Defined in SCTE 54[18] as being the same as MPEG-2 Video.
0x81	ATSC A/53 audio	Dolby AC-3 as defined in ATSC A/52A[8] and as constrained in ATSC A/53B[14] Annex B and Annex C.
0x82	SCTE Standard Subtitle	As specified in SCTE 54[18] and defined in SCTE 27 (formerly DVS 026)[24].
0x83	SCTE Isochronous Data	As specified in SCTE 54[18] and defined in ANSI/SCTE 19 (formerly DVS 132)[25].
0x84-0x85	Reserved by ATSC and SCTE for future use	
0x86	SMPTE Splice Info per SMPTE 312M[27]	
0x87-0x94	Reserved by ATSC and SCTE for future use	
0x95	ATSC Data Service Table, Network Resources Table	As specified in the ATSC A/90 *Data Broadcast Standard*[23].
0x96-0xC1	Reserved by ATSC and SCTE for future use	
0xC2	ATSC synchronous data stream	PES packets containing stream-synchronous data per the *ATSC A/90 Data Broadcast Standard*[23].
0xC3	SCTE asynchronous data stream	As defined in SCTE 53[26] (formerly DVS 051).
0xC4-0xFF	User private range	Requires use of an MPEG-2 Registration Descriptor to identify the entity quoting the user private value (see "User Private Stream Types" on page 301).

elementary_PID

Identifies the PID value associated with TS packets that carry the program element being described in this iteration of the program element "for" loop.

ES_info_length

Indicates the number of bytes of ES-level descriptors to follow. Even though it is a 12-bit field, the section length is limited to 1024 bytes; therefore ES_info_length is correspondingly limited.

ES info descriptors

Table 3.14 lists some types of descriptors that can appear at the Elementary Stream level in the PMT section.

TABLE 3.14 Elementary Stream-Level Descriptors in the PMT

Descriptor Tag	Descriptor Name	Where Defined	Description
0x05	MPEG-2 Registration	MPEG-2 *Systems*	Used to identify an entity supplying private data. In the ES-level descriptors loop of the PMT section, an MPEG-2 Registration Descriptor (MRD) identifies the entity that has supplied any private data that might appear in the program element. If the stream_type value is in the User Private range 0xC4-0xFF, the MRD indicates the private entity that has supplied that program element. For details, see Chapter 15, "Private Data" on page 295.
0x06	Data Stream Alignment	MPEG-2 *Systems*	Indicates the type of alignment present in the associated Elementary Stream. Types of alignment include slice or video access unit, group of pictures or sequence header.
0x09	Conditional Access	MPEG-2 *Systems*	The presence of the MPEG-2 Conditional Access (CA) Descriptor in the PMT section at the ES level indicates that the program element is scrambled; that is, subject to conditional access. The Conditional Access Descriptor indicates the PID values for TS packets where Entitlement Control Messages (ECMs) may be found. Note: typical CA systems treat all components of a service equally, so this descriptor does not normally appear at the ES level.
0x0A	ISO 639 Language	MPEG-2 *Systems*	Identifies the language pertinent to the associated program element.
0x80	Stuffing	A/65 PSIP	No function other than to make a placeholder of a given number of bytes. Stuffing bytes are disregarded by decoders.

TABLE 3.14 Elementary Stream-Level Descriptors in the PMT (continued)

Descriptor Tag	Descriptor Name	Where Defined	Description
0x81	AC-3 Audio	A/65 PSIP	Defines parameters relevant to AC-3 audio stream decoding, including the type of the audio stream.
0x86	Caption Service	A/65 PSIP	Appears in the ES information descriptor loop associated with the video service. Lists the caption services present.
0xA3	Component Name	A/65 PSIP	Provides a textual name for the given component. Required on audio Elementary Streams when there are multiple audio tracks for the same language and type, or when an audio track has no associated language.

PMT repetition rate

Acquisition of the Program Map Table section for a given program may be necessary for proper decoding of audio and video, therefore a fast repetition rate is recommended. We say "may be" because decoding may commence before the PMT section is recovered if the decoder has saved the PID values used the last time this Transport Stream was processed. These values may be used until the PMT section is recovered. If they are found to be in error, the decoders can be reset and the proper PID values used. Using cached PIDs often speeds up acquisition, and can only help—not hurt—performance.

Note that another source of audio and video PID values is the Service Location Descriptor found in the Terrestrial Virtual Channel Table. PID values from TVCT and PMT should be equivalent and whichever are found first can be used in acquisition.

MPEG-2 *Systems* does not directly specify minimum or maximum repetition rates for the TS_program_map_section(). Again, the average bitrate for System data (the combination of PAT, CAT and the selected PMT section) must not exceed 80,000 bps.

The ATSC A/53B[14] *Digital Television Standard* places no constraints on the repetition rate for the PMT section. For cable transport, SCTE 54[18] specifies that the TS_program_map_section() must be repeated every 400 milliseconds. The repetition rate of any given PMT section is therefore at least 2.5 per second on cable-compliant Transport Streams. As with the PAT, the upper limit on the repetition rate is set by the MPEG-2 *Systems* 80 kbps aggregate bit-rate limit for System data.

Example PMT section

Presented below in Table 3.15 is the bit-stream representation of an example PMT section for an MPEG-2 program with the following characteristics:

- the program_number value is 0xE007;
- the program consists of one video and two audio Elementary Streams;
- one audio stream is offered in English and the other is in French;
- the video is carried in TS packets with PID value 0x0031, the English audio ES is carried in TS packets with PID value 0x0034, and the French is carried in TS packets with PID value 0x0035.

A Content Advisory Descriptor is present indicating the program is rated for the US region (region_number 0x01) "TV-PG-V."

In the Figure, descriptors are shaded and heavier lines indicate the iterations of "for" loops and other significant data structure boundaries.

TABLE 3.15 Program Map Table Example Bit-stream Representation

Field Name	Value	Description
table_id	0x02	Identifies the table section as being a PMT section.
section_syntax_indicator	1b	The PMT section uses the MPEG "long-form" syntax.
private_indicator	0b	Set to 0 in the PMT section.
reserved	11b	Reserved bits are set to 1.
section_length	0x005D	Length of rest of this section is 93 bytes.
program_number	0xE007	As referenced in PAT.
reserved	11b	Reserved bits are set to 1.
version_number	00100b	Version number is 4.
current_next_indicator	1b	Data is applicable now (current).
section_number	0x00	First section.
last_section_number	0x00	Last is first (only one section).
reserved	111b	Reserved bits are set to 1.
PCR_PID	0x0031	PCR PID value is 0x0031.
reserved	1111b	Reserved bits are set to 1.
program_info_length	0x01F	31 bytes of program information descriptors.
descriptor_tag	0x05	Indicates descriptor is an MPEG-2 Registration Descriptor.
descriptor_length	0x04	Length is 4 bytes.
format_identifier	0x47413934	ISO Latin-1 code for "GA94".
descriptor_tag	0x87	Descriptor is a Content Advisory Descriptor.
descriptor_length	0x17	Length is 23 bytes following length byte itself.
reserved	11b	Reserved bits are set to 1.

TABLE 3.15 Program Map Table Example Bit-stream Representation (continued)

Field Name	Value	Description
rating_region_count	0x01	One region.
rating_region	0x01	Region 1, US + possessions.
rated_dimensions	0x02	Two dimensions in the US system are rated.
rating_dimension_j	0x00	First one is dimension 0 (TV rating).
reserved	1111b	Reserved bits are set to 1.
rating_value	0x3	Value is 3, meaning "TV-PG".
rating_dimension_j	0x04	Second one is dimension 4 (Violence).
reserved	1111b	Reserved bits are set to 1.
rating_value	0x1	Value is 1, meaning "V".
rating_description_length	0x0F	Fifteen bytes of MSS text describing the rating.
number_strings	0x01	One string.
ISO_639_language_code	0x656E67	Language is English (code "eng").
number_segments	0x01	One segment.
compression_type	0x00	No compression.
mode	0x00	Mode zero selects ISO Latin-1 coding.
number_bytes	0x07	Length of the string to follow.
compressed_string_byte[k]	0x54562D50 472D56	ISO Latin-1 encoding of "TV-PG-V".
stream_type	0x02	Stream type is MPEG-2 video.
reserved	111b	Reserved bits are set to 1.
elementary_PID	0x0031	Video is carried in TS packets with PID value 0x0031.
reserved	1111b	Reserved bits are set to 1.
ES_info_length	0x0000	No descriptors pertinent to the video ES are present.
stream_type	0x81	Stream type is AC-3 audio.
reserved	111b	Reserved bits are set to 1.
elementary_PID	0x0034	This audio ES is carried in TS packets with PID value 0x0034.
reserved	1111b	Reserved bits are set to 1.
ES_info_length	0x0011	17 bytes of descriptors pertinent to this audio ES.
descriptor_tag	0x05	Indicates descriptor is an MPEG-2 Registration Descriptor.
descriptor_length	0x04	Length is 4 bytes.
format_identifier	0x41432D33	ISO Latin-1 code for "AC-3".
descriptor_tag	0x81	Indicates AC-3 Audio Stream Descriptor.

TABLE 3.15 Program Map Table Example Bit-stream Representation (continued)

Field Name	Value	Description
descriptor_length	0x03	The descriptor is three bytes long.
sample_rate_code	000b	A/53 Annex B constrains the sample rate to 48 kHz, so this field must be set to zero.
bsid	01000b	Bit stream identification. Set to 8 for the current version of this standard.
bit_rate_code	000010b	Exact bitrate code of 48 kbps.
surround_mode	10b	Surround mode encoded.
bsmod	000b	Bit stream mode value zero indicates the audio stream is a Complete Main audio.
num_channels	1101b	Indicates 5.1 channel encoding.
full_svc	1b	Indicates the audio stream is a full service.
descriptor_tag	0x0A	Indicates ISO 639 Language Descriptor.
descriptor_length	0x04	The descriptor has four bytes following this field.
ISO_639_language_code	0x656E67	"eng" coded in ISO 8859-1 (Latin-1).
audio_type	0x00	Undefined audio type.
stream_type	0x81	Stream type is AC-3 audio.
reserved	111b	Reserved bits are set to 1.
elementary_PID	0x0035	This audio ES is carried in TS packets with PID value 0x0035.
reserved	1111b	Reserved bits are set to 1.
ES_info_length	0x0011	17 bytes of descriptors pertinent to this audio ES.
descriptor_tag	0x05	Indicates descriptor is an MPEG-2 Registration Descriptor.
descriptor_length	0x04	Length is 4 bytes.
format_identifier	0x41432D33	ISO Latin-1 code for "AC-3".
descriptor_tag	0x81	Indicates AC-3 Audio Stream Descriptor.
descriptor_length	0x03	The descriptor is three bytes long.
sample_rate_code	000b	A/53 Annex B constrains the sample rate to 48 kHz, so this field must be set to zero.
bsid	01000b	Bit stream identification. Set to 8 for the current version of this standard.
bit_rate_code	000010b	Exact bitrate code of 48 kbps.
surround_mode	10b	Surround mode encoded.
bsmod	000b	Bit stream mode value zero indicates the audio stream is a Complete Main audio.
num_channels	1010b	Used for 2.0 channel encoding.

TABLE 3.15 Program Map Table Example Bit-stream Representation (continued)

Field Name	Value	Description
full_svc	1b	Indicates the audio stream is a full service.
descriptor_tag	0x0A	Indicates ISO 639 Language Descriptor.
descriptor_length	0x04	The descriptor has four bytes following this field.
ISO_639_language_code	0x667265	"fre" coded in ISO 8859-1 (Latin-1).
audio_type	0x00	Undefined audio type.
CRC_32	0x1E8479FE	MPEG-2 CRC (example).

CAT Syntax and Semantics

Table 3.16 describes the syntax and semantics of the CAT table section, or CA_section().

TABLE 3.16 Syntax of the Conditional Access Table Section

Field Name	Number of Bits	Field Value	Description
CA_section() {			Start of the CA_section().
table_id	8	0x01	Identifies the table section as the Conditional Access Table.
section_syntax_indicator	1	1b	Indicates the section is formatted in MPEG "long-form" syntax.
'0'	1	0b	MPEG-2 *Systems* defines this bit as 0b for the CA_section().
reserved	2	11b	Reserved bits are set to 1.
section_length	12	<=1021	An unsigned integer that specifies the length, in bytes, of data following the section length field itself to the end of this table section. The maximum length of the full CA_section() is 1024 bytes.
reserved	18	0x3FFFF	Reserved bits are set to 1. Note that the table_id_extension field is reserved in the CA_section().
version_number	5		The version number reflects the version of a table section, and is incremented by one (modulo 32) when anything in the table changes.
current_next_indicator	1		Indicates, whether the table section is currently applicable (value 1) or is the next one to be applicable (value 0).
section_number	8		Gives the section number when the CAT is too big to fit into one table section. The first section is identified with section_number 0x00.

TABLE 3.16 Syntax of the Conditional Access Table Section (continued)

Field Name	Number of Bits	Field Value	Description
last_section_number	8		Gives the section number of the last section. For a CAT that fits into a single section, last_section_number must be 0x00.
for (i=0; i<N; i++) {			Start of the CA descriptors. The value of N is determined indirectly by section_length. Receivers are expected to process descriptors until the position within the section reaches the CRC_32 field.
descriptor()			A descriptor, of type-length-data format, giving CA information.
}			End of the CA descriptors.
CRC_32	32		A 32-bit CRC value that gives a zero result in the registers after processing the PMT section.
}			End of the CA_section().

A few things to note about the CA_section(): its length is limited to a total of 1024 bytes per section, but it is sectionable. It won't always be necessary for a decoder to handle a sectioned CAT or even the CAT itself. The CA provider indicates within the implementation details supplied with the license whether or not the CAT is required. As we have seen, it is possible in some systems to use hard-wired PID values for certain data elements. If the CAT is required, it may be that it is always short enough to fit into a single table section.

Conditional Access Table descriptors

Aside from possible private data, there is really only one descriptor used in the context of the CAT, and that is the Conditional Access Descriptor defined in MPEG-2 *Systems*. We look at the CA_descriptor() in detail below.

Note that CA_descriptor()s may be located within either the PMT or the CAT (or both). When it is located within a CAT, it contains EMM information. When it located within a PMT section, it contains ECM information.

Descriptors Defined in MPEG-2 *Systems*

The MPEG-2 *Systems*[1] Standard defines twenty-six descriptor types and partitions the remaining range of tag values between those that could be assigned in future versions of the Standard and those defined as "user private." In this context, user private means outside the scope of the MPEG-2 Standard. Different standards bodies throughout the world have extended MPEG in the user-private range.

Some of the descriptor types defined in MPEG-2 *Systems* are listed in Table 3.17.

TABLE 3.17 Some Descriptors Defined in MPEG-2 *Systems*

Descriptor Tag	Name or Note
0-1	Reserved for future use by MPEG
2	Video Stream Descriptor
3	Audio Stream Descriptor
4	Hierarchy Descriptor
5	Registration Descriptor
6	Data Stream Alignment Descriptor
7	Target Background Grid Descriptor
8	Video Window Descriptor
9	Conditional Access Descriptor
10	ISO 639 Language Descriptor
11	System Clock Descriptor
12	Multiplex Buffer Utilization Descriptor
13	Copyright Descriptor
14	Maximum Bit-rate Descriptor
15	Private Data Indicator Descriptor
16	Smoothing Buffer Descriptor
17	System Target Decoder (STD) Descriptor
18	IBP Descriptor
19-26	Defined in ISO/IEC 13818-6 DSM-CC
27-63	ISO/IEC 13818-1 Reserved
64-255	MPEG User Private

Only a very few of these are commonly used. The ones required or mentioned in the current terrestrial broadcast and cable standards for digital television are discussed below.

Video Stream Descriptor

Within the video_stream_descriptor() can be found various coding parameters used to enable the proper decoding of the video in the Elementary Stream associated with the descriptor. Parameters include the following:

- the video's frame rate code.

- an indication as to whether the video has a single frame rate, or may have multiple frame rates.

- a flag that can indicate the video is coded only according to MPEG-1 *Video*.

- a flag indicating whether only still picture data is present, or whether a mixture of still and moving picture data is present.

- the profile and level of the video coding (MPEG-2 *Video* defines profiles and levels to establish requirements for decoding of various resolutions and frame rates of video).

- flags providing information about color format and other decoding-related data.

Typically, video decoders operate just fine without the need for data provided in the Video Stream Descriptor. In fact, it probably would not be used anywhere were it not for the desire on the part of the SCTE Digital Video Subcommittee to allow a cable-compliant Transport Stream to include still picture data as well as regular video with moving pictures.

What is still picture data and why would anyone want to use it? Let's say a cable operator wishes to offer a number of music channels where the main component of interest on these channels was aural rather than visual. This operator certainly does not want to use a lot of Transport Stream bandwidth for the video component of these channels. Yet it would be nice if some kind of picture were to appear on-screen if a user's Digital TV were tuned to one of them.

Still pictures can be used like a "slide show." Once a frame has been decoded, in still picture mode the decoder holds it in view until the next picture is decoded. When not in still picture mode, a decoder may clear the screen to black if the video buffer underflows. However, in still picture mode a buffer underflow is not considered a decoding error or problem.

SCTE 54[18] specifies that cable Transport Streams may include still pictures. It states that if still pictures may be present in a video stream, the MPEG-2 Video Stream Descriptor shall be present to indicate that fact.

MPEG-2 Registration Descriptor (MRD)

MPEG recognized the need to support private data within a Transport Stream compliant with the MPEG standards. For every data element that might be present in the Transport Stream, a decoder must have a prior understanding as to that element's structure; in other words, its syntax and semantics. The purpose and function of the MPEG-2 Registration Descriptor (MRD) is to identify the entity that has provided data structures (tables or stream types for example) that are outside of those defined in the MPEG-2 standards.

ATSC and SCTE, being "user private" to MPEG, do conform to the MPEG-2 *Systems* guidelines and specify optional inclusion of an appropriate MRD in the Transport Stream.[*] The basic mechanism used in the MRD to identify the entity that has supplied the private data is the 32-bit "format identifier" code. ATSC uses the value 0x47413934, which is "GA94" in ASCII for this function. "GA" stands for Grand Alliance, and the "94" refers to the year it was formed, 1994. SCTE uses the value 0x53435445, or "SCTE" in ASCII coding.

Everything defined within the ATSC Digital Television system is, from the point of view of MPEG, "user private." But receivers deployed in the US will probably operate expecting that the Transport Streams they see comply with US standards. Multiplexes received by de-modulating an 8-VSB digital carrier are assumed to be compliant with the ATSC A/53[14] Standard for terrestrial broadcasting. Those received by de-modulating a 64- or 256-QAM signal on cable are expected to comply with SCTE's cable standards. So it is not clear that a receiver needs to care whether or not the "GA94" or "SCTE" MRD is found. The best course of action is to try to decode the components of the program in any case.

Another example of a standard use of the MRD is with Dolby AC-3 audio. Since the US standard doesn't use MPEG-2 audio coding but instead uses an audio coding that is "user private" from the point of view of MPEG, the ATSC and SCTE transport standards state that an MRD may be present in the audio ES_info loops with the format_identifier "AC-3" (value 0x41432D33) to indicate Dolby AC-3 audio coding. These audio streams also use a stream_type code (value 0x81) that is in the MPEG user-private range. Again, the MRD indicating AC-3 can be considered somewhat redundant as all the decoders in the US are built to recognize 0x81 as the Dolby AC-3 audio stream.[†]

Table 3.18 describes the syntax and semantics of the MPEG-2 registration_descriptor().

format_identifier

This 32-bit field is assigned by a Registration Authority designated by the ISO/IEC. The Society for Motion Picture and Television Engineers (SMPTE) in the US currently manages assignment of format_identifier codes to maintain worldwide uniqueness. You can find information at their web site: http://www.smpte-ra.org/mpegreg.html. As of this writing, thirteen organizations have registered format_identifier codes.

---

[*] The "B" revision of ATSC A/53 requires the use of the GA94 MRD, but as of this writing an amendment is being prepared for ballot to make this MRD optional.

[†] As with the GA94 MRD, the "B" revision of ATSC A/53 requires the AC-3 MRD, but an amendment is being prepared to make it optional.

TABLE 3.18 Syntax of the MPEG-2 Registration Descriptor

Field Name	Number of Bits	Field Value	Description
registration_descriptor() {			Start of the registration_descriptor().
descriptor_tag	8	0x05	Identifies the descriptor as an MRD.
descriptor_length	8		Gives the number of bytes following the descriptor_length field itself, to the end of the descriptor.
format_identifier	32		Identifies an entity that has registered with a Registration Authority (RA) sanctioned by ISO/IEC for MPEG specifications. SMPTE in the US is fulfilling the function of RA.
for (i=0; i<N; i++) {			Optional extra data bytes, often not used. The value of N is given by descriptor_length-4.
additional_identification_info			Some optional data defined by the assignee of the format_identifier value. According to MPEG-2 *Systems*, once assigned, these bytes must not change.
}			End of optional data.
}			End of the registration_descriptor().

Note that the Digital Video Broadcasting (DVB) standards use a different MPEG-2 *Systems* mechanism to identify private data. Instead of the MRD they use the MPEG-2 Private Data Indicator Descriptor for this purpose. It also has a 32-bit code for identification of the private entity, but MPEG-2 *Systems* states that MPEG will not manage assignment of values. DVB has managed assignment of private_data_indicator values in ETSI publication ETR 162[20].

Data Stream Alignment Descriptor

MPEG-2 *Systems* defines a Data Stream Alignment Descriptor that can be used in the inner loop of the PMT section to indicate what type of alignment is used for that particular program element. The type of alignment indicated in the descriptor reflects the data structure that is to be found in the PES packet just following a PES header containing a data_alignment_indicator flag whose value is set to '1.'

The types of alignment that can be identified with the Data Stream Alignment Descriptor include the video syntactic elements slice, video access unit, group of pictures header, and sequence header.

The "B" revision of the ATSC A/53[14] *Digital Television Standard* requires the Data Stream Alignment Descriptor to be present in the TS_program_map_section() inner loop for every video ES component. ATSC constrains the descriptor to indicate video access unit as the type of data stream alignment. Since A/53 also constrains each PES packet to begin with a video access unit, the requirement for the descriptor appears to be redundant (and not really useful to decoding equipment). A future revision to A/53 may make inclusion of the Data Stream Alignment Descriptor optional.

Conditional Access Descriptor

In our discussion of the Conditional Access Table (CAT) above we introduced the concepts of Entitlement Management Messages (EMMs) and Entitlement Control Messages (ECMs). We stated the CAT in PID 0x0001 is what tells a decoder where to find EMMs in the Transport Stream. As mentioned, the Conditional Access Descriptor is the only type of descriptor expected to appear in the CAT and its function is to point to the PID of the TS packets where EMM data can be found.

We also noted in the discussion of PMT descriptors that the Conditional Access Descriptor can be used to indicate the program is scrambled (i.e., subject to conditional access). When it appears in the Program Map Table, a Conditional Access Descriptor indicates the PID value of TS packets that carry ECM information for the program.

When descriptors are included in the PMT section, it is possible to place them at the program information (outer) loop or in one or more of the program element (inner) loops. At the program information level the descriptor applies to all program elements while a descriptor placed in the descriptor loop within one of the program elements applies only to that program element.

A Conditional Access Descriptor could appear in the program information loop if the entire program were to be subject to the conditional access provisions it described. Or it could be placed into one or more program element loops if certain program elements were subject to conditional access and others were not. But all of the CA systems in current use in the US, as far as is known, apply CA only to the program as a whole so the Conditional Access Descriptor is typically placed in the program information loop. Human factors considerations play a role in this choice—one expects to have access to all the audio, video, and data if one is a subscriber to that channel or is willing to accept the pay-per-view charges. Consumers would object if, after paying the $2.50 for a movie via PPV, another screen came up to say "Thank you for purchasing the video portion. If you would like audio, the charge is $2.00 more." Not quite so bad but also unpleasant would be a screen that said "Spanish language audio is available for an additional $1.00."

Now we look at the Conditional Access Descriptor in a bit more detail. Table 3.19 describes the syntax and semantics.

TABLE 3.19 Syntax of the Conditional Access Descriptor

Field Name	Number of Bits	Field Value	Description
CA_descriptor() {			Start of the registration_descriptor().
descriptor_tag	8	0x09	Identifies the descriptor as a CA_descriptor().
descriptor_length	8		Gives the number of bytes following the descriptor_length field itself, to the end of the descriptor.
CA_system_id	16		Identifies the CA system provider. See text.
reserved	3	111b	Reserved bits are set to 1.
CA_PID	13		When the CA_descriptor() appears in the PMT: the PID of TS packets which contain ECM data for the CA system identified in CA_system ID. When the CA_descriptor() appears in the CAT: the PID of TS packets which contain EMM data for the CA system identified in CA_system ID.
for (i=0; i<N; i++) {			Optional extra data bytes, often not used. The value of N is given by descriptor_length-4.
private_data_byte			Some optional data defined by the CA_system_id assignee.
}			End of optional bytes.
}			End of the CA_descriptor().

CA_system_id

This 16-bit number uniquely identifies a conditional access system provider. MPEG-2 *Systems* states that MPEG does not act as the Registration Authority for assignment of CA_system_id values. DVB has maintained a registry of CA_system_id values registered for use within the DVB system in ETSI *Technical Report 162*[20]. The 1995 release of ETR 162 listed about two dozen CA system operators, including US companies Scientific-Atlanta and General Instrument (now part of Motorola).

Practically speaking, whenever one wishes to build a decoder that is to be capable of descrambling services protected by conditional access, one must agree to the terms of a technology license. That license spells out the details of system operation, including the methods to be used to find and parse EMMs and ECMs. If MPEG-2 *Systems* methods such as the Conditional Access Descriptor are used, the license will indicate the value of CA_system_id to be used.

If at some time the ATSC A/70 conditional access standard for terrestrial broadcast is implemented, ATSC will likely coordinate registration of CA_system_id values. If they do, these values will almost certainly be harmonized with the assignments already made in ETR 162.

CA_PID

This 13-bit value identifies the PID of TS packets carrying EMM or ECM data. When the Conditional Access Descriptor is found in the PMT, the CA_PID indicates the PID value where ECMs may be found. When the descriptor is in the CAT, the CA_PID points to EMMs. Note that the data found in ECM and EMM TS packets is typically in the form of MPEG-2 private sections, but they could just as well be in any proprietary format. Again, the CA license gives the details needed to process these messages.

ISO 639 Language Descriptor

Audio tracks can be provided in any of a wide variety of spoken languages and even within a single language there are various types of audio. Two examples are audio for the visually impaired that includes descriptions of visual elements of the program, and audio for the hearing impaired in which all passages are available at increased volume. Some audio streams have no associated language at all. An example of this type of audio track is the ambient sounds of a sporting event.

The function of the ISO 639 Language Descriptor is to specify the language of the associated program element. For example, a language descriptor may be placed in the program element "for" loop in the PMT section to identify the language of an audio stream type. This descriptor's name comes from ISO 639-2[21], an international standard that associates three-letter codes with an exhaustive list of the world's languages. Table 3.20 describes the syntax and semantics of the ISO 639 Language Descriptor.

ISO_639_language_code

These three characters are defined in ISO 639-2[21] and consist of three ISO Latin-1[22] lowercase characters (note that all the characters used are in the ASCII range "a" to "z" or 0x61 to 0x7A). Inspection of the ISO 639-2 Standard shows that two different language code lists are defined, one for "Bibliographic" uses and one for "Terminology." The Bibliographic list is referred to as ISO 639-2/B while the Terminology list is ISO 639-2/T. Criteria used in derivation of the B list included:

- preference of the countries using the given language.

- established usage of the codes in national and international bibliographic databases.

- the vernacular or English form of the language.

All of the digital television standards use the bibliographic list. Some examples of ISO 639-2/B language codes are given in Table 3.21.

TABLE 3.20 Syntax of the ISO 639 Language Descriptor

Field Name	Number of Bits	Field Value	Description
ISO_639_language_descriptor() {			Start of the ISO_639_language_descriptor().
descriptor_tag	8	0x0A	Identifies the descriptor as an ISO_639_language_descriptor().
descriptor_length	8		Gives the number of bytes following the descriptor_length field itself, to the end of the descriptor.
for (i=0; i<N; i++) {			The value of N is given indirectly by descriptor_length; it is equal to descriptor_length divided by four.
ISO_639_language_code	24		Three lowercase ASCII characters, coded according to ISO 639-2[21], and representing a language.
audio_type	8		Indicates the type of an audio stream: 0x00 – Undefined 0x01 – Clean effects 0x02 – Hearing impaired 0x03 – Visual impaired commentary Note: in the US, the bsmod field in the Dolby AC-3 descriptor is used to indicate audio type and this field is set to zero.
}			
}			End of the ISO_639_language_descriptor().

TABLE 3.21 Some ISO 639-2/B Language Codes

Language	Language Code	Binary Encoding	Language	Language Code	Binary Encoding
Chinese	chi	0x636869	Polish	pol	0x706F6C
English	eng	0x656E67	Portuguese	por	0x706F72
French	fre	0x667265	Russian	rus	0x727573
German	ger	0x676572	Spanish	spa	0x737061
Greek	gre	0x677265	Swedish	swe	0x737765
Italian	ita	0x697461	Tagalog	tgl	0x74676C
Japanese	jpn	0x6A706E	Vietnamese	vie	0x766965

The table also shows the 24-bit field that results from encoding each of these three-character language codes. A receiving device can directly recognize the value 0x656E67 as representing English. The ISO 639-2 Standard states that the three-character codes were not intended to be an abbreviation of the language but instead a device to identify a given language or group of languages. Stated another way, the intended use for the codes was an identifier, not a text string to be displayed as-is.

Nevertheless, if the characters are displayed without interpretation, any of the hundreds of languages given in ISO 639 can be represented in on-screen displays.

audio_type

The ISO 639 Language Descriptor also includes the audio_type field to identify the type of audio track. MPEG-2 *Systems* defines four choices for audio_type: "clean effects" (the name for the case where no language applies), hearing impaired audio, visual impaired commentary, and a value indicating audio type is undefined. However, in the US system, the presence of a Dolby AC-3 Audio Stream Descriptor is required for terrestrial broadcast. The Dolby descriptor includes a "bit-stream mode" field that has a richer set of type values than the ISO 639 Language Descriptor offers. US cable standards specify the language of an audio track with the ISO 639 Language Descriptor's ISO_639_language_code field, and specify the audio type with the AC-3 descriptor's bit stream mode (bsmod) field. As of this writing, an amendment to Annex C of A/53B is being prepared that will track this approach for the terrestrial broadcast case.

Usage rules for cable

Rules for cable transport given in SCTE 54[18] state that an ISO 639 Language Descriptor must be present in the program element loop of an audio Elementary Stream when there is more than one audio track of the same type in the program. Audio type is given by the bsmod field in the AC-3 Audio Descriptor.

SCTE 54[18] states that the audio_type field in the ISO 639 Language Descriptor shall be set to zero (the undefined state) and that receivers are expected to use the AC-3 Audio Descriptor to determine the type of an audio track. We discuss the syntax and semantics of the AC-3 Audio Descriptor later in this book (see "AC-3 Audio Descriptor" on page 217).

Usage rules for terrestrial broadcast

ATSC A/52A[8] indicates that language identification is done with the ISO 639 Language Descriptor and that the language code previously defined within the AC-3 Audio Stream Descriptor is no longer to be used. As of this writing an update to the ATSC A/53B[14] Standard is being generated to reflect these changes. It echoes the rules adopted in cable for usage of the ISO 639 Language Descriptor.

References

1. ITU-T Recommendation H.222.0 | ISO/IEC 13818-1:2000, "Information Technology—Generic coding of moving pictures and associated audio information—Part 1: Systems."

2. Chernock, Richard S., Regis J. Crinon, Michael A. Dolan, and John R. Mick, *Data Broadcasting*, McGraw-Hill, 2001.

3. ISO/IEC 11172-1, "Information Technology—Coding of moving pictures and associated audio for digital storage media at up to about 1,5 Mbit/s—Part 1: Systems."

4. ISO/IEC 11172-2, "Information Technology—Coding of moving pictures and associated audio for digital storage media at up to about 1,5 Mbit/s—Part 2: Video."

5. ITU-T Recommendation H.262 | ISO/IEC 13818-2:1996, "Information Technology—Generic coding of moving pictures and associated audio information—Part 2: Video."

6. ISO/IEC 13818-6:1998, "Information Technology—Generic coding of moving pictures and associated audio information—Part 6: Digital Storage Media—Command and Control."

7. ISO/IEC 13818-3:1996, "Information Technology—Generic coding of moving pictures and associated audio information—Part 3: Audio."

8. ATSC Standard A/52A, "Digital Audio Compression Standard (AC-3)," Advanced Television Systems Committee, 20 August 2001.

9. ETSI EN 300 468 v1.4.1 (2000-11), "Digital Video Broadcasting (DVB); Specification for Service Information (SI) in DVB systems."

10. ATSC Standard A/58, "Harmonization with DVB SI in the Use of the ATSC Digital Television Standard," Advanced Television Systems Committee, 16 September 1996.

11. ATSC Standard A/55, "Program Guide for Digital Television," Advanced Television Systems Committee, 3 January 1996 (obsolete).

12. ATSC Standard A/56, "System Information for Digital Television," Advanced Television Systems Committee, 3 January 1996 (obsolete).

13. SP-RFIv2.0-I01-011231, "Data-Over-Cable Service Interface Specification, Radio Frequency Interface Specification," Cable Television Laboratories, 2001.

14. ATSC Standard A/53B, "ATSC Digital Television Standard," Advanced Television Systems Committee, 7 August 2001.

15. ATSC Standard A/70, "Conditional Access System for Terrestrial Broadcast," Advanced Television Systems Committee, 17 July 1999.

16. EIA 775.2, "Service selection information for digital storage media interoperability," Electronic Industries Alliance, 2000.

17. SCTE 18 2001 (formerly DVS 208), "Emergency Alert Message for Cable," Society of Cable Telecommunications Engineers.

18. SCTE 54 2002A (formerly DVS 241), "Digital Video Service Multiplex and Transport System Standard for Cable Television," Society of Cable Telecommunications Engineers.

19. ISO/IEC 11172-3, "Information Technology—Coding of moving pictures and associated audio for digital storage media at up to about 1,5 Mbit/s—Part 2: Audio."

20. ETSI ETR162, 1995, "Digital broadcasting systems for television, sound and data services; Allocation of Service Information (SI) codes for Digital Video Broadcasting (DVB) systems."

21. ISO 639-2:1998(E/F), "Codes for the representation of names of languages—Part 2:Alpha-3 code."

22. ISO/IEC 8859-1:1998, "Information technology – 8-bit single-byte coded graphic character sets—Part 1: Latin alphabet No. 1."

23. ATSC Standard A/90, "Data Broadcast Standard," Advanced Television Systems Committee, 26 July 2000.

24. SCTE 27 1996 (formerly DVS 026), "Subtitling Methods for Broadcast Cable," Society of Cable Telecommunications Engineers.

25. ANSI/SCTE 19 2001 (formerly DVS 132), "Standard Methods for Isochronous Data Services Transport," Society of Cable Telecommunications Engineers.

26. SCTE 53 2002 (formerly DVS 051), "Methods for Asynchronous Data Services Transport," Society of Cable Telecommunications Engineers.

27. SMPTE 312M, "Splice Points for MPEG-2 Transport Streams," Society of Motion Picture and Television Engineers.

Virtual Channels

Virtual channels are the cornerstone of ATSC's solution to the naming, numbering, and navigation problem for Digital Television. As we have described, when a viewer tunes a terrestrial broadcast or cable receiver to a digital channel, the channel number he or she uses to identify that channel is not tied to that channel's RF carrier frequency. Instead, the user's notion of the channel number is a parameter set by the broadcaster or cable operator in PSIP data. In this chapter we look into the virtual channel concept in detail.

What's in a Virtual Channel?

To see what the Virtual Channel Table adds, let's look for a moment at what the world would look like without it, or without any service information aside from that defined in the MPEG-2 *Systems* Standard. We consider a receiver built to only use the MPEG-2-defined PSI tables: the Program Association Table and the Program Map Table. We assume this receiver could find digital multiplexes on its own because it knows the frequency plans and modulation methods in use in a particular region of interest. Let's say a particular MPEG-2 Transport Stream was found at a certain carrier frequency. The receiver knows the following information from the PSI tables:

- the RF channel number of the carrier used to receive the TS.

- the Transport Stream's ID (TSID).

- that the TS includes some number of MPEG-2 programs.

- the MPEG-2 "program number" associated with each program.

- by inspection of the PMTs, the composition of each program in terms of audio and video streams.

What can the receiver display to the user to help in navigation? First of all, the RF channel of the carrier may be helpful. It could also somehow use the MPEG-2 program number associated with each service, but that number may not be user-friendly—it can range from 1 to 65,535. Furthermore, it could change over time. A receiver could number services in their order of appearance in the Program Association Table: the first one could be called program #1, the second #2, etc.

In contrast let's look at what the PSIP Virtual Channel Table adds to each service definition:

- a one- or two-part channel number that can be used for channel surfing or for reference to printed program guides.

- the channel's name, which typically indicates a broadcaster's call letters or for cable channels, the name of the service (for example "KNSD," "HBO," "C-SPAN," etc.).

- the type of channel, whether it's analog NTSC television, a digital television service, an audio-only service, or a data-only service.

The VCT gives additional information the receiver can use to help the user find and/or identify the service or to properly display data relating to it. This information includes:

- the channel's Source ID, a 16-bit number that links this channel with the program guide data.

- a flag indicating whether or not the channel is "hidden," meaning it is only available via an application or only to equipment owned by the broadcaster or cable operator.

- a flag indicating whether or not the channel should be included in Electronic Program Guide displays (note that some channels may be "inactive" meaning that they are currently unavailable [hidden], yet they may appear in EPG displays).

- a flag indicating whether the channel may require a subscription for viewing.

- an optional text string that also provides the name of the channel without the 7-character length limit.

- zero or more descriptors providing further information or attributes.

The VCT provided with terrestrial broadcast Transport Streams includes a list of the PID values associated with the audio and video elementary streams. This can help decrease the time it takes for a receiver to present audio and video when the channel is first tuned.

Terrestrial and Cable Virtual Channel Tables

A/65 actually defines two different Virtual Channel Tables. The one for use with terrestrial broadcasting is called the Terrestrial Virtual Channel Table or TVCT. The second one is the Cable Virtual Channel Table or CVCT and it is designed for use with cable signals when service information is carried in-band (within the same multiplex that carries the audio/video services). You may notice that the terrestrial broadcast and cable versions of the VCT are very similar to one another. Their structures are identical, and the only difference (aside from table ID) is that the cable version has a couple of extra flags defined. Throughout this book, when we refer to the "Virtual Channel Table" without mentioning terrestrial broadcast or cable we refer to the common aspects of the two types.

Service References

Each digital channel in the VCT is associated with a specific program within a particular Transport Stream by two numbers, the TSID and the MPEG-2 program_number. Figure 4.1 illustrates a Virtual Channel Table with five channels; four are digital and one is analog. Two of the digital services are carried on the same Transport Stream that carries the VCT. The TSID value for this TS is 0x10FD. The other two services are on a TS with TSID value 0x10FF.

Note that the MPEG program_number field is scoped to be unique only to the level of the Transport Stream. In this example, the services on TSID 0x10FD use program_number values 1 and 2 and so do the services on TSID 0x10FF. The example illustrates that the TSID value must be used together with the program number to resolve the reference. Channel TSID combined with program_number always results in a reference to just one MPEG-2 program because TSID is globally unique (managed by the ATSC for regions using the ATSC standard) and program_number is unique within the TS.

The figure shows that the fifth virtual channel is an analog service and the Transmission Signal Identifier is given as 0x1100. The referenced service is an analog broadcast signal carrying XDS packets in the VBI identifying it as TSID 0x1100. Program number values for analog services are set to 0xFFFF in the VCT.

One-part Numbers

So far, we've talked at length about two-part channel numbers. However, PSIP supports labeling a virtual channel transmitted on cable with either the major/minor (two-part) numbering scheme or a one-part channel number. Note that some cable systems use channel numbers in the range 1 to 999 for all of the services offered on the network although one-part numbers up to 16,383 can be represented.

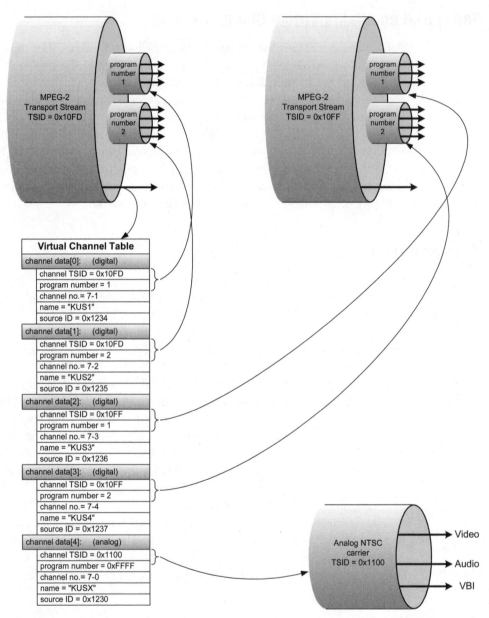

Figure 4.1 Virtual Channel Analog and Digital Service References

PSIP on Cable Transport Streams

Transport Streams received directly from cable may or may not include PSIP data. Whenever a service is provided in unscrambled format, PSIP data will be present in that Transport Stream if the cable operator expects that cable-ready devices that don't have conditional access support to be able to access it. If all of the services on a given TS are scrambled (that is, subject to conditional access), a cable operator is not required to include PSIP data at all within that particular Transport Stream.

The logic behind this is as follows: to view scrambled services requires use of a descrambling device. For US cable, that device is called the Point of Deployment (POD) module. Any device supporting the use of a POD module also must support an out-of-band (OOB) channel. An OOB channel involves a separate dedicated tuner/demodulator in the receiver for continuous communication with the cable headend. If a cable-ready device does not have a POD module (and hence has no access to the OOB channel), the only navigation data available will be the in-band PSIP data.

Cable operators prefer to send navigation data on the OOB channel because they avoid the need to duplicate the same data on all of the digital multiplexes. They feel the second dedicated (OOB) tuner is important because if a receiver only had one tuner and that tuner happened to be on an analog channel, the communication link to the headend would be lost.

So for all of the digital scrambled channels on cable, the virtual channel tables and EPG data (if sent) are typically delivered on the out-of-band channel. The standard that specifies SI on the out-of-band channel is called SCTE 65[1] (formerly known as DVS 234); we look at SCTE 65 in detail in Chapter 17. As we'll see there, parts of SCTE 65 are equivalent to PSIP while others reflect cable practice in the years before A/65 was developed.

Cable Virtual Channel Table

As mentioned, the Cable VCT is nearly identical to its Terrestrial VCT counterpart. Let's take a look at specifically how the two differ.

table_id

Obviously, the two are assigned separate table_id values. TVCT uses 0xC8 while CVCT uses 0xC9. This way, they can be distinguished from one another, and both can be sent in the same Transport Stream.

major_channel_number and minor_channel_number

As mentioned above, these fields have an extended meaning in the Cable VCT because they can indicate that the service is associated with either a one- or a two-part channel number. Also, the range restrictions for television services are eliminated: any service can have a major channel number in the 1 to 999 range, and any service can use a minor channel number in the 0 to 999 range.

carrier_frequency

Like the carrier frequency field in the TVCT, the Standard recommends that the carrier frequency field in the CVCT should be set to zero. That recommendation is for all services except those carried on the out-of-band multiplex. Virtual channels defined in the CVCT can point to services carried in the out-of-band channel. So far, it appears that no cable operator has enough spare bandwidth available to carry any kind of service on the out-of-band channel.

path_select

This flag indicates which of two physical input cables carry the associated service. Note that two-cable cable systems are not common in the US, and it's not likely that cable-compatible devices will support them.

out_of_band

This is a flag that indicates, when it is set, that the associated virtual channel is carried on the out-of-band multiplex. It appears unlikely that out-of-band virtual channels will be used, but if they do they will most certainly be for data rather than audio or video.

Cable VCT in the Terrestrial Multiplex

As mentioned, the VCT comes in two flavors, the Terrestrial VCT (TVCT) and the Cable VCT (CVCT). To comply with the ATSC standard for terrestrial broadcasting, a broadcaster must include a TVCT. Receivers that are able to pick up the 8-VSB-modulated broadcast signal use the TVCT for navigation.

A/65 states that a terrestrial broadcaster may include a Cable VCT in the broadcast multiplex in addition to the TVCT. Including the CVCT in a broadcast multiplex is permitted when a broadcaster has coordinated consistent channel numbering and labeling with all of the cable operators in the area who will carry that broadcaster's signal, and the desire is to use different numbering and/or labeling on cable as compared with terrestrial broadcast. A device accessing the TS via a terrestrial broadcast tuner and 8-VSB demodulator is expected to use the TVCT, while a device accessing it using a cable tuner and QAM demodulator is expected to use the CVCT.

In San Diego, for example, the local NBC affiliate is KNSD. This broadcaster obtained a license to broadcast an NTSC signal on RF channel 39, so receivers capturing KNSD's signal out of the air find it at channel 39. KNSD made arrangements with the cable operators in the San Diego area to have them all carry the KNSD signal on cable channel 7. So KNSD's logo and advertising use both "39" and "7" (see Figure 4.2). In printed advertising, they refer to themselves as "NBC 7/39."

Figure 4.2 KNSD Logo (used with permission)

KNSD has launched digital broadcasting using the RF channel assigned to them by the FCC, RF channel 40. The TVCT broadcast on channel 40 indicates major channel 39 for each of KNSD's programming services. When cable operators wish to include KNSD's digital services on cable, KNSD management may want to keep their association with channel 7 for the digital services. To facilitate this, they can include a Cable VCT labeling all their services with major channel number 7. A consumer's cable-ready DTV would then see the analog KNSD channel 7 as usual, and would see one or more digital channels on channel numbers such as 7-1, 7-2, etc.

In this example, KNSD included a CVCT in the broadcast multiplex because all the cable operators agreed to let them keep their services associated with channel 7. All of the operators benefited because they avoided the need to re-process KNSD's digital feed to alter the PSIP data, a step that would have been necessary if they had wanted to associate KNSD with a major channel number other than 39.

Figure 4.3 depicts this situation. In the example shown in the figure, broadcaster WGVH wishes to be known on cable as channel 8, but must be known to terrestrial broadcast receivers by the channel number corresponding to the original NTSC broadcast license, 35. A DTV receiving the signal off-air navigates to WGVH-DT via channel 35.1. The cable operators in the area have agreed to accept WGVH's signal as-is off the air (e.g. they won't touch PSIP), although each is free to modulate the Transport Stream within any 6-MHz spectral slot they choose.

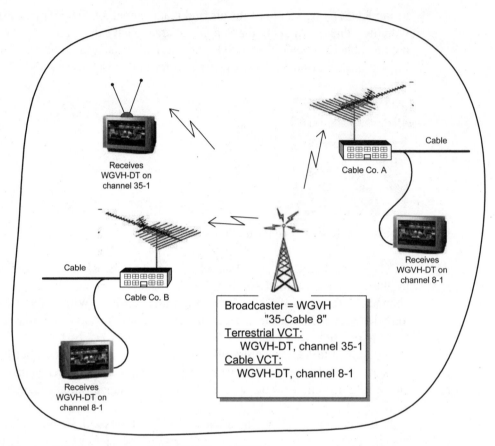

Figure 4.3 Broadcasting Both CVCT and TVCT

References

1.   SCTE 65 2002 (formerly DVS 234), "Service Information Delivered Out-of-band for Digital Cable Television," Society of Cable Telecommunications Engineers.

Two-Part Channel Numbers

As we saw in Chapter 4, ATSC A/65 introduces the concept of the two-part channel number, where transmitted PSIP data associates each broadcast programming service with both a "major" and a "minor" channel number. In this chapter we look at possible ways designers of Digital Television receivers can implement navigation when two-part channel numbers are involved. We also review the rules with which a broadcast station operator in the US must comply when assigning major and minor channel numbers to broadcast services.

As we have noted, channel numbers in analog television were tied explicitly to the FCC's channel numbering for 6-MHz spectral slots, and therefore in a sense were "hard-wired." A broadcaster was stuck with the channel number that corresponded to their broadcast license frequency. For those broadcasters lucky enough to have carriers assigned in the VHF band, these were nice easy-to-remember numbers in the 2-to-13 range. As digital broadcasting was phased in, some broadcasters said they did not want to have to lose the long-established brand identity they had built up with years of advertising. Many were comfortable with channel numbers in the "prime real estate" part of the number space, those numbers below 14, and they would much prefer to keep them.

In contrast to the old "hard-wired" channel numbering, PSIP offers "soft" channel numbering. Instead of being defined by the carrier frequency, the channel numbers associated with digital programming are defined by PSIP data transmitted with the digital services themselves.

A "Mental Model"

Ultimately, the systems engineering that has gone into definition of the PSIP protocol has to be boiled down into something that a consumer can actually use. In a scenario of interest here, a consumer picks up the DTV's remote control, hits a key to turn on the power, and begins searching for something good to watch. The viewer may have created a list of favorite channels or perhaps may view an Electronic Program Guide to see the program schedule for different channels. A common and traditional way to look for programs of interest is to "channel surf." Usually, this means hitting "channel up" or "channel down" until a program of interest appears on the screen, or until the viewer reaches a certain favorite channel. It can also involve cycling among channels in a list of favorites and stopping somewhere to take a look.

The mental model television viewers have used since the birth of NTSC is based on the idea that the channels are arranged in a list, sorted by channel number. When I hit "channel up," the TV tunes to the next highest channel/frequency. (It may also offer a mode in which "channel up" takes me to the next channel in my list of favorites.)

How might channel surfing work with ATSC terrestrial broadcast television? Each digital channel is identified by two numbers, a major channel number and a minor channel number (sometimes known as the sub-channel). So what happens when I hit "channel up" in this digital world?

Navigation Scenarios for Two-Part Channel Numbers

Various schemes are possible, of course, limited only by the imaginations of human factors engineers. One simple scheme is shown in Figure 5.1.

In the Figure the familiar channel up/down keys allow the viewer to move among all the available channels as usual, except that each minor channel is accessed in turn before moving to the next major channel group. In this example, the mental model is still a one-dimensional list of channels.

The channels in Figure 5.2 are linked as in the previous Figure but are arranged to emphasize the major channel number groupings.

Now we begin to see that perhaps a two-dimensional mental model can be made to work. In the vertical direction, one could choose among major channel groups while in the horizontal direction one could choose among channels within a given group.

Using channel up and down, a channel surfer would be able to move among channel groups. For example, with those keys one could jump from the set of channels operated by Fox to those operated by ABC to the NBC channels to the CBS ones, then to whatever independent stations were available.

Using the left and right arrow keys (or "channel left/right") one could choose among the various minor channels that might be available within a channel group. This scenario is depicted in Figure 5.3.

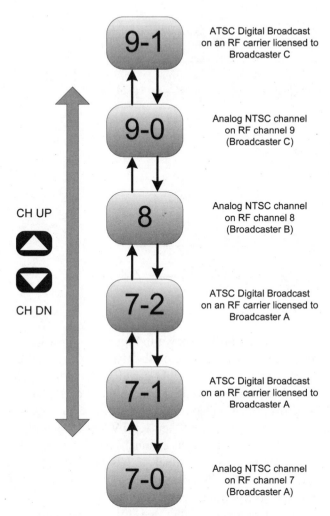

Figure 5.1 Linear Channel Surfing Example

Rules for Major and Minor Number Assignment

Broadcasters are not free to assign major and minor channel numbers to their digital programming in whatever manner they choose. To avoid conflicts, the rules and conventions for major and minor channel number assignments must be coordinated among broadcasters within any country using the ATSC Digital Television Standard. A conflict arises whenever one receiver can access, on two different RF carrier frequencies, digital multiplexes in which two different programming services are associated with the same major-minor channel number pair.

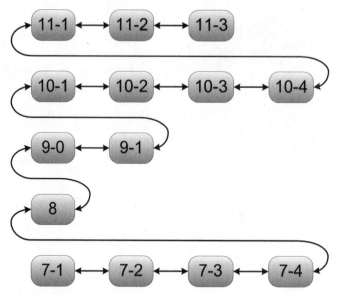

Figure 5.2 Major Channel Grouping

An example of this kind of clash is shown in Figure 5.4, in which two terrestrial broadcast transmitters are shown. The one on the left broadcasts a digital multiplex on RF channel 41, while the one on the right uses RF channel 36. The geographic area where the signal from each transmitter can be received is outlined[*]. Between the two is an area of overlap where one receiver can access both signals. In this example, both broadcasters have labeled their HDTV service with the same major/minor channel number: 10-1. The virtual channel names and TSIDs, though, are different. The broadcaster using RF channel 41 has labeled his channel WAAA, and uses TSID value 0x0401. The broadcaster using RF channel 36 has named the channel WBBB and tags his multiplex with TSID 0x0502.

In spite of the channel number conflict, a DTV receiver can (and should) be designed to allow a user to access both programming services. The two are clearly distinguishable from one another in the receiver because their TSID values differ. The problem really occurs with respect to human factors issues: if the user picks up the remote control unit and punches up channel 10-1, which signal should be acquired and displayed? A receiver can deal with the ambiguity in a number of ways. One possibility is to present the user with an on-screen display offering a choice between the two channels labeled 10-1. A very simple user interface could involve something like "Hit 1 for WAAA or 2 for WBBB."

---

[*] In practice, the situation illustrated will be quite rare and will typically only occur during propagation anomalies. Whenever signal areas are known to overlap, rules for assignment of virtual channel numbers ensure uniqueness.

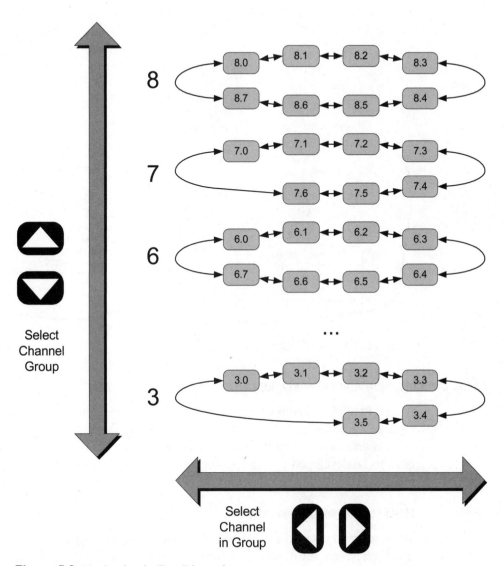

Figure 5.3 Navigation in Two Dimensions

Another difficulty arises if the receiver implements an Electronic Program Guide. Collected PSIP data may be presented on a channel-by-channel basis, but the appearance of two channels labeled 10-1 is troublesome and unexpected.

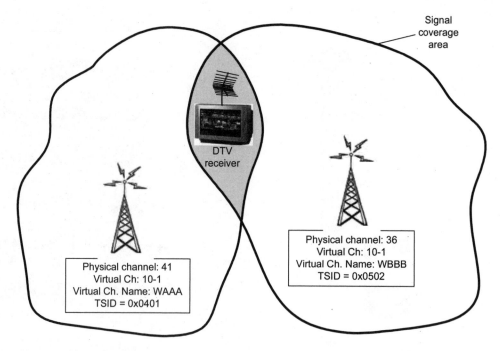

Figure 5.4 Channel Number Conflict Example

Assignment of major/minor channel numbers in the US

Annex B of A/65 includes specific rules for the assignment of major and minor channel numbers in the United States. We review those rules here. Note that for other countries of the world where the ATSC standard is used, analogous rules will need to be developed by the appropriate regulatory agencies in those countries in order to provide the needed coordination in number assignment.

Major channel numbers in the US

When a broadcaster in the US wishes to go on the air with a digital transmission, that broadcaster must adhere to the following rules in setting the major channel number:

- If the broadcaster has a pre-existing license to broadcast an analog NTSC signal, the major channel number for the digital TV channels is set to the RF channel number of the existing NTSC channel. This rule preserves channel branding.

- If the broadcaster does not have a pre-existing license for analog NTSC, the major channel number for the digital TV channels is set to the RF channel number of the digital broadcast, as assigned by the FCC.

These two rules effectively eliminate overlap of major channel numbers, because the FCC has assigned broadcast frequencies to minimize conflicts at the physical transmission level and these assignment rules build upon that foundation.

Another rule used in the US addresses the case wherein the FCC must ask a broadcaster to quit using the frequency first assigned to it and begin using another. When the FCC derived the DTV channel allotments, they avoided assignments in the 51-69 range because of the desire to re-allocate that portion of the spectrum to other uses at some time in the future. However, at the present time some broadcast stations have been assigned RF channels above 51.

The rule covering the case in which the FCC asks the operator of a digital television station to shift its RF transmission frequency is such that after the shift, that broadcaster must keep the major channel number it originally started with. Again, this rule preserves channel branding. The consumer benefits because most people will be unaware of the change as they recognize the digital channel by the virtual (not the physical) channel number.

Let's say Broadcaster A has an existing NTSC license and obtains a license to broadcast digital TV. This broadcaster operates both analog and digital broadcasts until the nationwide transition to digital broadcasting is complete. At that time, they shut down the analog transmitter. Now it may happen that a new broadcaster, Broadcaster B, comes into the area and is assigned the same RF channel as was previously used by the original broadcaster for the old analog service. What major channel number must Broadcaster B use?

The rule adopted by the United States for this situation says that the new digital broadcaster must use the RF channel number of the original analog broadcaster's assigned DTV RF carrier. This works because all of the digital services operated by the original broadcaster use the original NTSC RF channel as their major channel number, leaving the DTV service's RF channel number unused in the area. Figure 5.5 illustrates this scenario.

It may be that one broadcaster owns or controls broadcast licenses for more than one RF frequency and the service areas overlap. For example, there could be two Public Broadcasting stations serving the same community. The rules for the US state that such a broadcaster may use a common major channel number for of all the services on all the carriers on which that broadcaster transmits. In the PBS example, this rule means that all of the PBS stations are accessible as a group, and the channel logo and advertising for this broadcaster can feature just one number.

Minor channel numbers in the US

The general rule that applies to the assignment of minor channel numbers is that they must be assigned so that two-part channel numbers are unique in any given DTV service area. For broadcasters who operate just one DTV transmitter and use just one major channel number, this simply means that each service in the digital

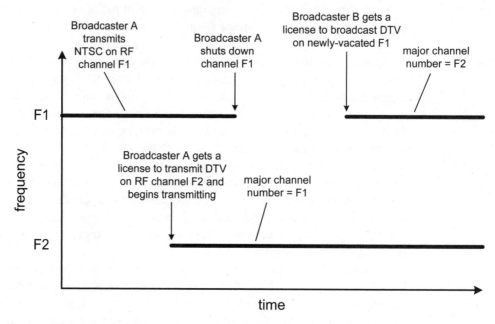

Figure 5.5 Re-assignment of Original NTSC RF Channel After the Digital Transition

multiplex must be assigned a different minor channel number. For broadcasters who operate more than one transmitter in one DTV service area and who choose to label the services in their digital multiplexes with common major channel numbers, the minor channel numbers must again be chosen so that no overlap occurs. One of the multiplexes, for example, could carry digital channels 15.1, 15.2 and 15.3, while a second multiplex carries 15.4, 15.5 and 15.6. Or, one could carry odd-numbered minor channels and the other even-numbered ones.

Other programming services .

In the US, some further flexibility is permitted with major/minor channel number assignments. The architects of the Standard realized that with the bandwidth available in a digital broadcast (nearly 20 Mbps), if a broadcaster chose to send standard-definition signals for some part of the day some extra bandwidth might be available that could be used by a programming service without a direct affiliation with that broadcaster. For example, a network affiliate might deliver, during the daytime hours, four standard-definition network-originated television channels plus one university or community service channel.

The rules state that services like these do not need to be labeled with the broadcaster's regular major channel number. Instead, they can be labeled with a major channel number in the 70-99 range. The choice of major channel number must be made such that all of the two-part channel numbers are unique in each potential receiving location. If they were not assigned uniquely, users would have difficulty selecting them, as we have seen.

So with this rule it is possible for a community college station in a certain service area to be associated with a certain major channel number, N. That station can continue to be associated with channel N even if it is shifted from one carrier to another.

Directional antennas

Coordination of channel number usage among broadcasters within countries using the ATSC standard is straightforward. As we have seen, the rules for assignment of major channel numbers in the US are tied to the assignment of their original NTSC carrier frequencies. The FCC has specified the frequency and power of analog carriers so that receivers in designated service areas receive at most one strong signal in any given 6-MHz band.

In the analog as well as the digital television system it is possible for a receiver to be situated in a location just between two stations. In the analog case, depending upon unusual weather-related propagation conditions or the orientation of a movable antenna, a receiver could pick up one station or the other at the same carrier frequency. In the digital case, the same thing could happen.

In Figure 5.6, a home is shown located between two digital broadcast transmitters. The two carriers are both modulated on RF channel 41. A rotatable rooftop antenna allows the homeowner to receive either one signal or the other at any given time.

A Digital Television receiver must "learn" what broadcasts are available to be received. The learning process involves tuning, one step at a time, through all the standard broadcast carrier frequencies and looking for either a digital or an analog signal. If a digital signal is found, an intelligently-designed receiver records that signal's TSID, the frequency used to tune it, and the Virtual Channel Table. If an analog NTSC signal is found, the receiver makes note of its frequency and attempts to discover the signal's Transmission Signal ID, also known as the analog TSID (for more on the analog TSID see "channel_TSID" on page 182).

Since a directional antenna may be involved, and since propagation conditions can change the set of signals available to the receiver at any given time, a receiver should retain the data associated with any signal it has received. It is also possible that the receiver itself might be moved to different locations: consider a DTV mounted in a recreational vehicle or motor home.

Looking again at Figure 5.6, we note that if the antenna is pointed west, WAAA can be received; if it's pointed east, the WBBB signal comes in. Whenever a digital signal is found at RF channel 41, the PSIP data should be collected. If the DTV

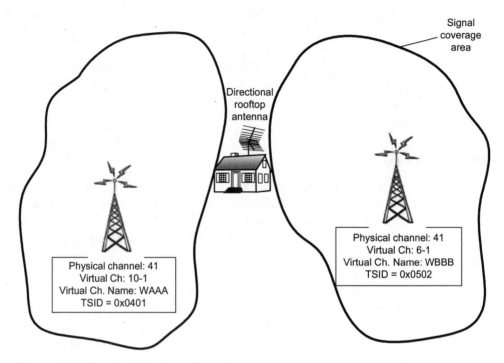

Figure 5.6 Directional Antenna Example

supports an Electronic Program Guide function, this data includes PSIP Event Information and Extended Text tables too.

The DTV might offer a list of channels and could let the user look at stored program guide data. In the channel list, both WAAA and WBBB appear along with whatever program titles and event descriptions might have been stored.

Whenever the user wants to watch WAAA, the receiver looks it up in memory and finds that WAAA was last found on RF channel 41. Tuning to RF channel 41 it might find a Transport Stream with ID 0x0401 (indicating WAAA) or a TSID of 0x0502 (that of WBBB). Here we can see the importance of TSID in identifying the digital multiplex. If a Transport Stream ID other than the desired 0x0401 is found, the DTV could respond with a message to the user such as "WBBB signal not available." Or, potentially it could be smart enough to operate the antenna rotor and acquire WBBB itself[*].

---

[*] A committee in the Consumer Electronics Association recently completed a standard interface specification for "smart" antenna control, EIA/CEA-909 *Antenna Control Interface*, that includes control of an antenna's rotational position.

Source IDs

The A/65 protocol introduces a data element called the Source ID. One way to look at Source ID is that it is a numeric representation of a service, but unlike a virtual channel number it is "not fit for human consumption." In other words, Source ID values should not be seen by users. More technically, the Source ID is the 16-bit number that links services listed in Virtual Channel Tables with program guide data given in the Event Information Tables and Extended Text Tables. Actually, Source ID embodies a concept that is more powerful than one might appreciate at first glance.

When we look at Source ID as the database linkage between entries in the Virtual Channel Table and program guide data given in Event Information Tables, the idea is simple and straightforward. Each channel in a Virtual Channel Table is associated with a unique value of Source ID. All of the Event Information Tables describing programming for a particular three-hour time slot are transported in Transport Stream packets with a common PID value. In each time slot where event data is delivered, there is one EIT for each programming service. Each EIT is tagged with the Source ID to which its data corresponds. Source ID is the data element that links the Event Information Tables to entries in the Virtual Channel Table.

Figure 6.1 shows a simplified Virtual Channel Table on the left, and Event Information Tables for the first three time slots to the right.

The Virtual Channel Table defines three programming services: channel 5.2, WCBA; channel 5.3, WCBA-D1; and channel 5.4, WCBA-D2. Each is associated with a unique Source ID, as shown. WCBA's Source ID is 50, so Event Information Tables labeled with a Source ID of 50 correspond to WCBA. When we look at Source ID in this way, it is seen as a simple linkage between two data structures.

Local vs. "Regional" Scope

We have described Source ID as the way events in the Event Information Table are tied to services defined in the Virtual Channel Table. In the ATSC terrestrial broad-

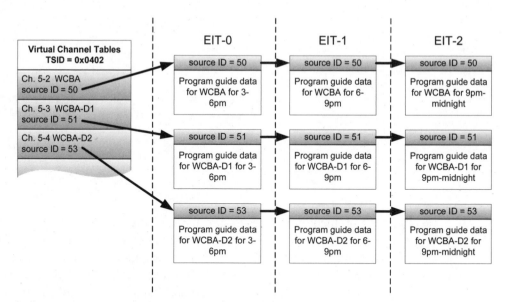

Figure 6.1 Source ID Example in One Transport Stream

cast multiplex, the VCT typically refers to values of Source ID that are to be found in Event Information Tables carried in that same multiplex.

When a Source ID is specified in PSIP, it can be scoped to the Transport Stream, or it can be "regionally" scoped. Values less than 4,096 are scoped to be unique only within the Transport Stream, while values 4,096 and above are regionally scoped. When scoped to the TS a Source ID can only be referred to by EITs carried in that same TS. When regionally scoped it can be referred to by EITs carried outside the TS, such as EITs on other Transport Streams, or even event information available via other means such as the Internet. Let's look at an example of a regionally-scoped Source ID.

Consider now a broadcast "super-station" whose signal can be received in its local metropolitan area as well as on cable and via satellite throughout the US. Figure 6.2 shows an example of a super-station called WTV originating in New York.

A portion of the Terrestrial Virtual Channel Table as broadcast in New York is shown at the upper left of the figure. WTV is carried on cable systems in Cleveland, Boston, and Washington DC. Portions of these Cable Virtual Channel Tables are shown at the left side of the figure.

Note that each Virtual Channel Table describes how and where viewers can access WTV in their local area. Viewers in New York, WTV's local broadcasting area, can pick up the signal by tuning the DTV to channel 17-1. Customers of Beltway Cable in Cleveland can find WTV on channel 7-2, while the customers of

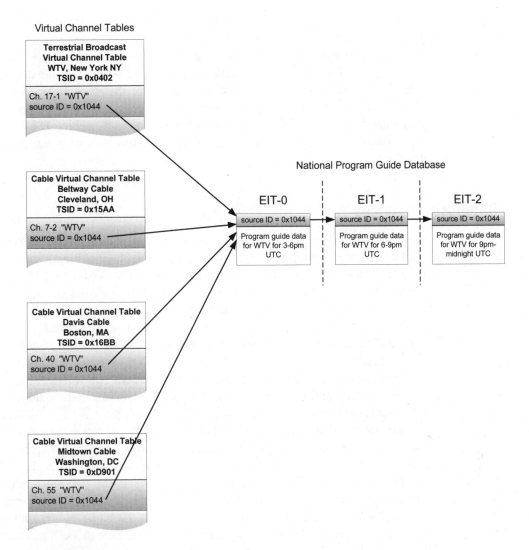

Figure 6.2 Source ID on a National Scale

Davis Cable in Boston can find it on channel 40. In Washington, Midtown Cable carries WTV on channel 55.

In each case, the Virtual Channel Table has listed a regionally-scoped value for Source ID, 0x1044. The figure shows one Event Information database that is equally applicable anywhere the WTV signal is carried. The EITs can be used to access the program schedule for WTV without regard for the specifics of what Transport Stream it is carried on. Any cable operator carrying WTV could even use a different text string name for the channel, and the Source ID value 0x1044 would still tie it unambiguously to WTV.

Cable operators and program guide providers use the concept described in Figure 6.2 today. These operators maintain one national program guide database. Each entry in the database is linked to the hundreds of different local systems by the Source ID. Building on regionally-scoped Source IDs, an Internet-based program guide database can be constructed to allow an Internet-enabled DTV access to a wide variety of programming information pertaining to any given programming service.

Composite Services

In some cable systems, one or more channels in the channel line-up switch "personalities" at some point in time each day. Perhaps the channel broadcasts educational programming until 9 p.m., but after that time it switches to an adult programming format. We call such a split-personality channel a "composite service" because its program schedule is composed of parts of two or more other services.

Whenever a composite service is created, it must be associated with a value of Source ID that is separate from any of the services from which it is made. As an example, let's say movie service A uses Source ID S_a and movie service B uses S_b. If a cable operator makes up a channel that broadcasts A's schedule during part of the day or week and B's schedule the other times, that channel would have to have a Source ID of its own.

Let's say another cable operator chooses to make a composite service from the same movie services, A and B. As long as the points where the switch-overs occur are at identical UTC times, the two can share a common Source ID. But if any switch-over point is different on any day of the week, the two must be considered separate programming services and must have distinct Source ID assignments.

Many different cable operators may broadcast the same programming service but substitute local advertising in certain designated time slots. Nationally-distributed cable channels are often reprocessed at cable headends to insert advertising pertinent to the area. As long as the *program schedule* remains the same, these services can (and should) share a common Source ID. The litmus test to determine whether two programming services can have the same Source ID is to compare the program schedules as they would appear in an Electronic Program Guide. Advertising does not factor into program schedules, so ads can differ between two services without requiring them to have separate Source IDs.

Registration of Source IDs

SMPTE has considered establishing and maintaining a database of Source IDs for use in the US. The effort may look at the problem from a global perspective, taking into account as well the needs of DVB in Europe and ARIB in Japan. In the US, General Instrument (now Motorola) created the first Source ID database in the mid-1990's for use with the company's digital satellite and cable products. This database was originally constructed by Prevue Inc. (now Gemstar-TV Guide International, Inc.); it included just over 7500 entries. Every broadcast licensee in the US, Canada and Mexico was represented, as well as every known cable and satellite programming service. SMPTE may use the GI/Prevue database as the starting point for their effort.

One of the reasons there were so many entries in the original Prevue database was that it had to include all of the composite channels in use in the US. As we transition to all-digital transport and virtual channels, the use of composite channels is expected to diminish significantly.

The SMPTE effort may do more than just maintain the numeric representation of the world's programming services. It could also maintain a registry of the textual representations of service names. In PSIP, services are identified with names of at most seven characters. SMPTE may also organize and maintain a database of the domain names associated with audio/video services, in support of the "tv:" URL scheme defined in RFC-2838[1].

References

1. IETF RFC-2838, "Uniform Resource Identifiers for Television Broadcasts," Internet Engineering Task Force, May 2000.

Program Content Advisories

A/65 standardizes the digital equivalent of the V-chip content advisory data now included in analog broadcasts in EIA/CEA-608-B[1] XDS packets. A content advisory is an indication provided with a given television program that tells something about that program's content, such as whether or not it contains scenes of a violent or sexual nature. Although the V-chip system has very specific requirements for content labeling for use in the US, PSIP was designed to accommodate the needs of any country in the world that may want to adopt the ATSC standard. PSIP defines a Rating Region Table (RRT) that specifies the structure of a multi-dimensional content advisory system for a specific region (e.g., country), and a Content Advisory Descriptor that can be used to associate specific program events with rating levels defined in the RRT.

This chapter describes the structure and capabilities of the ATSC content advisory system.

Dimensions and Levels

In ATSC PSIP, content advisory data is structured around concepts called dimensions and levels. A dimension is a particular aspect of content or a particular way of characterizing content. Commonly used dimensions include:

- Age: This dimension specifies the minimum age recommended for viewers of the given content.

- Violence: The violence dimension indicates the prevalence or intensity of violent content.

- Sexuality: The sexuality dimension can indicate the presence of sexual situations, the explicitness of sexual scenes, or the adult nature of the program.

- Language: This dimension provides a way to characterize the extent of use of crude or profane language in the program.

- Standard rating system: It is possible to tie the definition of a rating dimension to an established rating system such as the scheme supported by the Motion Picture Association in the US.

The term "level" may be used to indicate the amount or intensity of content, for a given dimension, the given program contains. For the Language dimension three simple states or levels could be defined: 1) not rated for language; 2) contains no offensive language; 3) contains some amount of bad language. Note that one can also distinguish between "no information provided" and "rated not to contain anything that could be considered harmful or offensive."

For a dimension such as Language, a more extensive definition of levels could be:

1. No bad language.

2. Some bad language.

3. Profane language.

Figure 7.1 shows an example rating system consisting of four dimensions. As shown, rating dimensions are laid out in the horizontal direction while levels within each dimension are drawn vertically. Virtually any rating system may be mapped into a two-dimensional table such as this. Each cell in the table represents a specific rating level for a particular dimension. Referencing any cell in the table involves specifying the dimension (column) and level within that dimension (row).

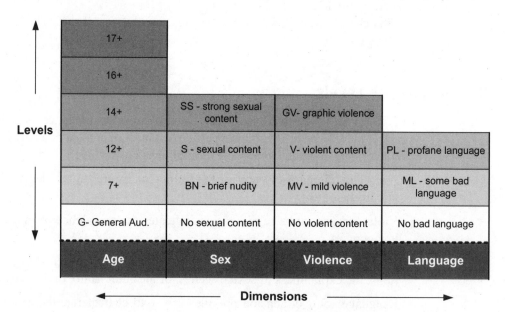

Figure 7.1 Rating Dimensions and Levels in an Example Rating System

Graduated Scales

Often, the definition of content ratings for a particular dimension are ordered in terms of increasing amount or intensity of the given content. The Language example given above is an example of a dimension defined on a "graduated scale."

In the PSIP system, it is possible to define both rating dimensions that are defined on a graduated scale as well as those that are not. The advantage of using a graduated scale is that a receiver can be set up to block programming at a given level of content and to automatically know that blocking of content at higher levels is appropriate. For instance, to continue with the example language dimension, if the user were to decide to block programs with mildly offensive language, programming with profanity of any kind would also be blocked.

In the following example, the "Not rated" level causes the content rating dimension not to work on a graduated scale:

1. All age levels.

2. Age 7 and up.

3. Age 13 and up.

4. Age 17 and up.

5. Not rated.

The designers may feel that programming rated "Not rated" is not necessarily designed only for viewers older than 17 years of age. Some programming may be "Not rated" simply because it has not been reviewed and given a rating, or perhaps it was produced before the rating system came into existence.

Any given program could be categorized into one of these five rating levels. A receiving device could be set up to accept or block programming rated for any of the five levels. It would be against the rating system designer's wishes, however, that a choice to block one level carries anything about preferences related to the others. The designers in this case wish to allow a parent to block programming at the "Age 17 and up" level without automatically blocking programming rated "Not rated." Hence, this system is not designated to conform to a graduated scale.

System Design Goals

As mentioned, the PSIP content advisory system was designed with flexibility in mind. An important design goal was the concept that a receiving device could be built that could adapt and accommodate any content advisory system in use within any digital Transport Stream it happened to receive. It was felt that no prior knowledge about the system in use in a particular region or country should be needed.

Another way to state this goal is to say that the definitions of the content rating system are "soft" rather than being hard-coded and inflexible.

Because the Rating Region Tables can be received and downloaded from within broadcast content, a receiver can be built to capture and to create from them an appropriate table-driven user interface. If an update, revision, or correction to an RRT were to be seen, a receiver's user interface could reflect that change automatically.

User preferences are typically established through user interaction with setup menus. A content advisory system plus the presence of content advisory data within broadcast programming means that any receiving device can offer the user a choice to conditionally accept or block programs containing certain rated content.

The user setup menus to support this feature could be organized such that an option from a setup screen first brings up a choice to select from among a number of named rating regions. Note that each Rating Region Table provides the textual name of the region it defines. If programs are rated for just one region, which will often be the case, this choice can be skipped.

Figure 7.2 shows an example menu tree. In the example at a), Parental Control is offered as a choice from the main user setup screen. Selection of that choice brings up a screen b) constructed from the names of Rating Regions encountered by or known to the receiver.

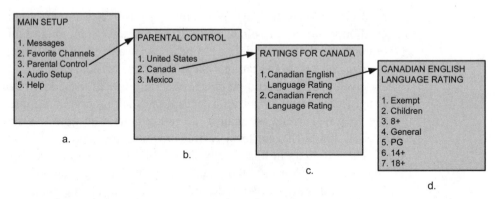

Figure 7.2 Example Menu Tree for Parental Control Setup

A choice of Canada brings up the screen c), where the Canadian rating dimensions are listed. Again, the number of dimensions and the names of each are derived from the downloaded RRT. If the English-language version is chosen, screen d) is generated, again using data from the Canadian RRT.

Rating Region Tables also accommodate the needs for text to be available in languages other than English. All of the text defining the content rating system can be delivered in one or more languages. For example, the rating system for Canada is defined in both English and French.

The US Rating Region Table

The US and Canadian PSIP Rating Region Tables are defined in EIA/CEA-766-A[2]. Two things become clear upon inspection of the RRT definitions for the United States and its possessions. First and most importantly, the interpretation of the definitions of some of the dimensions for the US region depend upon values in other dimensions. This will become clear as we look into the definition of the US RRT in detail. Secondly, EIA/CEA-766-A states that the definition of the RRT for the US region shall not change. Broadcasters and others in the United States chose to define and standardize a fixed rating system that could not be updated.

US rating dimensions

In the US Rating Region Table the dimensions are as follows:

1. **Entire Audience**. This dimension provides basic parental guidance about program content, and is applicable to programming suitable for any age group. Within the Entire Audience dimension, the levels are defined on a graduated scale:

   - "None" is defined to mean the program is rated for the entire audience and is not intended to be blocked.

   - "TV-G," means that many parents are likely to find the program's contents suitable for a general audience of any age group.

   - "TV-PG," denotes that the program may be unsuitable for younger children and that parental guidance is suggested.

   - "TV-14," indicates that parents are strongly cautioned that many would find the program unsuitable for children under 14 years of age.

   - "TV-MA," indicates the program is intended for adult viewing only and may not be suitable for those under 17 years of age.

2. **Dialogue**. This dimension provides an indication of the presence of sexually-suggestive dialogue in a program. Instead of being defined with a number of different levels of dialogue content, the Dialogue dimension is just a single flag abbreviated "D." The meaning of the D flag varies depending upon context:

   - D with TV-PG means "some suggestive dialogue."

   - D with TV-14 means "intensely suggestive dialogue."

- The Dialogue indication is not allowed to be used with TV-G or TV-MA rated programming.

3. **Language**. This dimension is used to signal that a program may contain objectionable language. The Language flag (L) is defined like the Dialogue flag such that it has context-dependent meaning:

- L with TV-PG means "infrequent coarse language."
- L with TV-14 means "strong coarse language."
- L with TV-MA means "crude indecent language."
- The Language indication is not allowed to be used with TV-G programming.

4. **Sex**. This dimension provides an indication that a program contains material of a sexual nature. It too is a flag, abbreviated "S," whose meaning depends upon the Entire Audience dimension:

- S with TV-PG means "some sexual situations."
- S with TV-14 means "intense sexual situations."
- S with TV-MA means "explicit sexual activity."
- The Sex indication cannot be used with programs rated TV-G.

5. **Violence**. This dimension provides an indication of a program's content of material of a violent nature. The context-dependent meaning of the Violence (V) flag is:

- L with TV-PG means "moderate violence."
- L with TV-14 means "intense violence."
- L with TV-MA means "graphic violence."
- The Violence indication cannot be used with programs rated TV-G.

6. **Children**. The "Children" dimension applies only to children's programming, and indicates the recommended age level for viewing. It is defined on a graduated scale:

- TV-Y indicates the program is expected to be suitable for all children.
- TV-Y7 indicates the program may include some violence or other element that may frighten children under the age of seven.

7. **Fantasy Violence**. This dimension, abbreviated FV, is applicable to children's programming to provide an indication that the program may contain a more intense or combative form of violence than other children's programs. It is a flag defined for use only in the context of TV-Y7 programming to signal the presence

of fantasy violence. Use of the FV flag is not allowed except with TV-Y7 programming.

8. **MPAA**. Finally, the US region includes a dimension reflecting the movie rating system adopted by the Motion Picture Association of America (MPAA), extending it one level beyond NC-17 to "X" and including levels to indicate "MPAA rating not applicable" and "Not rated by MPAA."

Clearly, the definition of the US region does not fit the model envisioned by the designers of the PSIP system. It is not possible to implement an appropriate user interface for the US solely based on the RRT because the Dialogue, Language, Sex, and Violence dimensions are flags whose meaning depends upon a program's rating in the Entire Audience dimension.

If a user's preferences for parental rating control were offered through a purely table-driven user interface, a situation like that depicted in Figure 7.3 could occur. In the figure, the user has chosen perhaps to block program viewing on the basis of Dialogue. But since the Dialogue dimension is defined only as a flag, the only thing that can be chosen is whether or not to block any program with the D flag set. It is not possible, then, to block "intensely suggestive dialogue" while allowing programs with "some suggestive dialogue."

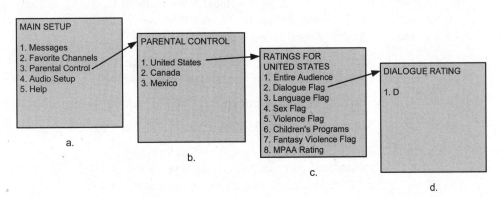

Figure 7.3 Example Table-Driven Menu for US Region

It should be clear by now that due to the way in which the US Rating Region Table is defined, the table-driven user interface model cannot be followed. Instead, a user interface and behavior based on EIA/CEA-766-A[2] and the CEA Recommended Practice document CEB1-A[3] are expected to be followed.

Transmission of the US RRT

As mentioned, broadcasters in the US have declared that the US Rating Region Table cannot be revised. The decision was made that the rating system for digital programming must exactly match the system that has been put in place for analog programming. For analog NTSC television, the "V-chip" system has been developed for content advisories. Rating data is placed into eXtended Data Service (XDS) packets in the signal's Vertical Blanking Interval (VBI). Televisions are required by the FCC to offer support for the V-chip system, and the implementation for analog sets is dependent upon a hard-coded design in these sets.

When one recognizes that the US RRT may not ever be updated, changed, or revised, one has to ask: why should it be required to transmit the US RRT within a digital broadcast signal? Indeed, ATSC A/65 has been revised to state that transmission of the US RRT is not required. If programs are rated for any region other than the US, it is required that the Transport Stream carry that region's current RRT.

For US cable, the standard defining requirements for cable transport is SCTE 54[4], published by the Society of Cable Telecommunications Engineers (SCTE). SCTE 54 states that transmission of the US RRT is optional because it is defined in EIA/CEA-766-A[2].

RRT Text

The Rating Region Table, whether downloadable from the digital multiplex or defined in a US standard such as EIA/CEA-766-A[2], defines text strings for the following aspects of each rating region:

- The name of the rating region. For the US region, the name is defined as "US (50 states + possessions)." For Canada, the name is simply "Canada."

- The name of each rating dimension, such as "Entire Audience" and "MPAA."

- The full name of each defined rating level, such as "Parental Guidance Suggested" or "8+."

- The abbreviated name of each rating level, such as "PG," "S," "L," or "V" in the US or "C" for Children in the Canadian system.

In some regions such as the US, one program can be rated for content in more than one dimension. In such a case, the program's rating can be displayed on-screen by concatenating the abbreviations for each dimension, separated by hyphens. For example, a program rated TV-PG and having the Dialogue, Sex, and Language flags set can be identified as being rated "TV-PG-D-L-S." For the US region, again, special rules apply, in that the display order for the flags is defined in the CEA standard.

Text in the RRT can be delivered multi-lingually. See "Multilingual text" on page 146 for a discussion of considerations for handling of multilingual text in receiving devices. The Canadian RRT, for example, is defined in EIA/CEA-766-A[2] with both English and French versions of each string.

The Canadian Rating Region Table

The Canadian RRT is an example of a rating system definition that does follow the intended PSIP design model. As shown in EIA/CEA-766-A[2], the Canadian RRT includes two rating dimensions:

1. Canadian English Language Rating. This dimension is defined on a graduated scale with the following seven levels:

- "Exempt" in English or "Exemptées" in French, abbreviated "E"

- "Children" in English or "Enfants" in French, abbreviated "C"

- "8+," meaning children 8 and above, abbreviated "C8+"

- "General" in English, or "Général" in French, abbreviated "G"

- "PG" in English, or "Surv. parentale" in French, abbreviated "PG"

- "14+," meaning children 14 and above, abbreviated "14+"

- "18+," meaning children 18 and above, abbreviated "18+"

2. Canadian French Language Rating, also defined on a graduated scale:

- "Exemptées" in French or "Exempt" in English, abbreviated "E"

- "Pour tous" in French or "For all" in English, abbreviated "G"

- "8+," meaning children 8 and above, abbreviated "8 ans+"

- "13+," meaning children 13 and above, abbreviated "13 ans+"

- "16+," meaning children 16 and above, abbreviated "16 ans+"

- "18+," meaning children 18 and above, abbreviated "18 ans+"

EIA/CEA-766-A[2] states that a given program may be rated with either the English or French dimension, but never both at once.

The Content Advisory Descriptor

Content advisory data pertinent to a particular program or event in PSIP is carried in a data structure called the Content Advisory Descriptor. We go into details in Chapter 11 (see "Content Advisory Descriptor" on page 231), but for now we consider the information conveyed in the descriptor. Each instance of a Content Advisory Descriptor can provide rating information for up to eight regions. Within a given region it can indicate the rating level for any dimension defined in that region.

Continuing with the example Rating Region Table definition given in Figure 7.4, let's look at some example Content Advisory data. A program's content advisory (parental rating) is defined in terms of a given region's RRT as the following Figure illustrates.

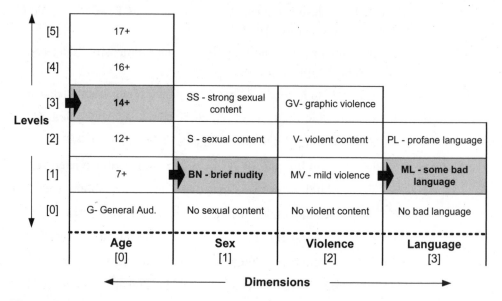

Figure 7.4 Example Program Content Advisory

This example shows the data for one region contained in an example Content Advisory Descriptor. Each Content Advisory Descriptor provides data pertinent to a given program's rated content. In the example, the program is rated "14+" for Age, BN for brief nudity, and NL because it contains some bad language. This program is not rated for violence content.

One program or event may be rated for more than one region. For example, a program may originate in the United States and be carried on Canadian or Mexican stations.

Data in the Content Advisory Descriptor is formatted as a set of parameter pairs for each identified region. The parameter pairs each consist of a dimension index and a rating level. As shown in the figure, one can consider the RRT to be a two-dimensional array, with Dimensions in the X-axis direction and Levels in the Y-axis direction.

Therefore, the Content Advisory Descriptor data in the descriptor would be encoded as: (0, 3), (1,1), and (3,1). This means dimension 0 is rated with a rating value of 3 (14+), dimension 1 is rated with a rating value of 1 (BN), and dimension 3 is rated with a rating value of 1 (ML).

References

1. ANSI/EIA/CEA-608-B, "Line 21 Data Services," American National Standards Institute, Electronics Industries Alliance and Consumer Electronics Association, 2001.

2. EIA/CEA-766-A, "U.S. and Canadian Rating Region Tables (RRT) and Content Advisory Descriptors for Transport of Content Advisory Information Using ATSC A/65A Program and System Information Protocol (PSIP)," Electronics Industries Alliance and Consumer Electronics Association, 2000 (http://www.ce.org/ or http://global.ihs.org/).

3. CEB1-A, "Recommended Practice for Content Advisories," Electronic Industries Alliance, December 1998.

4. SCTE 54 2002A (formerly DVS 241), "Digital Video Service Multiplex and Transport System Standard for Cable Television," Society of Cable Telecommunications Engineers.

Caption Services

In this chapter we take a brief look at caption services in the ATSC Digital Television System. PSIP plays a role in closed captioning in that the presence of caption services is announced within the Program Map Table or Event Information Table by the Caption Service Descriptor defined in A/65.

With regard to closed captioning, the A/53 ATSC Digital Television Standard refers the reader to EIA-708-B[1] *Digital Television (DTV) Closed Captioning*. Whereas the analog NTSC television system transports captions in VBI line 21 (on both field 1 and field 2), closed caption data for digital television is transported within the video data stream itself, in a portion of the video bit stream called Video User Data.

Transport of Closed Caption Data

Figure 8.1 illustrates the carriage of closed caption data within a Transport Stream. Captioning data that goes with a particular video service is carried within the video Elementary Stream inside transport stream packets of the same Packet Identifier as the associated MPEG-2 video stream. Extraction of closed caption data involves first extracting those packets in the TS that are identified with the video PID of interest. This PID filtering step yields the video Packetized Elementary Stream, shown in the figure as the pipe labeled "Video."

The MPEG-2 *Systems* Standard describes the method whereby the video PES can carry additional data related to video. This additional data, called Video User Bits, may be inserted into the video at any of three levels, the Sequence level, the Group of Pictures (GOP) level, or the Picture Data level. ATSC specifies that user_data() carrying closed caption data may only appear at the Picture Data level. The presence of a user_data() data structure within the video bit stream is signaled by a user data start code, value 0x0000 01B2. For ATSC user_data(), the start code is inserted into video within the extension_and_user_data(2) data structure.

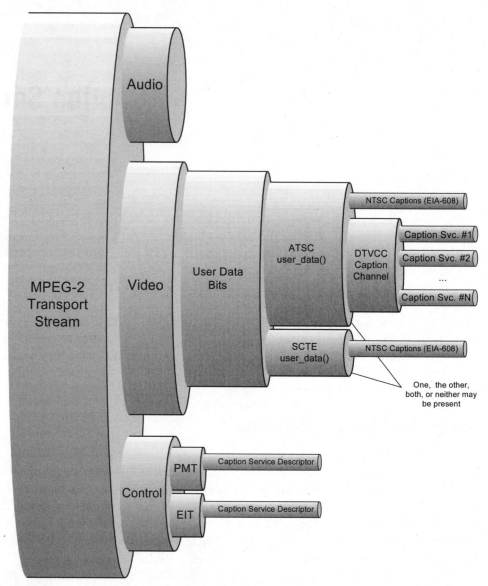

Figure 8.1 Caption Data in the MPEG-2 Transport Stream

Closed captioning for terrestrial broadcast

Annex A, Section 5.2.2 of the ATSC A/53B[2] Standard specifies the syntax and semantics of ATSC user_data(). An MPEG-2 user_data_start_code sequence followed by an ATSC_identifier (value is 0x4741 3934 or "GA94" in ASCII) signifies the presence of ATSC user data.

ATSC has included a user_data_type_code byte in the data structure and specified that value 0x03 for this byte indicates the syntax and semantics are as currently defined. Values 0x04 and 0x05 are defined in SCTE 21[3] to carry other types of VBI-related data for cable applications. Values other than these three may be used in the future to identify other ATSC- or SCTE-defined data structures. Receivers must always verify the user_data_type_code byte and discard the user_data() if the value does not indicate a supported type.

DTV closed captioning services

Figure 8.1 illustrates that one or more DTV closed captioning services may be present with any video service. Two services are defined with a special meaning that is analogous to the primary and secondary NTSC closed captions of the analog system. Service #1 is designated as the Primary Caption Service and is the caption service for the primary language being spoken in the accompanying program audio. Service #1 is expected to reflect the actual words being spoken. Service #2 is designated as the Secondary Language Service, and it contains captions in a secondary (translated) language.

Closed captioning for cable

Cable Transport Streams may also include closed captioning. When present, closed captioning on cable is also transported in the video user_data() portion of the video PES. A Transport Stream demodulated from cable may include captioning data in video user_data() compliant with the terrestrial broadcast formats, or it may include captioning data within a different user_data() data structure, as defined in SCTE 20[4]. The SCTE 20[4] Standard describes NTSC VBI data for line-21 and can describe other VBI lines as well. Equipment designed for consumer use, however, is expected only to reconstruct line-21 data so the other aspects of SCTE 20[4] may be disregarded. The only standard method for delivering DTV closed captioning (sometimes called "advanced" closed captioning) is via ATSC user_data() and EIA-708-B[1].

As shown in Figure 8.1, a given video PES may include either ATSC user_data(), SCTE user_data(), both ATSC and SCTE user data, or neither. A receiver that is designed to be compliant with both cable and terrestrial broadcast Transport Streams must handle each of these four cases. For the case that both ATSC and SCTE data are present, ANSI/EIA/CEA-608-B[5] data may be taken from either the ATSC or the SCTE data structures.

The receiver must handle dynamic changes in the closed caption data formats as well. One approach to receiver design is to accept the two bytes per field of ANSI/EIA/CEA-608-B[5] data from either the ATSC or the SCTE user_data() data structures. If ATSC user data provides the two bytes for a given field, those bytes are stored for use in reconstruction of that field. Likewise, if SCTE user data provides the two bytes, those are stored. If ATSC and SCTE user data both provide the two bytes for a given field, bytes in whichever data structure is received last are used.

This "take data from whichever source" strategy is likely to be more robust than a method whereby the receiver attempts to switch modes, moving back and forth between "use ATSC data" and "use SCTE data." If mode switching were to be used, when the type of data being used in the current mode stopped the receiver might lose some captioning data when switching to the other path.

Closed captioning data rates

In the NTSC closed captioning system defined in ANSI/EIA/CEA-608-B *Line 21 Data Services*[5], two bytes of data per video field can be sent. The data rate is thus $2*8*60 = 960$ bits per second. EIA-708-B[1] specifies that the total bit-rate for digital plus analog captioning is 9600 bits per second so that when analog captioning is present, 90% is dedicated to digital captioning and 10% to analog captioning. Eventually, NTSC closed captioning will cease and the full 9600 bits per second could then be used for DTV closed captioning.

NTSC closed captions in digital video

NTSC captions may be carried in a digital service so a decoder can reconstruct line-21 in the NTSC Vertical Blanking Interval. Receivers are not expected, under any circumstances, to derive NTSC line-21 caption data from the DTV closed captioning data provided in EIA-708-B[1].

Announcement of Closed Caption Services

Whenever a digital Transport Stream carries captioning services in either the ANSI/EIA/CEA-608-B[5]-style NTSC captions or the DTV closed captions format, information describing them must be included in the TS. Figure 8.1 shows a pipe in the Transport Stream labeled Control, from which emanates the Program Map Table (PMT) and the Event Information Table (EIT). Either of these tables can carry the Caption Service Descriptor defined in A/65.

The Caption Service Descriptor includes the following information about each caption service:

- its language.

- whether the caption service is a DTV closed caption service as defined in EIA-708-B[1], or an NTSC line-21 caption service as defined in ANSI/EIA/CEA-608-B[5].

- for NTSC line-21 services, whether the caption service is carried on field 1 or field 2 of the NTSC waveform.

- for DTV closed captions, the caption_service_number associated with the caption to link the announcement to the corresponding service number in the DTV closed caption packet stream.

- a flag indicating whether or not this service is tailored to the needs of beginning readers.

- a flag signaling whether the caption service is formatted for 16:9 displays.

EIA-708-B[1] specifies that there shall be no more than 16 simultaneous caption services described in the Caption Service Descriptor.

As mentioned, the Caption Service Descriptor may appear in either or both the PMT or the EIT when a given program is captioned. The usage rules for the descriptor state that, for terrestrial broadcast applications, the descriptor must be present in the EIT and may also be present in the PMT. For cable, the Caption Service Descriptor must appear in the PMT if the program is captioned and must also appear in EIT-0 if EIT-0 is present in the TS.

We go into the details of the syntax and semantics of the Caption Service Descriptor in Chapter 11 (see "Caption Service Descriptor" on page 226).

References

1. EIA-708-B, "Digital Television (DTV) Closed Captioning," December 1999.

2. ATSC Standard A/53B, "Digital Television Standard," 7 August 2001.

3. SCTE 21 2001 (formerly DVS 053), "Standard for Carriage of NTSC VBI Data in Cable Digital Transport Streams," Society of Cable Telecommunications Engineers.

4. SCTE 20 2001 (formerly DVS 157), "Standard Methods for Carriage of Closed Captions and non-Real Time Sampled Video," Society of Cable Telecommunications Engineers.

5. ANSI/EIA/CEA-608-B, "Line 21 Data Services," Electronic Industries Alliance and Consumer Electronics Association, 2001.

Data Representation

Before we dive into the detailed syntax and semantics of the PSIP tables, let's look at the different types of data the tables deliver and how each type is represented in the transmitted bit stream.

Unsigned Integers

PSIP table syntax includes unsigned integers of various lengths. When a range is not specified, any value that can be represented in a bit string of that length is valid. In accordance with MPEG-2 *Systems*, unsigned integers are placed into the data stream most significant bit first.

Sometimes the zero or all-ones value of an unsigned integer field has a special interpretation (in the VCT, for example, value 0x0000 for program_number is specified for use when the virtual channel is inactive, and value 0xFFFF is reserved for analog services). Some unsigned integer fields represent simple integers such as the number of data bytes to follow or the number of iterations of a "for" loop. Others are "enumerated types," which means that particular values have specific meanings. Typically for the latter case a table in the Standard defines the meaning assigned to a particular value or range for that parameter.

Binary Flags

Certain fields in PSIP tables are simply one-bit flags. Typically, the name of the field has been chosen so that the field name reflects the flag in its "true" or "set" state. For example the hidden bit in the Virtual Channel Table, when set, indicates that the channel is to be hidden from normal view.

Text Representation

Almost all of the textual data in the PSIP Standard is delivered in a structure called the Multiple String Structure (MSS). It is called a "multiple" string structure because it may include more than one text string. As we'll see, each of the N strings carried in one instance of the MSS is tagged with a language code. So it is possible to represent the same string in several languages with one MSS.

As an example, let's say the description of a program was "Homer is hired as a writer at the cookie factory." A Spanish version of that string could be included in the MSS as well: "Homer se emplea como un escritor en la fábrica de galleta." One receiver may choose to use the English version while another chooses the Spanish translation. Note that a receiver is expected to pick, based on language, just one of the strings for display and discard the other(s).

The coding for all text is based on the Unicode Standard, Version 3.0[1]. The UTF-16 representation of Unicode character data is in accordance with that defined by the Unicode Standard[1], which is identical to that defined by ISO/IEC 10646-1:2000[2] Annex C.

Figure 9.1 diagrams the structure of the MSS.

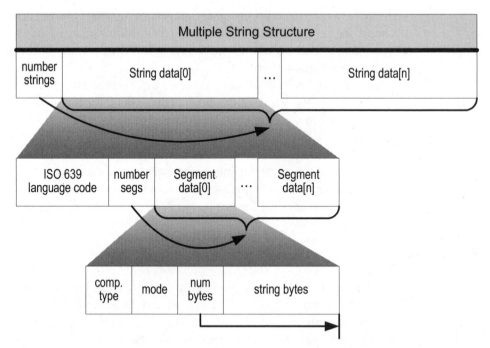

Figure 9.1 Structure of the Multiple String Structure

As shown in the Figure, the MSS consists of one or more blocks of "string data." The first byte of the MSS indicates the number of blocks. The first three bytes of each string data block identify the language of the text in that block

After the language tag, each string data block consists of one or more segments. Each segment includes a compression type byte, a mode byte, the number of bytes in the string, and then the bytes representing the string itself. The bytes representing the string may be uncompressed Unicode character codes or a Huffman-coded serial bit string. The compression type and mode bytes indicate the format of the string bytes.

Table 9.1 describes the syntax and semantics of the Multiple String Structure.

TABLE 9.1 Multiple String Structure Bit Stream Syntax

Field Name	Number of Bits	Description
multiple_string_structure() {		Start of the multiple_string_structure().
number_strings	8	Gives the total number of strings to follow in the MSS.
for (i=0; i<number_strings; i++) {		Start of the number_strings "for" loop.
ISO_639_language_code	24	Indicates the language of the string in this iteration of the "for" loop.
number_segments	8	Indicates the number of segments the string in this iteration of the number_strings "for" loop is broken into.
for (j=0; j<number_segments; j++) {		Start of number_segments "for" loop.
compression_type	8	Indicates the type of text compression, if any, used for this segment.
mode	8	Indicates the mode to be used in interpretation of the segment's contents. See text.
number_bytes	8	The size, in bytes, of the segment data to follow.
for (k=0; k<number_bytes; k++) {		Start of segment data bytes.
compressed_string_byte[k]	8	Bytes of the string segment (despite the name of this field, whether or not text compression is used depends upon the value of the compression_type field).
}		End of the segment data bytes.
}		End of the number_segments "for" loop.
}		End of the number_strings "for" loop.
}		End of the multiple_string_structure().

number_strings

This 8-bit unsigned integer indicates the number of different representations of the string that are contained in the MSS. Each is tagged with a language code, and the expectation is that whenever more than one string is present the MSS contains the string in the native language plus one or more translations of the string.

ISO_639_language_code

The MSS uses ISO 639-2/B[3] alpha-3 coding for languages, just like the MPEG-2 ISO 639 Language Descriptor. Each of the three characters is an ISO Latin-1 lower-case character in the "a" to "z" range. Codes for these characters are in the ASCII range and range from 0x61 to 0x7A. The coding of the ISO Latin-1 character set is defined in ISO/IEC 8859-1[4].

number_segments

A string representing the text item in one of the languages may be broken for delivery into pieces called segments. The reason one might split a string into segments is because the total length of the MSS may be smaller if segments are used. For example, the first ten words of a phrase might compress well using the Huffman tables but the last two words do not encode into short Huffman codes. That string could be broken into two segments, the first compressed using Huffman coding and the second sent uncompressed. In another example, a piece of text may have one or two non-ASCII character codes. The full string could be broken into segments to isolate these non-ASCII characters so the rest of the string can benefit from Huffman encoding.

Each segment can make use of a compression type and/or mode that differs from the preceding or the following segment.

compression_type

PSIP currently defines three compression types; other encodings of the compression type field are reserved for future use. Table 9.2 defines the encoding of the compression_type field.

mode

Each segment's "mode" reflects the method by which character data bytes in the MSS are to be converted to character codes for display. There are many different standards for character encodings. ASCII is one of the oldest and most common in the US. The Unicode Basic Multilingual Plane (BMP) represents a 16-bit mapping including more than 40,000 entries. PSIP references the BMP of Unicode Version 3.0[1] for representation of multilingual character data.

TABLE 9.2 Compression Types

Value	Compression Method	Notes
0x00	No compression	Indicates the bytes are uncompressed (except for mode 0x3E, see text).
0x01	Huffman coding using the Huffman "title" tables	These Huffman encode/decode tables are defined in Annex C of A/65 in Tables C.4 and C.5. They are optimized for use with program titles, where the first letter of words are often capitalized.
0x02	Huffman coding using the Huffman "description" tables	These Huffman encode/decode tables are defined in Annex C of A/65 in Tables C.6 and C.7. These are optimized for description text where capitalized words are less common.
0x03-0xAF	Reserved for future use by ATSC	
0xB0-0xFF	Used in private systems	ATSC will avoid defining future compression types in this range.

The mode mechanism in the Multiple String Structure takes advantage of the fact that sets of characters commonly used in various languages are grouped together in the definition of the Basic Multilingual Plane. For example, the characters making up the Greek alphabet (both upper- and lower-case characters) are found in the BMP in the 0x0300 to 0x3FF range. Russian characters are in the 0x0400 to 0x04FF range, Arabic characters are in the 0x0600 to 0x06FF range, and Thai is in the 0x0E00 to 0x0EFF range.

Table 9.3 describes the encoding of the mode field and indicates some of the languages covered in some of the 256-character ranges.

TABLE 9.3 Mode Byte Encoding

Value	Meaning	Notes
0x00	Selects ISO Latin-1 character set	ISO Latin-1 is equivalent to UTF-16 characters in the 0x0000 to 0x00FF range.
0x01	Selects Unicode code range 0x0100 to 0x01FF	Includes many European Latin characters.
0x02	Selects Unicode code range 0x0200 to 0x02FF	Includes the standard phonetic alphabet.
0x03	Selects Unicode code range 0x0300 to 0x03FF	Includes the Greek alphabet.
0x04	Selects Unicode code range 0x0400 to 0x04FF	Includes Russian characters and characters used in Slavic languages.
0x05	Selects Unicode code range 0x0500 to 0x05FF	Many characters in this range are used in Armenian and Hebrew languages.
0x06	Selects Unicode code range 0x0600 to 0x06FF	Includes Arabic characters.
0x07-0x08	Reserved	
0x09	Selects Unicode code range 0x0900 to 0x09FF	Includes characters used in Devanagari and Bengali languages.

TABLE 9.3 Mode Byte Encoding (continued)

Value	Meaning	Notes
0x0A	Selects Unicode code range 0x0A00 to 0x0AFF	Includes characters used in Punjabi and Gujarati languages.
0x0B	Selects Unicode code range 0x0B00 to 0x0BFF	Includes characters used in Oriya and Tamil languages.
0x0C	Selects Unicode code range 0x0C00 to 0x0CFF	Includes characters used in Telugu and Kannada languages.
0x0D	Selects Unicode code range 0x0D00 to 0x0DFF	Includes characters used in the Malayalam language.
0x0E	Selects Unicode code range 0x0E00 to 0x0EFF	Includes characters used in Thai and Lao.
0x0F	Selects Unicode code range 0x0F00 to 0x0FFF	Includes characters used in Tibetan.
0x10	Selects Unicode code range 0x1000 to 0x1FFF	Includes characters used in Georgian.
0x11-0x1F	Reserved for future ATSC use	
0x20-27	Selects appropriate Unicode code page	These Unicode code pages include characters used for mathematics, typesetting, and other special uses.
0x28-0x2F	Reserved for future ATSC use	
0x30	Selects Unicode code range 0x3000 to 0x30FF	Includes characters of the Hiragana and Katakana Japanese alphabets.
0x31	Selects Unicode code range 0x3100 to 0x31FF	Includes Bopomopho characters.
0x32	Selects Unicode code range 0x3200 to 0x32FF	Enclosed Chinese-Japanese-Korean characters, some ideographic characters.
0x33	Selects Unicode code range 0x3300 to 0x33FF	Enclosed Chinese-Japanese-Korean characters, some ideographic characters.
0x34-0x3D	Reserved for future ATSC use	
0x3E	Selects the standard compression scheme for Unicode	Defined in SCSU[5].
0x3F	Selects Unicode, UTF-16 Form	Standard 16-bit Unicode character encoding.
0x40	Character is defined by an included bitmap	Used in Taiwan, defined in A/68[6].
0x41	Character is defined by an included bitmap	Used in Taiwan, defined in A/68[6]; includes a 16-bit Unicode UTF-16 character preceding.
0x42-0xDF	Reserved for future ATSC use	
0xE0-0xFE	Used in other systems	Will not be defined by ATSC in the future.
0xFF	Reserved	

Many of the defined values for mode indicate that each of the character codes in the segment is a 16-bit Unicode character formed by performing the calculation character code byte + (256 * mode). Note that for non-Latin languages it won't always be possible for a given string to use only characters within one of the 256-

character ranges. For example, a Greek string that includes Arabic numerals must be segmented because the numbers 0 to 9 use mode zero.

number_bytes

This integer simply indicates the length of the compressed_string_byte[] array to follow.

compressed_string_byte[k]

These are the bytes comprising the string data itself. For some values of compression_type and mode, these bytes are interpreted as a bit string. In other cases, each byte is interpreted as a character code, while in other modes, pairs of bytes are interpreted as 16-bit character codes.

Huffman text compression

The A/65 PSIP Standard includes two sets of Huffman encode/decode tables for use with English-language text. One is optimized for program title text (Figures C.4 and C.5 in the A/65 Standard), in which the first letter of words is often capitalized. The other is optimized for use with program descriptions (Figures C.6 and C.7 in A/65). Annex F of the PSIP Standard describes Huffman-based text compression and gives a general background and some simple examples.

Huffman coding, in general, is cleverly designed such that shorter symbols represent more common objects. As applied in the ATSC PSIP Standard, each transmitted symbol represents the next character in the text string from the context of the previous character in the string. This so-called "second-order" Huffman coding creates shorter symbols for the more commonly occurring character pairs. To handle the beginnings and ends of the string, the first and last characters in the string are assumed to be "termination" characters.

The Huffman tables included in the ATSC Standard were derived from actual program title and description data and hence perform reasonably well. Although compression of short strings is not quite as efficient as that of longer ones, the tables can reduce the size of typical transmitted text data by approximately 50%.

Let's take a look at the Huffman symbols representing some letter pairs. In the Huffman table for program titles, the letter sequence "Ho" has a one-bit Huffman code (value 0). The one-bit length indicates that a capital "H" was followed by lowercase "o" more often than any other letter in the set of program titles analyzed to create this coding table.

Some other letter pairs are used infrequently. For example, a space character followed by an uppercase "X" is encoded with the 10-bit Huffman code value 0b0000000010. A space followed by an X would only occur in a multi-word program title where a word other than the first word in the title started with the letter X—obviously a possibility but not a likely one.

Some letter pairs are so infrequently used that they do not have Huffman codes at all. How can these pairs these be represented? In our coding scheme, such pairs involve the use of an "escape" sequence, where temporarily the string switches from Huffman coding to literal, not compressed, character coding. In the tables the escape code is decimal value 27, corresponding to the ASCII code for the <esc> character. Looking again at the Huffman table for title text, we can see an entry for uppercase J followed by <esc>, coded as 000. For "J" followed by the vowels "a," "e," "i," or "u," the table defines Huffman codes. A "J" followed by any other letter generates an escape sequence where that letter is placed into the bit-string without compression.

Let's first look at an example of how the Huffman title tables are used to compress and de-compress the program title "Moonlighting." The termination character, shown as "<term>" in Table 9.4 and coded as 0 in the Huffman encode table definitions, is assumed to precede the first character of the string.

TABLE 9.4 Huffman Encoding—Example 1

Prior Symbol	Symbol		Bit String
<term>	M	→	1010
M	o	→	00
o	o	→	0011
o	n	→	111
n	l	→	111010110
l	i	→	001
i	g	→	1100
g	h	→	00
h	t	→	1111
t	i	→	001
i	n	→	10
n	g	→	101
g	<term>	→	110

As shown, the bit-string encoding for the title "Moonlighting" is "1010 00 0011 111 111010110 001 1100 00 1111 001 10 101 110." In hexadecimal coding, the bytes are 0xA0, 0xFF, 0x58, 0xE1, 0xE6, and 0xB8. Zero-bits are used to pad out the last byte. In this example, the 12-character program title has been compressed to six bytes, a compression ratio of 50%.

For a second example, we pick a string that includes an unusual letter pair, the title "CatDog" which has a lowercase "t" followed by a capital "D." Table 9.5 shows the coding.

TABLE 9.5 Huffman Encoding—Example 2

Prior Symbol	Symbol		Bit String		
<term>	C	→	1011		
C	a	→	100		
a	t	→	011		
t	D	→	<esc>	=	11000010
<esc>	D	→	ASCII "D"	=	01000100
D	o	→	00		
o	g	→	00010		
g	<term>	→	110		

No Huffman code is defined for "D" preceded by "t," so we must use the code for <esc> (value 27) preceded by "t." Next, the code for "D" preceded by <esc> is used, but the <esc> character has a special meaning: the eight bits in the compressed string following <esc> are always an uncompressed byte. The <esc> character is actually an "escape" from compression into uncompressed mode. So the bit-stream representation for "D" preceded by "t" is <esc>-D.

The coding for "CatDog" turns out to be 1011 100 011 11000010 01000100 00 00010 110, or in hexadecimal, 0xB8 0xF0 0x91 0x01 0x60. The length of the encoded string is six bytes, the same as the length of the uncompressed string. In cases like these compression is not helpful. When the length after compression is equal to or greater than the length without any compression, Huffman coding gives no benefit and is typically not used.

The decoding process involves the use of the corresponding Huffman decode table, in this case the one for titles, to undo the encoding. For the first-order Huffman coding used in PSIP, each and every character code in the range 0 to 127 has its own decode tree, each of which represents all of the symbols that have defined Huffman codes when starting from that character.

Huffman decoding tables may be represented in computer memory or in standards documents in various ways, but the PSIP Standard uses the following format: The first 256 bytes are a list of 128 16-bit pointers to decode tables for individual characters in the range 0 to 127. Let's call this initial block of data the "decode tree offset list" because each entry represents the offset, or number of bytes, from the top of the data to where that particular decode tree begins.

Following the decode tree offset list are the 128 decode trees themselves. Nodes in each tree are defined by two data bytes, one to indicate the "0" branch (also known as the "left child") and the other to indicate the "1" branch (the right child). Each branch either terminates in a "leaf" or points (indirectly through the initial table) to the decode table for the next character. Leaves in our application are the ASCII-coded values of the character resulting from the decoding process.

Whether the table entry represents a leaf or another node is indicated by the value of the most-significant bit (MSB) in the byte. When the MSB is value 1, the entry is a leaf, and the lower seven bits are the ASCII character code. When the MSB is a zero, the entry is an index into the decode tree offset list where the child node may be found. The table entry for each node, then, is a byte for the "0" branch and a byte for the "1" branch. The PSIP Huffman decode table for titles can be viewed as being 970 by 16 bits, or as defining a tree having a total of 970 nodes (not counting the root nodes in each tree). The Huffman decode tree for description text is somewhat smaller, with 891 nodes and leaves.

As an exercise, we can reconstruct the decode tree when we start from the letter "D." The letter D in ASCII is 0x44 or 68 decimal. Therefore, the byte offset to the "D" decode tree is the 68th entry in the decode tree offset list. Since each entry is two bytes, the byte offset for the "D" decode tree can be found at byte 68*2=136. The offset to the "D" tree is found to be (2*256) + 172 = 684. At byte offset 684 we see data bytes representing the "D" tree: 5, 6, 249, 155, 1, 245, 2, 242, 233, 229, 239, 3, 225, 4.

Interpreting the most significant bit to indicate either a leaf character or an index into the decode tree offset list, we can reconstruct the decode table for the "D" tree. Starting at the top (root) node, the "5" entry means that the definition of the left child node (0) starts at a word offset of 5 from the start of the table. Data for this branch is at offset 684 + (2*5) = 694. At byte offset 694 we find value 239 for the left child node (second 0), which has bit 7 set and equals 128 + 111. The ASCII code value 111 is character "o." So we have found the first leaf, and the bit string "00" is the Huffman symbol for "D" followed by "o." The same procedure can be continued to construct the full tree, shown in Figure 9.2.

We can follow a general procedure that can be used with the decode tables to decompress a Huffman encoded string. First we define the following notation:

- Tc – the byte offset from the top of the decode tree. We start decoding the string at the root node, at byte offset 256 in Table C.4 of the A/65 Standard.

- Bn – the next bit of input data, either 0 or 1.

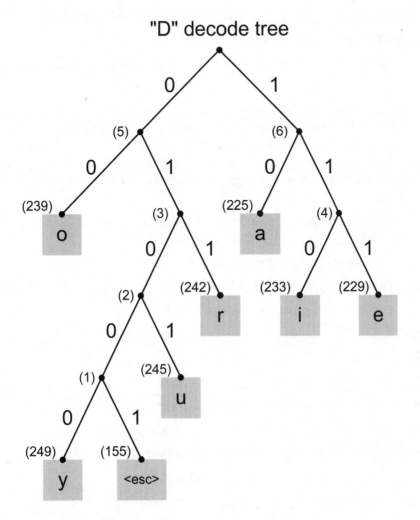

Figure 9.2 Huffman Decode Tree for Letter "D"

The steps that can be used to decode strings using the data structures as they are defined in A/65 are:

1. Look up the value at byte offset (Tc + Bn).

2. If the value is greater than or equal to 128:

- subtract 128 and use the result as an ASCII code for an output character. If the output character is <esc> (value 27), take instead the eight following bits of input data as the output character. If the output character is 0, we have found the end of the string. If it is not zero, record the character and look up the offset of that character's decode table to use as the new Tc, and continue at Step 1.

- Otherwise (if the value is less than 128), continue to the next step.

3. Multiply the value by 2 to get a word offset.

4. Add this offset to Tc to get the address of the data for the next node.

5. Loop back to Step 1.

As a last exercise, let's manually decode our second example of the bit string 10111000111100001001000100000000101100000. We don't know where one symbol ends and the next starts until we begin to trace through the decode tables. We start at the root node, address 256. The square brackets notation [x] is shorthand for "look up the value at offset x into the data." The input data bit string is highlighted in bold.

1.   [256+**1**]=28; 28*2=56 → 56+256 = 312, the address of the next node we need to use.

2.   [312+**0**]=25; 25*2=50 → 50+256 = 306

3.   [306+**1**]=20; 20*2=40 → 40+256 = 296

4.   [296+**1**]=195 → Bit 7 + ASCII(67) = "**C**"

5.   The "C" decode table is at 67*2 = 134 → 2*256 + 146 = 658

6.   [658+**1**]=12; 12*2=24 → 24+658 = 682

7.   [682+**0**]=10; 10*2=20 → 20+658 = 678

8.   [678+**0**]=225 → Bit 7 + ASCII(97) = "**a**"

9.   The "a" decode table is at 97*2 = 194 → 4*256 + 28 = 1052

10. [1052+**0**]=25; 25*2=50 → 50+1052 = 1102

11. [1102+**1**]=22; 22*2=44 → 44+1052 = 1096

12. [1096+**1**]=244 → Bit 7 + ASCII(116) = "**t**"

13. The "t" decode table is at 116*2 = 232 → 6*256 +208 = 1744

14. [1744+**1**]=21; 21*2=42 → 42+1744 = 1786

15. [1786+**1**]=19; 19*2=38 → 38+1744 = 1782

16. [1782+**0**]=15; 15*2=30 → 30+1744 = 1774

17. [1774+**0**]=11; 11*2=22 → 22+1744 = 1766

18. [1766+**0**]=8; 8*2=16 → 16+1744 = 1760

19. [1760+**0**]=5; 5*2=10 → 10+1744 = 1754

20. [1754+**1**]=3; 3*2=6 → 6+1744 = 1750

21. [1750+**0**]=155 → Bit 7 + ASCII(27) = <esc>. Take next 8 bits as output: 01000100 = "**D**"
The "D" decode table is at 68*2=136 → 2*256+172 = 684

22. [684+**0**]=5; 5*2=10 → 10+684 = 694

23. [694+**0**]=239 → Bit 7 + ASCII(111) = "**o**"

24. The "o" decode table is at 111*2 = 222 → 6*256 + 10 = 1546

25. [1546+**0**]=27; 27*2=54 → 54+1546 = 1600

26. [1600+**0**]=23; 23*2=46 → 46+1546 = 1592

27. [1592+**0**]=17; 17*2=34 → 34+1546 = 1580

28. [1580+**1**]=12; 12*2=24 → 24+1546 = 1570

29. [1570+**0**]=231→ Bit 7 + ASCII(103) = "**g**"

30. The "g" decode table is at 103*2=206 → 5*256 + 6 = 1286

31. [1286+**1**]=16; 16*2=32 → 32+1286 = 1318

32. [1318+**1**]=14; 14*2=28 → 28+1286 = 1314

33. [1314+**0**]=128 → Bit 7 + 0 = **<term>**

Once the terminator character is reached, any remaining bits are simply discarded.

Compression for Unicode

Unicode Transformation Format-16 (UTF-16) is a 16-bit character encoding including nearly all the characters used in languages throughout the world. Delivery of uncompressed 16-bit characters can consume considerable bandwidth, so PSIP offers two methods to reduce the size of transmitted data when UTF-16 is used.

A simple run-length encoding is available for any segment in which a run of Unicode characters share the same value for the upper byte of the first 64K code points. Unicode is organized such that alphabets for many languages are coded within one 256-byte block in the 16-bit code space, so this run-length method may be helpful.

For example, the Greek word for "keyboard" is the 12-letter word "πληκτρολόγιο." The UTF-16 representation is: 0x03A0, 0x03BB, 0x03AE, 0x03BA, 0x03C4, 0x03C1, 0x03BF, 0x03BB, 0x03CC, 0x03B3, 0x03B9, and 0x03BF. All of the upper- and lower-case Greek characters are in the range 0x0370 through 0x03FF in UTF-16 encoding, and one of the methods used in the Multiple String Structure takes advantage of this fact. An MSS representation of the word "πληκτρολόγιο" is 20 bytes long using the run-length encoding mode, whereas it would be 32 bytes if uncompressed UTF-16 encoding were to be used.

The second option for Unicode-encoded characters is to use a compression method standardized by the Unicode Consortium in Unicode Technical Standard #6, A Standard Compression Scheme for Unicode (SCSU)[5]. The latest version of this report can be found on the web at http://www.unicode.org/unicode/reports/tr6/. Note that one MSS instance may include, for a string representing a single language, many segments. It may have segments that are uncompressed, segments that are compressed using one of the Huffman tables, and/or segments that are compressed with Unicode compression. Choice as to which method is best for a given segment or string is based on which one results in the shortest representation.

SCSU uses some clever techniques involving dynamically positioned windows, wherein once a window position is set, character code values are represented as byte-value offsets relative to the position of the currently active window. One example shown in the report compressed 116 16-bit UTF-16 characters (232 bytes) down to 178 bytes, a compression ratio of 1.3 to 1.

Null strings

Sometimes in the PSIP data no text is available for a particular string. There are at least four possible ways to indicate a null string, each using a different number of transmitted bytes:

1. Preceding each instance of the Multiple String Structure is a length field, indicating the length of the MSS to follow. A zero value for this length field indicates a null string (no string data available for any language). This is the most compact way to indicate a null string and is the preferred method when no strings are defined in any language.

2. In the MSS itself, one can indicate zero for the number of strings. This method leaves the MSS with just one byte.

3. Again in the MSS, any string with zero indicated as the number of segments in that string is a null string. This method should be avoided because it wastes bandwidth and is no different than simply not including a string for that language in the MSS.

4. An even more wasteful method in the MSS is to specify a zero byte-length for each of the segments in the string.

Multilingual text

With just one exception, all textual data delivered in PSIP data is carried within instances of the Multiple String Structure. The exception is the short channel name, which is represented as seven uncompressed 16-bit Unicode (UTF-16) characters. Whenever text is available in more than one language, a receiving device must choose one and only one for display. Often, a user setup option is provided so the user can indicate a language preference. For products designed for a certain country or region, common practice among manufacturers seems to be to set the default language setting as the official or most commonly used language in that region.

It may be that a receiver has been set up with one language as a preference. Say that choice is French. Let's say that text for a program title is only available in one language, say English. The receiver could display nothing, or (probably a better choice) display the English text anyway. Whenever a string is available in just one language, a reasonable rule to follow is to display it as-is without regard to language.

If a string is offered multi-lingually, the receiver can make a choice and select the one that best fits the user's preferences. If text is available in more than one language but neither matches the one set as the user's preference, another method must be used to make the choice. To completely cover all of the cases, a receiver would have to offer the user a way to specify a prioritized list of language preferences.

Example Multiple String Structure

Table 9.6 shows an example MSS in which an English string and its Spanish equivalent are represented. The strings are "Homer is hired as a writer at the cookie factory." and its Spanish translation "Homer se emplea como un escritor en la fábrica de galleta."

Languages

Language tags are present in A/65 in several places to associate an object with a written or spoken language. In the Caption Service Descriptor, the language of each caption service is indicated. In the Service Location Descriptor, the language of each audio track is given. The Multiple String Structure discussed above tags each of its text strings with a language code. In each of these instances, the language code is a 24-bit field coded according to the ISO 639-2/B[3] Standard.

MPEG-2 *Systems* also uses the ISO 639-2/B[3] Standard for coding the names of languages, in a descriptor aptly named the ISO 639 Language Descriptor. We covered this descriptor in Chapter 3 (see "ISO 639 Language Descriptor" on page 86).

Representation of Time and Date

Time and date fields appear in various PSIP and ATSC data broadcast tables, including the System Time Table where the current date and time of day is provided, and the Event Information Table where the scheduled start times of future

TABLE 9.6 Multiple String Structure Example

Field Name	Value (hex)	Description
number_strings	02	This MSS includes two strings.
ISO_639_language_code	454E47	"eng" (English) in ASCII.
number_segments	01	This string has one segment.
compression_type	00	No compression (in this example).
mode	00	ISO-Latin-1 character set.
number_bytes	2F	The segment data to follow is 47 bytes long.
compressed_string_byte[k]	48 6F 6D 65 72 20 69 73 20 68 69 72 65 64 20 61 73 20 61 20 77 72 69 74 65 72 20 69 6E 20 61 20 63 6F 6F 6B 69 65 20 66 61 63 74 6F 72 79 2E	ISO-Latin-1 coding of "Homer is hired as a writer at the cookie factory."
ISO_639_language_code	737061	"spa" (Spanish) in ASCII.
number_segments	01	This string has one segment.
compression_type	00	No compression (in this example).
mode	00	ISO-Latin-1 character set.
number_bytes	3A	The segment data to follow is 58 bytes long.
compressed_string_byte[k]	48 6F 6D 65 72 20 73 65 20 65 6D 70 6C 65 61 20 63 6F 6D 6F 20 65 6E 20 65 73 63 72 69 74 6F 72 20 65 6E 20 6C 61 20 66 E1 62 72 69 63 61 20 64 65 20 67 61 6C 6C 65 74 61 2E	ISO-Latin-1 coding of "Homer se emplea como un escritor en la fábrica de galleta."

programs are announced. Wherever a time/date is indicated, it is given as the number of seconds since the beginning of "GPS time," 00:00:00 UTC January 6th, 1980. The US Naval Observatory operates the Master Clock that serves as the reference point for time-of-day clocks, Coordinated Universal Time (known as UTC). Time zones throughout the world are often specified relative to UTC, which in the past was known as Greenwich Mean Time (GMT). GPS time is synchronized to within 1 microsecond of the US Naval Observatory's Master Clock.

Simply counting seconds from a given point in the past is not sufficient by itself for keeping track of time of day, because date and time of day must be kept aligned with the position of the Earth in the heavens. The Earth's annual trip around the Sun is not exactly 365 days. Actually, it is close to 365.2422 days, the period of time known as the "cycle of the seasons" or the tropical year. We introduce the concept of leap years, which are the years when we add an extra day (February 29th). Leap years usually occur every fourth year.

If we were to simply add a day every four years our average year would be 365.25 days in length, which is a bit long. Therefore as a further refinement, we skip the leap years on centennial years not divisible by 400. Thus, there are 97 leap years in each 400-year span. With this additional adjustment, the average year is 365.2425 days long, which matches the duration of the tropical year to within 0.0003 days (26 seconds) per year.[*]

Leap years help keep our calendars aligned with the Earth's orbit around the Sun. Another aspect of the Earth's motion is its rotation about its axis. We need to make sure our clock time stays aligned with the cycle of each day as well. The time period known as the "second" was defined based on astronomical observations made during the eighteenth and nineteenth centuries. The Earth's rotational rate is slowly decreasing over time due to the braking action of the tides. Whereas the number of seconds in a day was exactly 86,400 in 1820, careful observations of the Earth's rotation using sensitive equipment has shown that the number of seconds in one complete revolution is now approximately 2 milliseconds longer than that.

Over a period of time, this difference accumulates. When that difference begins to approach a whole second (every eighteen months or so), a future date is chosen for insertion of a "leap second." On a day it is scheduled to occur, which by convention is always a June 30th or December 31st, the extra second is inserted just following 23:59:59 UTC.

So for purposes of keeping very accurate track of time of day, UTC is the world standard. GPS time tracks it to within a microsecond, but GPS time is a linear count of seconds since January 6th 1980 while UTC is at times discontinuous. In order to convert GPS time to UTC, all that is needed is knowledge of the number of leap seconds that have occurred since the start of GPS time. This leap second count is provided in the PSIP System Time Table, and it is also available from the Global Positioning Satellites themselves as a part of their broadcast signal. You can learn more information about leap seconds by visiting the website operated by the Time Service Department of the US Naval Observatory at http://tycho.usno.navy.mil/leapsec.html. The international agency responsible for scheduling leap second events is the International Earth Rotation Service (IERS). You can visit their website at http://www.iers.org/ and download the latest "Bulletin C" to find out the current count. UTC, then, can be computed from the count of GPS seconds by subtracting the number of leap seconds that have occurred since GPS time began.

The designers of the PSIP Standard chose to base the representation of time on GPS seconds rather than UTC for several reasons. For uses such as setting the clock in a DTV, making timed recordings, or displaying program guides, one second here or there does not matter much. But various events occurring within the distribution chain of digital television such as insertion of commercials or synchronization of

---

[*] In about 3000 years, we'll need to skip another leap year to continue to stay on track.

data with video, may need precision far more accurate than one second. GPS receivers are now common and relatively inexpensive, and the output of such a receiver can provide time references for use by switching or data insertion equipment at microsecond accuracy.

Although it would happen rarely if ever, use of GPS seconds as a time reference allows an event to occur right at the point at which a leap second is inserted. These points in time typically cannot be referenced using UTC time because UTC is discontinuous at those moments.

PSIP-generating equipment may or may not anticipate the occurrence of a leap second. To be absolutely precise, generating equipment has the capability to recognize the scheduled time at which a leap second is to be inserted and can adjust the GPS time for future events accordingly. For certain types of events, such precision may be necessary.

We can look at the two possible cases for treatment of leap seconds by PSIP-generating equipment. In method A, no accounting is made for future leap seconds. If and when UTC jumps, all the future times are re-computed and re-transmitted. In method B, the point in time at which a leap second is scheduled to occur is known in advance, and announcements for the time of future events on the other side of that boundary already take the leap second into account.

The PSIP Standard does not mandate which method shall be used. Practically speaking, receivers should accommodate the one-second errors that can occur due to the occurrence of leap seconds.

The difference between method A and method B is illustrated in the following example. For simplicity, times are given relative to UTC so that local time zones and Daylight Saving Time issues do not need to be accounted for:

- current time of day (UTC):1:00 p.m., December 30th, 2001;

- event start time (UTC):2:00 p.m., January 2nd, 2002;

- let's say a leap second is scheduled to occur just after 12:59:59 p.m. on December 31, 2001;

- the leap seconds count on December 31st is 13.

The System Time Table indicates the following parameters for the current time of day:

- GPS seconds = 693,752,413, or 0x2959D25D;

- GPS to UTC offset (count of leap seconds) = 13.

Using Method A, where the upcoming leap second is not accounted for:

- event start time in EIT:694,015,213, or 0x295DD4ED;

- converted to UTC that's 2:00:00 p.m., January 2nd, 2002;

- the number of seconds until the event is $694,015,213 - 693,752,413 = 262,800$;

- the number of hours, minutes, and seconds to the event: 73:00:00.

With method A, all event start times are recomputed just after the leap second event occurs and all EITs updated. In receiving equipment, the number of seconds to the event must be adjusted when the EIT update is processed and the event start time is found to have changed.

- Event start time in EIT (after re-computation):694,015,214, or 0x295DD4EE

Using method B, where the upcoming leap second is anticipated and it is known the given event will occur after the next leap second is to be inserted, PSIP-generating equipment can send the following information:

- Event start time in EIT:694,015,214, or 0x295DD4EE

- Converted to UTC: 2:00:01 p.m., January 2nd, 2002

- Number of seconds to event: $694,015,214 - 693,752,413 = 262,801$

- Hours, minutes, seconds to event: 73:00:01

When method B is used, the number of seconds until the event is to occur is exactly correct. The indicated event start time does not change as time moves across the leap second boundary, hence there is no need for receiving devices to recompute the time until the event is to occur.

Receiving equipment, in general, does not know whether Method A or Method B is used at leap second boundaries. Consumer equipment may not do anything special with regard to leap seconds, since the error is just one second at most and occurs very rarely. Of course, the GPS to UTC offset must always be used to convert GPS time to UTC. Equipment for commercial use can take advantage of the precision offered by the GPS system, take careful account of the leap seconds, and perform timing functions with very high precision.

Time zones

As mentioned, the counts of GPS seconds delivered in PSIP tables reflect the number of seconds that have elapsed since the beginning of GPS time, 00:00:00 January 6th, 1980. GPS time identifies time relative to UTC, the time zone local to Greenwich, England and previously known as Greenwich Mean Time (GMT). To convert to local time of day for time zones east or west of GMT, one must adjust by adding or subtracting the number of hours corresponding to the time zone offset.

For example, Pacific Standard Time (PST) is behind GMT by eight hours. To convert GPS seconds to Pacific Standard Time, eight hours must be subtracted from the count of GPS seconds.

The following formula may be used to convert GPS seconds into local time of day:

```
Local time in seconds =
    GPS sec. – leap sec. count + time zone offset (in seconds) +
                         daylight saving offset (in seconds, if applicable)
```

Daylight Saving Time

Daylight Saving Time applies in most communities in the US in the summer months. The PSIP System Time Table delivers information that can automate the process of handling Daylight Saving Time in receiving equipment. Again, all of the times reflected in PSIP table parameters are relative to UTC, so local fluctuations in time zone or Daylight Saving Time have no impact on these values. Daylight Saving Time and time zone adjustment relate only to the interpretation and display of the time data delivered in the PSIP tables; in other words, only to the user interface.

A consumer device receiving PSIP data cannot typically know by itself its physical location (which would identify its time zone) or whether or not Daylight Saving Time is observed in its community. The user must supply these parameters when the device is configured for use. An exception occurs in some cable set-top boxes in which the cable operator knows these parameters, and knowing the box location, can set its time zone. In some other cases, location can be determined from other consumer-supplied data such as postal zip code or telephone area code and prefix.

The PSIP System Time Table delivers these pieces of data relative to Daylight Saving Time:

- a flag indicating whether a transition into or out of Daylight Saving Time is imminent (will occur within the next thirty days).

- an indication of the day of the month and hour in the day the transition is to take place.

The System Time Table indicates one of the following four states related to Daylight Saving Time:

1. Daylight Saving Time is not in effect, and will not go into effect for at least another 30 days.

2. Daylight Saving Time will go into effect on the day of the month and time indicated.

3. Daylight Saving Time is in effect and will stay in effect for at least another 30 days.

4. The transition out of Daylight Saving Time will occur on the day of the month and time indicated.

Please see "Interpretation of the Daylight Saving Time fields" on page 161 for details regarding daylight saving time signaling.

References

1. The Unicode Standard, Version 3.0, The Unicode Consortium, Addison-Wesley Pub. ISBN 0201616335, 2000.

2. ISO/IEC 10646-1:2000, "Information technology — Universal Multiple-Octet Coded Character Set (UCS) — Part 1: Architecture and Basic Multilingual Plane."

3. ISO 639-2:1998, "Codes for the representation of languages – Part 2: Alpha-3 code."

4. ISO/IEC 8859-1:1998, "Information technology – 8-bit single-byte coded graphic character sets – Part 1: Latin alphabet No. 1."

5. Unicode Technical Standard #6, "A Standard Compression Scheme for Unicode," Version 3.4, The Unicode Consortium, 2002.

6. ATSC Standard A/68, "Use of ATSC A/65A PSIP Standard in Taiwan," 11 July 2001.

The Main PSIP Tables

In this chapter we explore the core set of SI tables specified in ATSC A/65: the System Time Table, Master Guide Table, Cable and Terrestrial Virtual Channel Tables, the Rating Region Table, the Event Information Table, and the Extended Text Table (we leave the Directed Channel Change tables for Chapter 13). Before we start, we look at the common structure of all of the PSIP tables.

Common Structure of PSIP Tables

All of the table sections defined in the A/65 Standard are derived from the "long form" syntax of the MPEG-2 private section syntax. MPEG uses the term "private" in this context to mean "defined outside MPEG." Each of the PSIP tables includes one additional 8-bit field after the MPEG-defined header bytes, the protocol_version field. Table 10.1 shows this common structure.

TABLE 10.1 PSIP Table Section Common Syntax

Field Name	Number of Bits	FIeld Value	Description
PSIP_table_section() {			Start of the PSIP_table_section().
table_id	8		Identifies the type of PSIP table section.
section_syntax_indicator	1	1b	All PSIP table sections are formatted in MPEG "long-form" syntax.
private_indicator	1	1b	This flag is not defined in MPEG and is set to 1 in PSIP tables.
reserved	2	11b	Reserved bits are set to 1.

TABLE 10.1 PSIP Table Section Common Syntax (continued)

Field Name	Number of Bits	FIeld Value	Description
section_length	12		An unsigned integer that specifies the length, in bytes, of data following the section_length field itself to the end of this table section. PSIP table sections are either limited to 1024 total bytes (STT, VCT, RRT) or 4096 bytes (MGT, EIT, ETT) corresponding to 1021 or 4093 byte limits for the section_length field.
table_id_extension	16		Used to differentiate multiple instances of a table section having a common table ID. Various PSIP tables define this field in different ways.
reserved	2	11b	Reserved bits are set to 1
version_number	5		Typically, the version number reflects the version of a table section, and is incremented when anything in the table changes.
current_next_indicator	1	1b	Indicates, in general, whether the table section is currently applicable (value 1) or is the next one to be applicable (value 0).
section_number	8	0	For tables segmented into sections, section_number indicates which part this section is.
last_section_number	8	0	Indicates the section number of the last section in a segmented table.
protocol_version	8	0	Indicates the protocol version of this table section.
PSIP table data()			The fields and data structures specific to this type of table section that make up the table section. The syntax and semantics for PSIP table data() are established by the table ID and protocol version fields.
CRC_32	32		A 32-bit checksum designed to produce a zero output from the decoder defined in the MPEG-2 *Systems* Standard.
}			End of table section.

Protocol version

The protocol_version offers yet another extension for the table ID in that it allows, in the future, ATSC to define a type of table section entirely independent of an existing table section definition while it shares the same table ID value. For the currently defined tables, protocol_version is zero.

For example, the System Time Table uses table ID value 0xCD. At some time in the future, a new type of table with syntax and semantics entirely different than the System Time Table but using the table ID value of 0xCD could be defined. It would use a value other than zero for protocol_version to distinguish it from the original System Time Table.

A receiver may only recognize the System Time Table (table ID 0xCD and protocol_version zero). In that case, a table section with table ID 0xCD and a non-zero protocol_version field would be summarily discarded. Another receiver, built to support a new protocol, could recognize both the System Time Table and the new table using the protocol_version field to distinguish between the two.

System Time Table (STT)

The System Time Table is a small data structure that is sent once per second to synchronize the clocks in all receiving devices to the current time of day. It also indicates whether or not Daylight Saving Time is in effect, and signals the day and hour for transitions into or out of Daylight Saving Time. Time in the STT is represented as the count of GPS seconds that have occurred since 00:00:00 January 6th, 1980. To convert this count into time of day, one must account for the number of leap seconds that have occurred since the start of GPS time and the present. The STT provides that number in a parameter called "GPS-to-UTC offset." GPS time and leap seconds were discussed in detail in Chapter 9 (see "Representation of Time and Date" on page 147).

ATSC A/65 stipulates that the count of GPS seconds and the leap seconds count delivered in the STT must be correct to within plus or minus one second. This accuracy tolerance was arrived at after much debate. Achieving this level of accuracy is straightforward, but there are several sources of potential error within the system:

- Accuracy of the reference clock used to feed the PSIP generator. If a GPS receiver is used, accuracy to one microsecond is possible, as such receivers often produce output pulses to that level of precision. If other clock sources are used, for example a PC's internal clock, accuracy could be much worse and could drift over time.

- Latency between the clock reference and the output of the STT from the generator.

- Latency between the moment PSIP generator data appears at the input of the multiplexer and the moment that data appears in the output of the multiplexer. Equipment capable of multiplexing audio/video data with PSIP data must buffer the PSIP data, generating a buffering delay. Depending upon the output data rate, the multiplexer must wait until an appropriate time before placing a transport packet into the output Transport Stream. Processing time in the multiplexer may be a factor as well.

Very good precision can be achieved by pre-compensating for some of these delays.

Structure of the System Time Table

Figure 10.1 illustrates the structure of the System Time Table section.

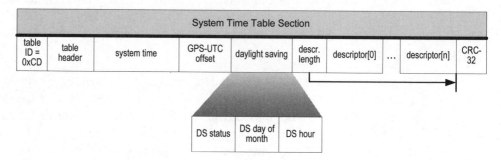

Figure 10.1 System Time Table Section Structure

Following the standard table header is the 32-bit system_time field, followed by the count of leap seconds since the beginning of GPS time, and then Daylight Saving Time information. At the end of the table section, descriptors pertinent to the System Time Table function may be placed.

STT transport

Transport of the System Time Table involves these considerations:

- An STT section, when carried in the Transport Stream, are placed into Transport Stream packets with PID value 0x1FFB, the SI base_PID.

- TS packets containing an STT section must not be scrambled (the transport_scrambling_control bits are set to zero).

- As with all PSIP tables, an adaptation field must not be present in TS packets carrying the VCT.

- The STT may not be sectioned for delivery

- Each STT section is limited to a total length of 1024 bytes.

- Unlike all other tables, the STT does not use the MPEG-2 versioning technique because in this case it is not helpful; every instance has new data and hence would be a new version.

STT syntax and semantics

Table 10.2 describes the syntax and semantics of the System Time Table section.

TABLE 10.2 System Time Table Section Syntax

Field Name	Number of Bits	Field Value	Description
system_time_table_section() {			Start of the system_time_table_section().
table_id	8	0xCD	Identifies the table section as the System Time Table.
section_syntax_indicator	1	1b	Indicates the section is formatted in MPEG "long-form" syntax.
private_indicator	1	1b	This flag is not defined in MPEG and is set to 1 in PSIP tables.
reserved	2	11b	Reserved bits are set to 1.
section_length	12		An unsigned integer that specifies the length, in bytes, of data following the section length field itself to the end of this table section.
table_id_extension	16	0x0000	Used to differentiate multiple instances of a table section having a common table_id. For the STT, this field is not used and is set to zero.
reserved	2	11b	Reserved bits are set to 1.
version_number	5	0	Typically, the version number reflects the version of a table section, and is incremented when anything in the table changes. For the STT, versioning is not used since the table is different every time it is sent.
current_next_indicator	1	1b	Indicates, in general, whether the table section is currently applicable (value 1) or is the next one to be applicable (value 0). For the STT, only the current table is sent.
section_number	8	0	Indicates which part this section is for tables segmented into sections. For the STT, this parameter is set to zero, as the STT is only one section long.
last_section_number	8	0	Indicates the section number of the last section in a segmented table. For the STT, this parameter is set to zero.
protocol_version	8	0	Indicates the version of this STT. At present, only protocol version zero is defined.
system_time	32		An unsigned integer that represents the count of the number of GPS seconds since 00:00:00 UTC, January 6th, 1980.
GPS_UTC_offset	8		An unsigned integer count of the number of leap seconds that have occurred between the beginning of GPS time and the present time.

TABLE 10.2 System Time Table Section Syntax (continued)

Field Name	Number of Bits	Field Value	Description
DS_status	1		A flag that indicates, when set, that all parts of the country that observe Daylight Saving Time are currently doing so. When clear, Standard Time is being observed throughout the country.
reserved	2	11b	Reserved bits are set to 1.
DS_day_of_month	5		Indicates the day of the month that a transition into or out of Daylight Saving Time is scheduled to occur. The field is set to zero when no transition is imminent.
DS_hour	8		Indicates the hour of the day that a transition into or out of Daylight Saving Time is scheduled. The field is set to zero when no transition is imminent.
for (i=0; i<N; i++) {			Start of the additional descriptors "for" loop. The value of N is given indirectly by section_length. Receivers are expected to process successive descriptors until the CRC_32 field is reached. If the value of section_length is 17, no descriptors are present.
descriptor()	var.		Zero or more descriptors, formatted as type-length-data.
}			End of the additional descriptors "for" loop.
CRC_32	32		A 32-bit checksum designed to produce a zero output from the decoder defined in the MPEG-2 *Systems* Standard.
}			End of the system_time_table_section().

system_time

This represents the number of GPS seconds that have occurred since 00:00:00 UTC January 6, 1980. A/65 states that broadcasters must maintain the accuracy of system_time to plus or minus one second. The clock in the consumer digital TV will likely be one of the most accurate in the household.

GPS_UTC_offset

An unsigned integer that is a count of the number of leap seconds that have been inserted since the start of GPS time. The GPS_UTC_offset is used to convert GPS seconds into time of day.

DS_status

This is a flag that indicates, when set to 1 (true) that either all of the time zones within the span of the network are in Daylight Saving Time, or they are in the process of making the transition out of Daylight Saving Time. When DS_status is set to zero (false), all of the time zones in the network are not observing Daylight Saving Time, or they are in the process of making the transition into it.

DS_day_of_month

When non-zero, the DS_day_of_month field represents the day of the month (relative to local time) that a transition into or out of Daylight Saving Time is to occur. If the DS_status field is zero when a nonzero DS_day_of_month is specified, the day represents the day in the month at which entry into Daylight Saving Time will occur. If the DS_status field is one when a nonzero DS_day_of_month is specified, the day represents the day in the month when the return to Standard Time will be made. DS_day_of_month is set to zero if the next transition will not occur for another month or more, and after all time zones in the network have made a transition (either into or out of Daylight Saving Time).

Note that the day of the month indicated in DS_day_of_month refers to the next date for which the day of the month matches the indicated value. That day may occur within the current month or it may occur in the next month.

DS_hour

When non-zero, DS_hour gives the hour in the day, relative to local time, that a Daylight Saving Time transition is to occur. In the US, the hour of the transition is typically 2:00 a.m., so the value of DS_hour is set to 2. If the DS_status field is zero when a nonzero DS_hour is specified, the hour represents the hour in the day at which entry into Daylight Saving Time will occur. If the DS_status field is one when a nonzero DS_hour is specified, the hour represents the hour in the day when the return to Standard Time will be made. DS_hour is set to zero if the next transition is more than a month away, and after all time zones in the network have made a transition.

STT descriptors

Optionally, the System Time Table can include descriptors. Currently, none are defined for use in the STT.

Interpretation of the Daylight Saving Time fields

The basic scenario of Daylight Saving Time is described in Table A.2 of the A/65 Standard. Figure 10.2 illustrates a time line representing one year-long cycle of transitions into and out of Daylight Saving Time. At the top of the Figure, the four phases of the cycle are shown.

1. In Standard Time: All time zones in the network are observing Standard Time, and the transition into Daylight Saving Time will not be complete in all time

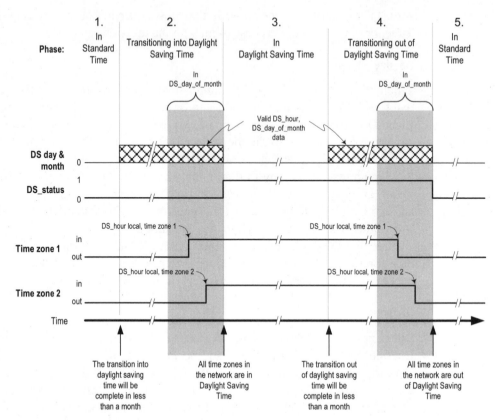

Figure 10.2 Transitions Into and Out of Daylight Saving Time

zones for at least another month. The day of the transition is given as a day in the month (range 1 to 31). The idea here is that if the transition is to occur on day N of a given month, the transition cannot begin any earlier than day N+1 of the prior month (or some receivers would make the transition one month early).

2. **Making the transition into Daylight Saving Time**: The transition into Daylight Saving Time begins at the point in time one month before the instant the first time zone in the signal's coverage area will make the transition to Daylight Saving Time. At this point, DS_day_of_month and DS_hour are transmitted with valid (non-zero) values. These settings are maintained until all of the time zones in the network have reached DS_hour local to each time zone. At that point, all of the time zones in the network have moved into Daylight Saving Time, and DS_day_of_month and DS_hour are set to zero. DS_status is set to 1 to indicate Daylight Saving Time is in effect.

The figure expands the time axis in the shaded regions to show that, on the day of the transitions, entry into or exit from Daylight Saving Time occurs when the local time reaches DS_hour. Therefore, more easterly time zones will transition before those in the west even though each makes the switch at the same local-time hour of the day.

3. **In Daylight Saving Time**: Throughout the summer, DS_status stays at 1 and DS_day_of_month and DS_hour stay at zero, indicating all time zones are in Daylight Saving Time and there is at least a month before the transition back to Standard Time will be completed.

4. **Moving out of Daylight Saving Time.**: The transition out of Daylight Saving Time begins one month before the instant the first time zone in the coverage area will change back to Standard Time. When the transition back to Standard Time will occur next month, and the day of the month is at least one greater than the day of the month that the transition is to occur, DS_day_of_month and DS_hour are set to non-zero values. These settings are maintained until all of the time zones in the network have reached DS_hour local to each time zone. At that point, all of the time zones in the network have reverted back to Standard Time, and DS_day_of_month and DS_hour are set to zero. DS_status is set to 0 to indicate Standard Time is in effect. The cycle is completed.

STT transport rate and cycle time

A/65 states that the maximum cycle time for the System Time Table in a terrestrial broadcast stream is 1000 milliseconds. Therefore, at least one TS packet per second is required to carry the STT.

Example STT

Table 10.3 shows an example bit-stream representation of a System Time Table section. In this example, the time of day is 3:41:01 a.m. December 16, 2001, UTC. There have been 13 leap second events in GPS time since January 6th, 1980. Daylight Saving Time is not in effect and will not be in effect for more than 30 days.

TABLE 10.3 System Time Table Example Bit-stream Representation

Field Name	Value	Description
table_id	0xCD	Identifies the table section as being an STT.
section_syntax_indicator	1b	The STT uses the MPEG "long-form" syntax.
private_indicator	1b	Set to 1 in PSIP tables.
reserved	11b	Reserved bits are set to 1.
section_length	0x0011	Length of the rest of this section is 17 bytes.
table_id_extension	0x0000	Set to zero in the STT.

TABLE 10.3 System Time Table Example Bit-stream Representation (continued)

Field Name	Value	Description
reserved	11b	Reserved bits are set to 1.
version_number	00000b	Version number is 0.
current_next_indicator	1b	Section is current STT.
section_number	0x00	First section.
last_section_number	0x00	Last is first (only one section).
protocol_version	0x00	Zero for the present protocol.
system_time	0x2946DA4D	12/16/2001 3:41:01 a.m. UTC.
GPS_UTC_offset	0x0D	13 leap seconds.
DS_status	0b	Not in Daylight Saving Time.
reserved	11b	Reserved bits are set to 1.
DS_day_of_month	00000b	Set to zero, as no transition is imminent.
DS_hour	0x00	Set to zero, as no transition is imminent.
CRC_32	0xDD28F315	MPEG-2 32-bit table section CRC.

Master Guide Table (MGT)

The PSIP Master Guide Table, or MGT, is called "master" because it refers to all other PSIP tables (except the System Time Table). The purpose of the MGT is to provide a convenient reference for receiving devices to the various instances of PSIP tables present in the Transport Stream. The MGT offers the following information for the benefit of receivers:

- A list of all PSIP tables present in the Transport Stream. Tables other than those defined in the A/65 Standard may be listed as well. For example, the ATSC Data Broadcast Standard A/90[2] uses the MGT to announce the presence of the Data Event Table (DET) and the text associated with it, and the Long Term Service Table (LTST).

- For each type of table listed, the PID value for the Transport Stream packets that are used to carry it.

- The version number of each type of table. In some cases, one MGT table type represents more than one table section or even more than one instance. For that case, all tables and table sections must share the common version number indicated in the MGT.

- The total number of bytes transmitted in all transmitted sections of tables of each type. The size parameter is helpful to the receiver so that it can determine beforehand whether or not necessary storage resources are available, and if available, how much storage must be allocated.

Structure of the Master Guide Table

Figure 10.3 depicts the structure of one section of the Master Guide Table.

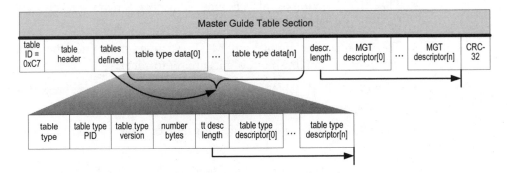

Figure 10.3 Master Guide Table Section Structure

As the figure shows, the MGT is structured as a field giving the number of table types to be defined, followed by that many blocks of table type data. After the table type data, the MGT can have zero or more descriptors pertinent to all of the table types described in the MGT section. Each block of table type data consists of an identifier as to the type of table, its version number, the size of all table sections of that type, and zero or more descriptors pertinent to this type of table.

References to private tables

The MGT may be used to announce the presence of private tables in the Transport Stream. Private tables are indicated by a "table type" value in a range reserved for private use. Whenever a private table type is used, an MPEG-2 Registration Descriptor must be present to identify the entity that has defined the private table type. Please see "MGT references to private tables" on page 298 for a detailed discussion.

MGT transport

The MGT is unlike any other table defined in A/65 in that it must be placed into the Transport Stream such that the first byte of the table (the table_id field) is the first byte of the packet payload. This rule means that receiving equipment, when searching for an MGT in the multiplex, can discard any packet that does not have the Packet Unit Start Indicator bit set and the pointer_field value set to zero. When a packet matching these conditions is found, the byte following the pointer field is a table_id byte that can be matched against the table_id of the MGT.

An advantage of requiring the start of the MGT to be aligned with the packet payload is that often, the full MGT fits into one Transport Stream packet. Acquisition time is thus minimized.

Transport of the Master Guide Table involves these transport considerations:

- The MGT section is placed into Transport Stream packets with PID value 0x1FFB, the SI base_PID.

- TS packets containing an MGT section must not be scrambled (the transport_scrambling_control bits are set to zero).

- As with all PSIP tables, an adaptation field must not be present in TS packets carrying the VCT.

- The MGT may not be sectioned for delivery; the whole thing must fit into one section.

- The MGT section is limited to a total length of 4096 bytes.

- As mentioned, the MGT must start at the first byte of a packet payload.t

MGT syntax and semantics

Syntax and semantics for the Master Guide Table section are given in Table 10.4.

TABLE 10.4 MGT Section Syntax and Semantics

Field Name	Number of Bits	FIeld Value	Description
master_guide_table_section() {			Start of the master_guide_table_section().
table_id	8	0xC7	Identifies the table section as being a master_guide_table_section().
section_syntax_indicator	1	1b	The MGT uses the MPEG "long-form" syntax.
private_indicator	1	1b	Set to 1 in PSIP tables.
reserved	2	11b	Reserved bits are set to 1.
section_length	12		The MGT section is limited to 4096 total bytes, so section_length is limited to 4093.
table_id_extension	16	0x0000	Set to zero for the MGT. At most one MGT table instance can appear in the SI base PID.
reserved	2	11b	Reserved bits are set to 1.

TABLE 10.4 MGT Section Syntax and Semantics (continued)

Field Name	Number of Bits	FIeld Value	Description
version_number	5		Typically, the version number reflects the version of a table section, and is incremented when anything in the table changes.
current_next_indicator	1	1b	Indicates, in general, whether the table section is currently applicable (value 1) or is the next one to be applicable (value 0).
section_number	8	0	The MGT must fit into a single table section.
last_section_number	8	0	Must be zero.
protocol_version	8	0	Indicates the protocol version of this table section. The only type of master_guide_table_section() currently defined is protocol_version zero.
tables_defined	16		An unsigned integer that gives the number of table types defined in this MGT table section. The maximum allowable value is limited by the section length to 370.
for (i=0; i<tables_defined; i++) {			Start of the tables defined "for" loop.
table_type	16		A 16-bit unsigned integer that defines the type of table being described in this iteration of the "for" loop. See Table 10.5 for the coding of table_type.
reserved	3	111b	Reserved bits are set to 1.
table_type_PID	13		Specifies the PID value that is used to transport sections of the table type indicated in table_type.
reserved	3	111b	Reserved bits are set to 1.
table_type_version_number	5		Specifies the version number of the tables of type table_type being defined in this iteration of the "for" loop. All the table instances of this table type must share the same table_type_version_number, so that if any one must be updated, all other instances of that type must be updated to bump their version number fields up by one.
number_bytes	32		Specifies the total size, in bytes, of all the sections of the type indicated in table_type.
reserved	4	1111b	Reserved bits are set to 1.
table_type_descriptors_length	12		This field indicates the total length of any descriptors included within this iteration of the "for" loop. If no descriptors are included, the field may be set to zero.

TABLE 10.4 MGT Section Syntax and Semantics (continued)

Field Name	Number of Bits	FIeld Value	Description
for (i=0; i<N; i++) {			Start of the table type descriptors "for" loop. The value of N (number of descriptors present) is given indirectly by table_type_descriptors_length. Receivers are expected to process successive descriptors until the total number of descriptor bytes processed equals table_type_descriptors_length. A value of zero for table_type_descriptors_length indicates no descriptors are present.
descriptor()	var.		Zero or more descriptors, formatted as type-length-data.
}			End of the table type descriptors "for" loop.
}			End of the tables defined "for" loop.
reserved	4	1111b	Reserved bits are set to 1.
descriptors_length	12		This field indicates the length of any additional descriptors.
for (i=0; i<N; i++) {			Start of the additional descriptors "for" loop. The value of N is given indirectly by descriptors_length. Descriptors should be processed until the total number of bytes processed equals the value given in descriptors_length (it may be zero).
descriptor()	var.		Zero or more descriptors, formatted as type-length-data.
}			End of the additional descriptors "for" loop.
CRC_32	32		A 32-bit checksum designed to produce a zero output from the decoder defined in the MPEG-2 *Systems* Standard.
}			End of the master_guide_table_section().

We now take a detailed look now at some of the more important fields in the MGT.

table_type

This 16-bit field indicates the type of table being described in this iteration of the table types "for" loop. We explore the details of the table_type field below and list the assigned values.

table_type_PID

One of the things the MGT indicates for each type of table is the PID of the TS packets carrying that table, and it is the table_type_PID that provides this data. Of course, for tables like the TVCT or CVCT, the PID is unchangeable (it must be 0x1FFB), but for tables like the Event Information Tables, inspection of the MGT is the only way a receiving device can know where to find them in the Transport Stream.

table_type_version_number

The table_type_version_number field indicates the version_number field in the table headers of all tables of the type indicated by table_type. Often, one table_type corresponds to many instances of the indicated table. For example, for each three-hour time slot there will be one EIT instance for each programming service (identified by its Source ID). Let's say there are six services (virtual channels) in a multiplex. In the PID value associated with the EITs for the current time slot, EIT-0, there are six separate EITs, one for each of the six Source IDs.

The table_type_version_number reflects the version_number field common to all of the tables of the type indicated in table_type. For the EIT example, all six EIT-0 table instances must share the same value for version_number so that table_type_version_number is accurate and correct for each. Since the granularity of the table_type definition requires that all tables of a given table_type value must have the same version_number, if a change needs to be made to any one of the instances, all tables of that type have to be updated with the new version.

number_bytes

This 32-bit field gives the total number of bytes needed to store all of the table sections corresponding to the type of table indicated in table_type. By processing the number_bytes field, a receiving device could determine in advance, for example, if it has sufficient RAM to store a certain type of table (maybe the EITs corresponding to a certain future time slot) and not waste time trying to collect it if it has insufficient storage space.

table_type_descriptors_length

The MGT can carry descriptors pertinent to the particular type of table indicated in table_type. It is the table_type_descriptors_length that indicates how many bytes of descriptors are present, and hence the location of the end of the "for" loop.

Currently, the only descriptor defined for use in the MGT is the MPEG-2 Registration Descriptor (MRD). Use of the MRD in the MGT to reference user private data tables is discussed below.

MGT table types

The MGT is called "master" because this table gives information about all of the other PSIP tables, except as noted, the STT. As extensions to the digital television standards for terrestrial broadcast, cable, and satellite continue to evolve, new types of tables are standardized. Typically, designers of the new tables wish to take advantage of the MGT's ability to announce the presence, associated PID values, version, and size of the new tables in the Transport Stream. This happened, for example, with tables invented for ATSC Data Broadcast Standard A/90, and it also occurred when the development of System Information for the cable out-of-band channel required the definition of some new table types.

A pointer mechanism is used in the MGT to refer to the tables it describes. This mechanism is the table_type field in the MGT. The table_type indicates the type of table, and in some cases an instance or group of table instances, of the table. Table 10.5 lists the currently-defined values for table_type.

TABLE 10.5 MGT Table Type Values and Ranges

Value or Range	Meaning	Where Defined
0x0000	Terrestrial Virtual Channel Table (current)	A/65 PSIP
0x0001	Terrestrial Virtual Channel Table (next)	A/65 PSIP
0x0002	Cable Virtual Channel Table (current)	A/65 PSIP
0x0003	Cable Virtual Channel Table (next)	A/65 PSIP
0x0004	Channel Extended Text Table	A/65 PSIP
0x0005	Directed Channel Change Table	A/65 PSIP
0x0006	Directed Channel Change Genre Code Table (DCGST)	A/65 PSIP
0x0007-0x000F	Reserved for future ATSC/SCTE use	
0x0010	Short-form Virtual Channel Table (VCM subtype)	SCTE 65 (out-of-band SI)
0x0011	Short-form Virtual Channel Table (DCM subtype)	SCTE 65
0x0012	Short-form Virtual Channel Table (ICM subtype)	SCTE 65
0x0013-0x001F	Reserved for future ATSC/SCTE use	
0x0020	Network Information Table (CDS subtype)	SCTE 65
0x0021	Network Information Table (MMS subtype)	SCTE 65
0x0021-0x002F	Reserved for future ATSC/SCTE use	
0x0030	Network Text Table (SNS subtype)	SCTE 65
0x0031-0x00FF	Reserved for future ATSC/SCTE use	
0x0100-0x017F	EIT-0 to EIT-127	A/65 PSIP
0x0180-0x01FF	Reserved for future ATSC/SCTE use	
0x0200-0x027F	Event ETT-0 to event ETT-127	A/65 PSIP

TABLE 10.5 MGT Table Type Values and Ranges (continued)

Value or Range	Meaning	Where Defined
0x0280-0x0300	Reserved for future ATSC/SCTE use	
0x0301-0x3FF	RRT with rating region 1-255	A/65 PSIP
0x0400-0xFFF	User private	
0x1000-0x10FF	Aggregate Event Information Table with MGT_tag 0 to 255	SCTE 65
0x1100-0x11FF	Aggregate Extended Text Table with MGT_tag 0 to 255	SCTE 65
0x1100-0x117F	Data Event Table-0 to –127	A/90 Data Broadcast
0x1180	Long Term Service Table	A/90 Data Broadcast
0x1181-0x11FF	Reserved for future ATSC/SCTE use	A/90 Data Broadcast
0x1200-0x127F	Extended Text Table associated with DET	A/90 Data Broadcast
0x1280-0xFFFF	Reserved for future ATSC/SCTE use	

References to private tables

The MGT may be used to announce the presence of private tables in the Transport Stream. Private tables are indicated by a table_type value in the range reserved for private use: 0x0400 to 0x0FFF. Whenever a private table type is used, an MPEG-2 Registration Descriptor must be present to signal the registration authority that has defined the private table type.

MGT transport rate and cycle time

ATSC A/65 specifies in Sec. 7 that for the terrestrial broadcast application the maximum cycle time of the MGT is 150 milliseconds. That means the MGT must be repeated at a rate not less than 6-2/3 repetitions per second.

A typical smaller-sized MGT might have references to a Terrestrial VCT, a channel ETT, and four EITs and ETTs each. Assuming no descriptors are present, that MGT would be a total of 127 bytes in length. If it were possible to send only the MGT, it would consume a bandwidth in the multiplex of 127 * 8 * 6.667 = 6773 bps. But, since one cannot put just 127 bytes into the multiplex by itself (bytes must be placed into TS packets), we must include TS packets at a rate of 6.667 per second.

Including the packet overhead yields a bandwidth usage of 10,027 bps, with 56 bytes per packet available for transport of other types of tables (for example, the STT).

Currently, there are no specifications related to transport rate requirements for cable.

Example MGT

Table 10.6 lists an example set of seven tables and their related data.

TABLE 10.6 Example MGT Table Set

Table Type	PID	Table Version	Number of Bytes
TVCT	0x1FFB	14	292
EIT-0	0x1D02	16	2005
EIT-1	0x1D03	2	1607
EIT-2	0x1D04	12	1479
EIT-3	0x1D05	16	686
ETT-0	0x1E02	10	530
Channel ETT	0x1E80	14	2557

Table 10.7 shows the bit-stream representation of the MGT table section that represents this data set. Note that the order in which the table types appear in the MGT has no significance.

TABLE 10.7 Master Guide Table Example Bit-stream Representation

Field Name	Value	Description
table_id	0xC7	Identifies the table section as being an MGT.
section_syntax_indicator	1b	The MGT uses the MPEG "long-form" syntax.
private_indicator	1b	Set to 1 in PSIP tables.
reserved	11b	Reserved bits are set to 1.
section_length	0x005B	Length of the rest of this section is 91 bytes.
table_id_extension	0x0000	Set to zero in the MGT.
reserved	11b	Reserved bits are set to 1.
version_number	10001b	Version number is 17.
current_next_indicator	1b	Section must be current MGT.
section_number	0x00	First section.
last_section_number	0x00	Last is first (only one section).
protocol_version	0x00	Zero for the present protocol.
tables_defined	0x0007	Seven tables are defined.
table_type	0x0000	Table type is TVCT.
reserved	111b	Reserved bits are set to 1.

TABLE 10.7 Master Guide Table Example Bit-stream Representation (continued)

Field Name	Value	Description
table_type_PID	0x1FFB	SI_base PID.
reserved	111b	Reserved bits are set to 1.
table_type_version_number	01110b	TVCT is at version 14.
number_bytes	0x00000124	Total size of TVCT is 292 bytes.
reserved	1111b	Reserved bits are set to 1.
table_type_descriptors_length	0x000	No additional descriptors.
table_type	0x0200	Table type is ETT-0.
reserved	111b	Reserved bits are set to 1.
table_type_PID	0x1E02	ETT-0 is carried in PID 0x1E02.
reserved	111b	Reserved bits are set to 1.
table_type_version_number	01010b	ETT-0 tables are at version 10.
number_bytes	0x00000212	Total size of all ETT-0 table sections is 530 bytes.
reserved	1111b	Reserved bits are set to 1.
table_type_descriptors_length	0x000	No additional descriptors.
table_type	0x0004	Table type is Channel ETT.
reserved	111b	Reserved bits are set to 1.
table_type_PID	0x1E80	Channel ETT is carried in PID 0x1E80.
reserved	111b	Reserved bits are set to 1.
table_type_version_number	01110b	Channel ETT tables are at version 14.
number_bytes	0x000009FD	Total size of Channel ETT table sections: 2557 bytes.
reserved	1111b	Reserved bits are set to 1.
table_type_descriptors_length	0x000	No additional descriptors.
table_type	0x0100	Table type is EIT-0.
reserved	111b	Reserved bits are set to 1.
table_type_PID	0x1D02	EIT-0 is carried in PID 0x1D02.
reserved	111b	Reserved bits are set to 1.
table_type_version_number	10000b	EIT-0 tables are at version 16.
number_bytes	0x000007D5	Total size of EIT-0 table sections is 2005 bytes.
reserved	1111b	Reserved bits are set to 1.
table_type_descriptors_length	0x000	No additional descriptors.
table_type	0x0101	Table type is EIT-1.
reserved	111b	Reserved bits are set to 1.
table_type_PID	0x1D03	EIT-1 is carried in PID 0x1D03.

TABLE 10.7 Master Guide Table Example Bit-stream Representation (continued)

Field Name	Value	Description
reserved	111b	Reserved bits are set to 1.
table_type_version_number	00010b	EIT-1 tables are at version 2.
number_bytes	0x00000647	Total size of EIT-1 table sections is 1607 bytes.
reserved	1111b	Reserved bits are set to 1.
table_type_descriptors_length	0x000	No additional descriptors.
table_type	0x0102	Table type is EIT-2.
reserved	111b	Reserved bits are set to 1.
table_type_PID	0x1D04	EIT-2 is carried in PID 0x1D04.
reserved	111b	Reserved bits are set to 1.
table_type_version_number	01100b	EIT-2 tables are at version 12.
number_bytes	0x000005C7	Total size of EIT-2 table sections is 1479 bytes.
reserved	1111b	Reserved bits are set to 1.
table_type_descriptors_length	0x000	No additional descriptors.
table_type	0x0103	Table type is EIT-3.
reserved	111b	Reserved bits are set to 1.
table_type_PID	0x1D05	EIT-3 is carried in PID 0x1D05.
reserved	111b	Reserved bits are set to 1.
table_type_version_number	10000b	EIT-3 tables are at version 16.
number_bytes	0x000002AE	Total size of EIT-3 table sections is 686 bytes.
reserved	1111b	Reserved bits are set to 1.
table_type_descriptors_length	0x000	No additional descriptors.
reserved	1111b	Reserved bits are set to 1.
descriptors_length	0x000	No additional MGT descriptors.
CRC_32	0xEEADB002	MPEG-2 32-bit table section CRC (example).

Virtual Channel Table (VCT)

One of the fundamental pieces of data needed by receivers is a list of available services, including information about each, such as the service's name and the channel number it is identified with in printed guides. Such a list is an essential part of the construction of an Electronic Program Guide. As we've seen, the Virtual Channel Table in PSIP fulfills this function. Knowledge of the VCT is essential even for a receiver not supporting an EPG feature since it is the VCT that enables all receivers to identify digital services in a consistent and user-friendly manner.

We introduced the basic concept of virtual channels in Chapter 4. Here, we look in detail into the data structure of the VCT and its syntax and semantics. A/65 defines two separate types of Virtual Channel Tables, one specifically for terrestrial broadcasting and the other for cable. Since the structure, syntax, and semantics of these two are nearly identical, we combine the two for purposes of our initial discussion. Later we explore those aspects that are specific to each transmission medium.

Structure of the Virtual Channel Table

Figure 10.4 depicts the structure of one section of the Virtual Channel Table.

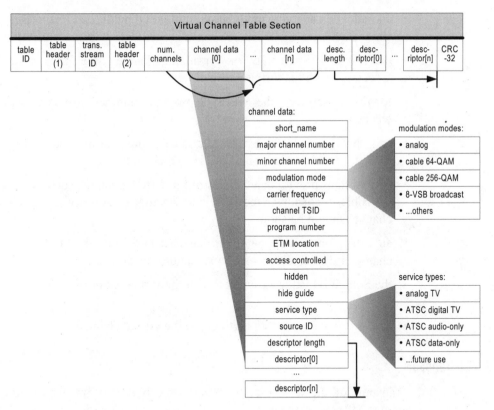

Figure 10.4 Virtual Channel Table Section Structure

The transport_stream_id field in the Virtual Channel Table section is at the position of the table_id_extension. If VCT sections from several Transport Streams are saved in memory together, the TSID is a convenient way to identify the separate instances.

Beyond the header bytes common to all of the PSIP table sections, the structure consists of a count of the number of virtual channels to be defined in the section, followed by each definition itself. Following the channel definitions is a field indicating the length of any descriptors that might be included in the section, followed by the descriptors themselves. Finally, the 32-bit CRC terminates the section.

The heart of the VCT structure is the definition of each virtual channel, the data structure called channel_data[n] in the Figure. Parameters comprising each virtual channel definition include:

- the channel's "short" name, an uncompressed text string of up to seven characters.

- the channel's major and minor channel numbers.

- an indication as to the modulation mode used by the carrier delivering the virtual channel.

- the Transport Stream ID (or, for analog channels, the Transmission Signal ID) associated with the virtual channel.

- for digital services, the MPEG-2 program_number associated with the virtual channel.

- an indication as to the existence and location of an Extended Text Table giving further textual description of the virtual channel.

- several flags providing attributes of the virtual channel, such as whether or not it is visible to consumer receivers and whether or not program guide data for the channel should be shown.

- an indication of the type of service associated with the virtual channel, such as analog or digital television service, audio- or data-only service.

- the Source ID value associated with the virtual channel (see Chapter 6, "Source IDs")

- zero or more descriptors pertinent to the virtual channel.

VCT transport

Transport of the terrestrial broadcast or cable VCT involves a few constraints against MPEG-2 *Systems*:

- a TVCT or CVCT section, when carried in the Transport Stream, are placed into Transport Stream packets with PID value 0x1FFB, the SI base_PID.

- TS packets containing a VCT section must not be scrambled (the transport_scrambling_control bits are set to zero).

- as with all PSIP tables, an adaptation field must not be present in TS packets carrying the VCT.

- although the VCT may be sectioned for delivery, each VCT section is limited to a total length of 1024 bytes.

Terrestrial Broadcast VCT (TVCT)

The Terrestrial Virtual Channel Table defines virtual channels available to terrestrial broadcast receivers. A/65 lists a number of requirements specific to the terrestrial broadcast application, including the requirement to transmit a TVCT to describe digital programming services delivered via terrestrial broadcast.

VCT syntax and semantics

Table 10.8 describes the syntax and semantics of the VCT, valid for both terrestrial broadcast and cable forms. Differences between the TVCT and CVCT are noted.

TABLE 10.8 Virtual Channel Table Syntax and Semantics

Field Name	Number of Bits	Field Value	Description
virtual_channel_table_section() {			Start of the virtual_channel_table_section().
table_id	8	0xC8 (TVCT) 0xC9 (CVCT)	Identifies the table section as being virtual_channel_table_section() (either TVCT or CVCT type).
section_syntax_indicator	1	1b	The VCT uses the MPEG "long-form" syntax.
private_indicator	1	1b	Set to 1 in PSIP tables.
reserved	2	11b	Reserved bits are set to 1.
section_length	12		The VCT section is limited to 1024 total bytes, so section_length is limited to 1021.
transport_stream_id	16		Set to the value of the TSID given in the PAT carried on PID 0 of the same TS carrying this VCT.
reserved	2	11b	Reserved bits are set to 1.
version_number	5		The version number reflects the version of a table section, and is incremented when anything in the table changes.
current_next_indicator	1	1b	Indicates whether the table section is currently applicable (value 1) or is the next one to be applicable (value 0).

TABLE 10.8 Virtual Channel Table Syntax and Semantics (continued)

Field Name	Number of Bits	Field Value	Description
section_number	8		The TVCT may be segmented into sections for delivery. When it is segmented, section_number indicates which part this section is. When the whole thing fits into one table section, section_number is set to zero.
last_section_number	8		Indicates the section number of the last section in a segmented table. The last_section_number field is set to zero if the whole table fits into one section.
protocol_version	8	0	Indicates the protocol version of this table section. The only type of virtual_channel_table_section() currently defined is for protocol_version zero.
num_channels_in_section	8		Indicates the number of virtual channels being defined in this section of the VCT.
for (i=0; i<num_channels_in_section; i++) {			Start of channels "for" loop.
short_name	7*16		Seven Unicode UTF-16 encoded characters, representing the "short name" of the virtual channel.
reserved	4	1111b	Reserved bits are set to 1.
major_channel_number	10		The "major" channel number to be associated with the virtual channel being defined in this iteration of the channels "for" loop.
minor_channel_number	10		The "minor" channel number to be associated with this virtual channel.
modulation_mode	8		This is an enumerated type field that indicates the modulation mode for the carrier transmitting this virtual channel. See text for coding.
carrier_frequency	32	0	Earlier versions of A/65 specified this field to be the carrier frequency, but the field has now been deprecated. The current recommendation is to set this field to zero.
channel_TSID	16		The TSID associated with the Transport Stream carrying this virtual channel. If the virtual channel is an analog NTSC service, channel_TSID indicates that signal's Transmission Signal ID.
program_number	16		Associates this virtual channel with one MPEG-2 program within the TS identified by channel_TSID.
ETM_location	2		Indicates whether or not an ETT exists to further describe this virtual channel, and if so, whether it is carried on this TS or the TS indicated by channel_TSID.

TABLE 10.8 Virtual Channel Table Syntax and Semantics (continued)

Field Name	Number of Bits	Field Value	Description
access_controlled	1		A flag indicating, when set, that programming on this virtual channel may be scrambled (requires CA to view). When clear, all programming is sent in the clear on this virtual channel.
hidden	1		A flag indicating, when set, that the channel is not accessible by consumer direct entry of the channel number. Program guide data for hidden channels can appear in the EPG if the channel has the hide_guide bit set to zero.
path_select	1		(CVCT only; reserved for TVCT) For cable systems involving two physical cables, indicates which of the two this virtual channel is associated with. Rarely used.
out_of_band	1		(CVCT only; reserved for TVCT) Indicates the virtual channel is carried on the cable out-of-band channel.
hide_guide	1		This flag indicates whether a given hidden channel can be included in program guide displays. When a hidden channel has a hide_guide flag set to zero, the channel is an inactive channel and its program schedule can be displayed. When a hidden channel has its hide_guide flag set to one, it is inaccessible and invisible to consumer devices.
reserved	3	111b	Reserved bits are set to 1
service_type	6		This field indicates the type of service associated with the virtual channel. Service types include analog and digital televisions, data-only, and audio-only services. See text.
source_id	16		This 16-bit field acts as the database linkage between the VCT and the EITs. It can also link VCTs to Internet-based EPG databases. We discuss the Source ID concept in detail in Chapter 6.
reserved	6	111111b	Reserved bits are set to 1.
descriptors_length	10		Indicates the extent of the channel descriptors "for" loop.
for (i=0; i<N; i++) {			Start of channel descriptors "for" loop. As before, the value of N is determined from descriptors_length.
descriptor()			
}			End of channel descriptors "for" loop.

TABLE 10.8 Virtual Channel Table Syntax and Semantics (continued)

Field Name	Number of Bits	Field Value	Description
}			End of channels "for" loop.
reserved	6	111111b	Reserved bits are set to 1.
additional_descriptors_length	10		Indicates the extent of the additional descriptors "for" loop.
for (i=0; i<N; i++) {			Start of additional descriptors "for" loop. The value of N is determined indirectly by additional_descriptors_length. Descriptors are processed until the position reaches the CRC_32 field.
additional_descriptor()			
}			End of additional descriptors "for" loop.
CRC_32	32		A 32-bit checksum designed to produce a zero output from the decoder defined in the MPEG-2 *Systems* Standard.
}			End of the virtual_channel_table_section() .

num_channels_in_section

This is an integer number that gives the number of virtual channel definitions to follow in the section. The largest number of channels that can be defined in one section is limited by the maximum length of a section.

short_name

The "short name" is a text string of one to seven characters identifying the channel. When channel names are listed in a program guide grid format on-screen, screen area is precious. This short channel name is length-limited to ensure that all channel names can fit within a limited display area. If the text to be represented is less than seven characters, the short_name field is padded with 16-bit fields of value 0x0000.

major_channel_number

The major and minor channel numbers together define the channel's familiar "channel number" in either a two-part or (for the cable application only) a one-part format. We discuss one-part channel numbers used in some cable systems shortly. For terrestrial broadcast television programs, the major channel number is in the 1 to 99 range. For cable, when two-part channel numbers are used the major channel number can range from 1 to 999. Data-only channels can be assigned a major channel number from 1 to 999 for either cable or terrestrial broadcast.

minor_channel_number

Minor channel numbers for digital television or ATSC audio services delivered via terrestrial broadcast can range from 1 to 99. For cable two-part numbers, minor channel numbers can range from 0 to 999. Other types of services such as data broadcasting can use minor channel numbers in the 1 to 999 range on terrestrial broadcast or 0 to 999 on cable.

One-part channel numbers for cable

It is possible for a cable operator to create a CVCT that associates a one-part channel number with any given virtual channel. If the six most-significant bits of the major_channel_number are set to ones, a one-part number is specified. Both the major and minor channel number fields are combined mathematically to form the one-part number according to the following equation, written in C syntax:

one_part_number = ((major_channel_number & 0x00F)<<10) + minor_channel_number

In English, this means take the least significant four bits of the major channel number field, multiply it by 1024 and add it to the value of the minor channel number. The result is a 14-bit one-part channel number. With this coding method, it is possible to create one-part channel numbers with values anywhere from zero to 16,383.

modulation_mode

The modulation mode field indicates the modulation mode for the transmitted carrier associated with the virtual channel. For channels carried on the same TS as the Virtual Channel Table itself, this field has no practical value. It can, however, be helpful if the virtual channel is delivered on some other Transport Stream. Modulation mode is an enumerated type field that can indicate one of a number of different modulation modes including NTSC analog, cable digital transport in 64- or 256- QAM modulation, or ATSC 8- or 16- VSB modes defined in the ATSC standards.
Table 10.9 defines the coding for modulation_mode.

carrier_frequency

It may not be obvious at first glance, but specification of the carrier frequency of a virtual channel is really not necessary. Complicating matters is the fact that in many cases it is not even practical for the transmitted signal to accurately reflect the true carrier frequency. The A/65 Standard has deprecated (phased out) the use of the carrier frequency field and receiving equipment should be built to disregard it. The Standard states that after January 1, 2010, the field reverts to a reserved field

TABLE 10.9 Modulation Mode

modulation_mode	Meaning	Notes
0x00	Reserved	
0x01	Analog	This modulation mode indicates the virtual channel is an analog NTSC television service modulated with standard analog methods on terrestrial broadcast or cable.
0x02	SCTE mode 1	This is the modulation mode used for carriage of Transport Streams on cable. Mode 1 is defined in ANSI/SCTE 07[3] and is typically used for 64-QAM modulation on cable.
0x03	SCTE mode 2	This is the modulation mode used for carriage of Transport Streams on cable. Mode 2 is defined in ANSI/SCTE 07[3] and is typically used for 256-QAM modulation on cable.
0x04	ATSC 8-VSB	The signal is modulated in accordance with ATSC A/53 using 8-VSB modulation (see "8-VSB modulation on cable" on page 327).
0x05	ATSC 16-VSB	The signal is modulated in accordance with ATSC A/53 using 16-VSB modulation. Note: 16-VSB modulation is not expected to be used and its use in the cable application is not specified in any current standards.
0x06-0xFF	Reserved for future ATSC/SCTE use	

(except where carrier frequency indicates the frequency of an out-of-band carrier on cable; more about that later). Later in this book we discuss the methods whereby a receiver may be built without the need for PSIP-supplied carrier frequency data.

channel_TSID

Every virtual channel defined in the VCT points to one and only one programming service. That service may be an MPEG-2 digital service, or it may be an analog NTSC service. For digital services, the channel TSID identifies the digital multiplex that carries that particular virtual channel. In many cases, the channel TSID indicates the TSID of the Transport Stream carrying the VCT itself. In some cases the channel TSID may indicate the Transport Stream ID of a different MPEG-2 multiplex.

If the type of service indicated by the modulation mode field is analog, channel TSID indicates the Transmission Signal Identifier associated with the analog NTSC signal. The Transmission Signal Identifier is also known as the "analog TSID" and is carried in line-21 Extended Data Service packets in the signal's VBI. Refer to Sec. 9.5.2.4 of EIA/CEA-608-B[4] for the specification of the carriage of Transmission Signal Identifier in NTSC VBI eXtended Data Service (XDS) packets.

To ensure the proper operation of receivers that rely on it (and to comply with the standards), broadcasters in North America must use unique digital and analog TSID values. As of this writing, the Association for Maximum Service Television (MSTV) has proposed TSID assignments for both digital and analog carriers. The MSTV TSID assignments can be found on the web at http://www.mstv.org/downloads/TSIDASGN.doc.

program_number

As mentioned, the program_number field is used in conjunction with the channel TSID to uniquely reference one digital service. While channel TSID identifies the Transport Stream, program_number identifies the MPEG-2 program (service) within that Transport Stream. The term "program" might be misleading here. The MPEG Standards use the term "program" in the way we normally use the term "service." In MPEG terminology, a program is a set of related elementary stream components, which are the audio, video, and data streams that make up the program.

ETM_location

Any virtual channel can have multilingual text associated with it if a broadcaster wishes to describe the channel in some way other than the short 7-character name or the extended channel name that can be given in the Extended Channel Name Descriptor. Such text could be a description of the kind of programming carried on the channel or some other information the operator wishes to include.

The ETM location field indicates whether or not a text record for this channel exists. If it does exist, the ETM location indicates whether it is carried on the same Transport Stream that carried this VCT section or it is on the TS identified by channel TSID.

ETM coding is shown in Table 10.10.

TABLE 10.10 ETM_location Coding

ETM_location	Meaning
0x00	No ETM—This virtual channel does not have an ETT record.
0x01	An ETM describing this virtual channel is located in the same Transport Stream as this Virtual Channel Table (the one given in the transport_stream_id field of this VCT).
0x02	An ETM describing this virtual channel is located in the Transport Stream whose ID is given by channel_TSID.
0x03	Reserved for future ATSC use.

access_controlled

Some digital services are scrambled and therefore only accessible to viewers who have a suitable descrambler and have paid for a subscription. Other digital services are free. The "access_controlled" bit is a simple flag that indicates, when set, that the service may include programs that require a descrambler and a subscription to view them.

hidden

When the "hidden" flag is set for a virtual channel, that channel should not be accessible by direct entry of the virtual channel number in a consumer receiver. Furthermore, receivers should skip over hidden channels when the user channel-surfs, and the hidden channels are to be treated as if undefined. A hidden channel may be used for test signals. Equipment owned by the broadcaster or cable operator is able to access hidden channels. In addition, hidden channels may also be accessible to applications.

hide_guide

When the "hide_guide" bit is set to one the channel is not to be included in Electronic Program Guide displays. When "hide_guide" is set to zero the channel is to be included in EPG displays. The following table illustrates the relationship between the hidden and hide_guide bits. The "x" indicates that the value of the bit is irrelevant.

The case of hidden channels that are nonetheless visible in the Electronic Program Guide deserves special mention. Broadcasters realized that they might want to broadcast High Definition programming during some part of the broadcast schedule and multi-channel Standard Definition programming during the rest of the time. HD programming takes nearly all of the 19.4 Mbps of digital bandwidth. Therefore, when the HD programming started up, one of the SD channels would switch formats to high-definition and the others would have to go off the air temporarily.

These channels are also known as "inactive" channels, meaning that they are currently not on the air (hence cannot currently be accessed) yet are scheduled to be on the air at one or more times in the future. Including inactive channels in the program guide allows viewers to see what programs will be broadcast when the channel comes on the air again, and possibly to set up a recorder or a reminder to capture the program for later viewing.

service_type

The service type field indicates the type of service of a particular virtual channel. Four service types are currently defined: analog television, digital television, digital

TABLE 10.11 Hidden and Hide Guide Flags

hidden	hide_guide	Channel Access	EPG Display
0	x (irrelevant)	Yes—Available by direct entry of channel number, available when channel surfing.	Yes—Included in EPG displays.
1	0	No—Skipped when channel surfing; not accessible by direct entry of the channel number.	Yes—Included in EPG displays. The channel is inactive (not currently on the air).
1	1	No—Skipped when channel surfing; not accessible by direct entry of the channel number.	No—Not included in EPG displays. The channel may be a test signal or a channel only accessible via an application

audio only, and an ATSC data-only service. The digital television and audio-only services may have optional data transmitted as well. Table 10.12 shows the coding of the values for service_type as they are currently defined.

TABLE 10.12 Service Types

service_type	Meaning	Notes
0x00	Reserved value	
0x01	Analog television.	Indicates the channel is a traditional NTSC analog service.
0x02	ATSC digital television	Indicates the service is digital television conforming to the ATSC A/63 Digital Television Standard. The service must include video and audio, and it can optionally include one or more data components as well.
0x03	ATSC audio-only	This type of service has no video, but includes digital audio conforming to the ATSC standard, as well as optional data.
0x04	ATSC data-only	This type of service has no video or audio, but includes one or more data components.
0x05-0xFF	Reserved for future ATSC use	

source_id

Each virtual channel that has associated Event Information Table data must specify a Source ID, a 16-bit number that uniquely identifies the service. The Source ID concept was discussed in detail in Chapter 6. Source IDs are the database linkage between virtual channels and the EPG data carried in the Event Information Tables. They may also be used to tie the services to EPG databases on the Internet or elsewhere.

Source_id values in the 0 to 0x0FFF range are required to be unique only within the Transport Stream, while those 0x1000 and above must be unique at a "regional" level. The ATSC Standard does not give a definition of the term "regional," but the PSIP architects had in mind a concept like the one the International Telecommunications Union (ITU) uses. For some of its standardization work, the ITU views the world as composed of three geographic regions. Region 1 includes Europe, Africa and the Middle East, Region 2 includes North and South America, while Region 3 covers Asia, Australia and the Pacific Rim.

As long as one receiver cannot access signals originating from multiple regions, values of accessible regionally-scoped Source IDs are unique. SMPTE has considered establishing a national and international registry for Source IDs.

descriptors

One or more descriptors can be placed into the Virtual Channel Table as needed. A descriptor that pertains only to a particular virtual channel is placed within the channel data, following the Source ID. If a descriptor pertains to all of the virtual channels being defined in the table section, it is placed in the descriptor area beyond the last array of channel data. Table 10.13 lists the three types of descriptors currently defined for use with the VCT.

TABLE 10.13 Descriptors in the Virtual Channel Table

Descriptor	Location	Function	Usage Rule
Extended Channel Name	Channel data	Long form of channel name.	Optional.
Service Location	Channel data	Gives PID values for various stream types including audio and video.	Required for terrestrial broadcast VCT, optional for cable VCT.
Time Shifted Service	Channel data	Indicates the channel has the same program schedule as a different specified service but is shifted in time by an indicated amount.	Optional.

TVCT transport rate and cycle time

ATSC A/65 states that the TVCT must be sent at a cycle time not to exceed 400 milliseconds, corresponding to a minimum rate of 2.5 repetitions per second. The maximum rate that can be used for the TVCT in a terrestrial broadcast multiplex is set by the maximum bitrate allowed for data transported in the SI base_PID 0x1FFB, which is 250,000 bps.

If a certain broadcaster carried five digital television services in a Transport Stream, the corresponding TVCT would be 392 bytes in length, assuming no descriptors other than the required Service Location Descriptors. If this TVCT were to be placed into three Transport Stream packets (the third would have 158 bytes left over that could be used for another type of table), and the TVCT repeated at a rate of 2.5 per second, the amount of Transport Stream bandwidth used would be 3 packets times 188 bytes per packet times 8 bits/byte times 2.5 repetitions per second, or 11,280 bps.

CVCT transport rate and cycle time

Minimum cycle times (highest allowed rates) for cable VCTs are currently not directly specified in any SCTE standard, although SCTE SCTE 54[5] limits the bitrate of the SI_base PID to 250,000 bits per second. For the maximum length of time between repetitions of the CVCT on cable (slowest allowed rates), SCTE 54[5] states that it must be no greater than 400 milliseconds.

Example VCT

We now present the bit-stream representation for a Terrestrial Virtual Channel Table describing four programming services. The first is an analog television station called WBCD and transmitting on UHF channel 39. This broadcaster also operates a digital transmitter and broadcasts three services, called WBCD-D1, WBCD-D2 and WBCD-D3. These services use minor channels 1, 2 and 3 respectively. Table 10.14 summarizes the four programming services to be defined in the TVCT.

TABLE 10.14 Example Analog and Digital Services

Short Name	Type	Major Channel	Minor Channel	Source ID	Extended Name
WBCD	Analog	39	0	4	WBCD Newark.
WBCD-D1	Digital	39	1	1	WBCD-D1 Newark.
WBCD-D2	Digital	39	2	2	WBCD-D2 Sports.
WBCD-D3	Digital	39	3	3	WBCD-D3 News.

Table 10.15 shows the bit-stream representation of the Terrestrial VCT table section that encodes these parameters. In this example, WBCD-D3 is broadcasting a multi-lingual program in which both an English and a French audio track are available. The Service Location Descriptor reflects this service structure. Descriptors are highlighted with shaded rows in the table.

Rating Region Table (RRT)

As we saw in Chapter 7, the Rating Region Table is what defines the rating parameters, called rating dimensions and levels, that are used by program content adviso-

TABLE 10.15 Virtual Channel Table Example Bit-stream Representation

Field Name	Value	Description
table_id	0xC8	Identifies the table section as being a TVCT.
section_syntax_indicator	1b	The VCT uses the MPEG "long-form" syntax.
private_indicator	1b	Set to 1 in PSIP tables.
reserved	11b	Reserved bits are set to 1.
section_length	0x0121	Length of rest of this section is 289 bytes.
transport_stream_id	0x0E21	Corresponds to TSID in PAT.
reserved	11b	Reserved bits are set to 1.
version_number	01110b	Version number is 14.
current_next_indicator	1b	Section is current TVCT.
section_number	0x00	First section.
last_section_number	0x00	Last is first (only one section).
protocol_version	0x00	Zero for the present protocol.
num_channels_in_section	0x04	This TVCT section defines four channels.
short_name	0x0057 0042 0043 0044 002D 0044 0032	Unicode UTF-16 of "WBCD-D2".
reserved	1111b	Reserved bits are set to 1.
major_channel_number	00 0010 0111b	39 decimal.
minor_channel_number	00 0000 0010	2 decimal.
modulation_mode	0x04	ATSC (8-VSB).
carrier_frequency	0x00000000	Deprecated (must be set to zero).
channel_TSID	0x0E21	Same as TSID in PAT; service carried here.
program_number	0x1001	Linkage to PAT and PMT.
ETM_location	01b	Indicates ETM located in this TS.
access_controlled	0b	Not access-controlled.
hidden	0b	Not hidden.
reserved	11b	Reserved bits are set to 1.
hide_guide	1b	hidden=0 so this bit is ignored (irrelevant).
reserved	111b	Reserved bits are set to 1.
service_type	000010b	ATSC digital television service.
source_id	0x0002	Source ID for WBCD-D2 is 2.
reserved	111111b	Reserved bits are set to 1.
descriptors_length	00 0010 1001b	41 bytes of descriptors.
descriptor_tag	0xA0	Extended Channel Name Descriptor.
descriptor_length	0x16	Length is 22 bytes.

TABLE 10.15 Virtual Channel Table Example Bit-stream Representation (continued)

Field Name	Value	Description
number_strings	0x01	MSS structure, one string.
ISO_639_language_code	0x65 6E 67	"eng" in ASCII, signifying English language.
number_segments	0x01	One segment.
compression_type	0x00	No compression.
mode	0x00	Selects ISO Latin-1 code page in UTF-16.
number_bytes	0x0E	14 byte string.
compressed_string_byte	0x57 42 43 44 2D-44 32 20 53 70 6F 72 74 73	"WBCD-D2 Sports"
descriptor_tag	0xA1	Service Location Descriptor.
descriptor_length	0x0F	Length is 15 bytes.
reserved	111b	Reserved bits are set to 1.
PCR_PID	0x0011	PCR PID is 0x0011 (13-bits).
number_elements	0x02	Two program elements.
stream_type	0x02	MPEG-2 video.
reserved	111b	Reserved bits are set to 1.
elementary_PID	0x0011	Video PID is 0x0011 (13-bits).
ISO_639_language_code	0x000000	No language specified.
stream_type	0x81	AC-3 Audio.
reserved	111b	Reserved bits are set to 1.
elementary_PID	0x0014	Audio PID is 0x0014 (13-bits).
ISO_639_language_code	0x65 6E 67	"eng" in ASCII, signifying English language.
short_name	0x0057 0042 0043 0044 002D 0044 0031	Unicode UTF-16 of "WBCD-D1".
reserved	1111b	Reserved bits are set to 1.
major_channel_number	00 0010 0111b	39 decimal.
minor_channel_number	00 0000 0001	1 decimal.
modulation_mode	0x04	ATSC (8-VSB).
carrier_frequency	0x00000000	Deprecated (must be set to zero).
channel_TSID	0x0E21	Same as TSID in PAT; service carried here.
program_number	0x1000	Linkage to PAT and PMT.
ETM_location	01b	Indicates ETM located in this TS.
access_controlled	0b	Not access-controlled.
hidden	0b	Not hidden.
reserved	11b	Reserved bits are set to 1.

TABLE 10.15 Virtual Channel Table Example Bit-stream Representation (continued)

Field Name	Value	Description
hide_guide	1b	hidden=0 so this bit is ignored (irrelevant).
reserved	111b	Reserved bits are set to 1.
service_type	000010b	ATSC digital television service.
source_id	0x0001	Source ID for WBCD-D1 is 1.
reserved	111111b	Reserved bits are set to 1.
descriptors_length	00 0010 1001b	41 bytes of descriptors.
descriptor_tag	0xA0	Extended Channel Name Descriptor.
descriptor_length	0x16	Length is 22 bytes.
number_strings	0x01	MSS structure, one string.
ISO_639_language_code	0x65 6E 67	"eng" in ASCII, signifying English language.
number_segments	0x01	One segment.
compression_type	0x00	No compression.
mode	0x00	Selects ISO Latin-1 code page in UTF-16.
number_bytes	0x0E	14 byte string.
compressed_string_byte	0x57 42 43 44 2D-44 31 20 4E 65 77 61 72 6B	"WBCD-D1 Newark".
descriptor_tag	0xA1	Service Location Descriptor.
descriptor_length	0x0F	Length is 15 bytes.
reserved	111b	Reserved bits are set to 1.
PCR_PID	0x0041	PCR PID is 0x0041 (13-bits).
number_elements	0x02	Two program elements.
stream_type	0x02	MPEG-2 video.
reserved	111b	Reserved bits are set to 1.
elementary_PID	0x0041	Video PID is 0x0041 (13-bits).
ISO_639_language_code	0x000000	No language specified.
stream_type	0x81	AC-3 Audio.
reserved	111b	Reserved bits are set to 1.
elementary_PID	0x0044	Audio PID is 0x0044 (13-bits).
ISO_639_language_code	0x65 6E 67	"eng" in ASCII, signifying English language.
short_name	0x0057 0042 0043 0044 002D 0044 0033	Unicode UTF-16 of "WBCD-D3".
reserved	1111b	Reserved bits are set to 1.
major_channel_number	00 0010 0111b	39 decimal.
minor_channel_number	00 0000 0011	3 decimal.

TABLE 10.15 Virtual Channel Table Example Bit-stream Representation (continued)

Field Name	Value	Description
modulation_mode	0x04	ATSC (8-VSB).
carrier_frequency	0x00000000	Deprecated (must be set to zero).
channel_TSID	0x0E21	Same as TSID in PAT; service carried here.
program_number	0x1002	Linkage to PAT and PMT.
ETM_location	01b	Indicates ETM located in this TS.
access_controlled	0b	Not access-controlled.
hidden	0b	Not hidden.
reserved	11b	Reserved bits are set to 1.
hide_guide	1b	hidden=0 so this bit is ignored (irrelevant).
reserved	111b	Reserved bits are set to 1.
service_type	000010b	ATSC digital television service.
source_id	0x0003	Source ID for WBCD-D3 is 3.
reserved	111111b	Reserved bits are set to 1.
descriptors_length	00 0010 1101b	45 bytes of descriptors
descriptor_tag	0xA0	Extended Channel Name Descriptor.
descriptor_length	0x14	Length is 20 bytes.
number_strings	0x01	MSS structure, one string.
ISO_639_language_code	0x65 6E 67	"eng" in ASCII, signifying English language.
number_segments	0x01	One segment.
compression_type	0x00	No compression.
mode	0x00	Selects ISO Latin-1 code page in UTF-16.
number_bytes	0x0C	12 byte string.
compressed_string_byte	0x57 42 43 44 2D 44 33 20 4E 65 77 73	"WBCD-D3 News".
descriptor_tag	0xA1	Service Location Descriptor.
descriptor_length	0x15	Length is 21 bytes.
reserved	111b	Reserved bits are set to 1.
PCR_PID	0x0031	PCR PID is 0x0031 (13-bits).
number_elements	0x03	Three program elements.
stream_type	0x02	MPEG-2 video.
reserved	111b	Reserved bits are set to 1.
elementary_PID	0x0031	Video PID is 0x0031 (13-bits).
ISO_639_language_code	0x000000	No language specified.
stream_type	0x81	AC-3 Audio.

TABLE 10.15 Virtual Channel Table Example Bit-stream Representation (continued)

Field Name	Value	Description
reserved	111b	Reserved bits are set to 1.
elementary_PID	0x0034	Audio PID is 0x0034 (13-bits).
ISO_639_language_code	0x65 6E 67	"eng" in ASCII, signifying English language.
stream_type	0x81	AC-3 Audio.
reserved	111b	Reserved bits are set to 1.
elementary_PID	0x0035	Audio PID is 0x0035 (13-bits).
ISO_639_language_code	0x66 72 65	"fre" in ASCII, signifying French language.
short_name	0x0057 0042 0043 0044 0000 0000 0000	Unicode UTF-16 for "WBCD".
reserved	1111b	Reserved bits are set to 1.
major_channel_number	00 0010 0111b	39 decimal.
minor_channel_number	00 0000 0000	0 decimal.
modulation_mode	0x01	Analog modulation.
carrier_frequency	0x00000000	Deprecated (must be set to zero).
channel_TSID	0x0E20	Analog Transmission Signal ID (analog TSID).
program_number	0xFFFF	Indicates analog service.
ETM_location	01b	Indicates ETM located in this TS.
access_controlled	0b	Not access-controlled.
hidden	0b	Not hidden.
reserved	11b	Reserved bits are set to 1.
hide_guide	1b	hidden=0 so this bit is ignored (irrelevant).
reserved	111b	Reserved bits are set to 1.
service_type	000001b	Analog television service.
source_id	0x0004	Source ID for WBCD analog is 4.
reserved	111111b	Reserved bits are set to 1.
descriptors_length	00 0001 0101b	21 bytes of descriptors.
descriptor_tag	0xA0	Extended Channel Name Descriptor.
descriptor_length	0x13	Length is 19 bytes.
number_strings	0x01	MSS structure, one string.
ISO_639_language_code	0x65 6E 67	"eng" in ASCII, signifying English language.
number_segments	0x01	One segment.
compression_type	0x00	No compression.
mode	0x00	Selects ISO Latin-1 code page in UTF-16.
number_bytes	0x0B	11 byte string.

TABLE 10.15 Virtual Channel Table Example Bit-stream Representation (continued)

Field Name	Value	Description
compressed_string_byte	0x57 42 43 44 20 4E 65 77 61 72 6B	"WBCD Newark".
reserved	111111b	Reserved bits are set to 1.
additional_descriptors_length	00 0000 0100b	No additional descriptors.
CRC_32	0xDD458FA9	MPEG-2 32-bit table section CRC.

ries. A Transport Stream can carry as many Rating Region Tables as there are ratings regions in use.

In some cases such as the United States, a certain region's RRT is unchangeable and defined outside the ATSC standard. In such cases, that region's RRT need not be sent. Receiving devices can be built using the standard definition available for that part of the world. The Rating Region Table for rating_region 0x01 (US plus possessions) is fully defined in EIA/CEA-766-A[1].

Note that the EIA/CEA-766-A[1] Standard also defines the Canadian RRT (rating_region 0x02), but unlike the one for the US, the Canadian table is specified as changeable. That means at some time in the future, an updated Canadian RRT can be sent. The new table would be identifiable as an update because the version_number field would have been incremented.

Now we explore in detail the structure, syntax, and semantics of the PSIP Rating Region Table.

Structure of the Rating Region Table

Any Transport Stream can carry zero or more Rating Region Table sections. Each RRT instance corresponds to one rating region, the value of which is identified within the table section header in order to distinguish one RRT instance from another in the TS.

Figure 10.9 illustrates the structure of the Rating Region Table section as it is formatted within transmitted PSIP data.

As shown, the region associated with each RRT instance is included in the header portion of the table. At the topmost level following the standard table section header, the RRT is structured in these parts:

- a string-length and text string, in MSS format, giving the name of the region in textual (and possibly multi-lingual) form,

- a count of the number of rating dimensions this RRT includes, followed by that many data blocks defining dimension data. In the Figure, dimension data is indicated as an indexed array of data blocks, with the "dimensions defined" field identifying the total number.

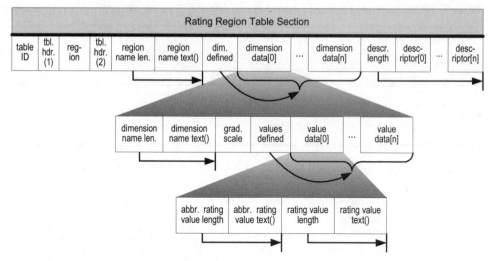

Figure 10.5 Structure of the Rating Region Table Section

- an optional descriptor loop to accommodate any descriptors pertinent to this RRT.

  Dimension data itself is structured into these parts:

- a string-length and text string giving the name of the dimension in textual (and possibly multi-lingual) form.

- a bit indicating whether or not the dimension is to be interpreted as a "graduated scale"

- a count of the number of values defined for this dimension, followed by that many blocks of value data. Value data is shown in the figure as an indexed array.

  Each value data block includes the following information pertinent to each rating level:

- a (possibly multi-lingual) text string indicating the name of the level in abbreviated form.

- a (possibly multi-lingual) text string indicating the full name of the level.

RRT transport

Transport of the RRT section involves a few constraints against MPEG-2 *Systems*:

- a Rating Region Table section, if carried in the Transport Stream, isplaced into Transport Stream packets with PID value 0x1FFB, the SI base_PID.

- TS packets containing the RRT section are not to be scrambled (the transport_scrambling_control bits must be set to zero).

- as with all PSIP tables, an adaptation field cannot be present in TS packets carrying the RRT.

- each RRT instance is limited to a total length of 1024 bytes.

- the RRT for a given rating region may not be sectioned; it must fit into a single 1024-byte section.

RRT syntax and semantics

Table 10.16 describes the syntax and semantics of the RRT section.

TABLE 10.16 RRT Syntax and Semantics

Field Name	Number of Bits	Field Value	Description
rating_region_table_section() {			Start of the rating_region_table_section().
table_id	8	0xCA	Identifies the table section as being a rating_region_table_section().
section_syntax_indicator	1	1b	The RRT uses the MPEG "long-form" syntax.
private_indicator	1	1b	Set to 1 in PSIP tables.
reserved	2	11b	Reserved bits are set to 1.
section_length	12	<=1021	The RRT section is limited to 1024 total bytes, so section_length is limited to 1021.
reserved	8	0xFF	Reserved bits are set to 1.
rating_region	8		Indicates the Rating Region being defined by this instance of the RRT.
reserved	2	11b	Reserved bits are set to 1.
version_number	5		The version number reflects the version of a table section, and is incremented when anything in the table changes.
current_next_indicator	1	1b	Set to 1, indicating the table section is currently applicable.
section_number	8	0	The RRT cannot be segmented.
last_section_number	8	0	The RRT cannot be segmented.
protocol_version	8	0	Indicates the protocol version of this table section. The only type of rating_region_table_section() currently defined is for protocol_version zero.

TABLE 10.16 RRT Syntax and Semantics (continued)

Field Name	Number of Bits	Field Value	Description
rating_region_name_length	8		An unsigned integer that gives the number of bytes in the rating_region_name_text() field to follow.
rating_region_name_text()	var.		Indicates the textual name of the region formatted as a Multiple String Structure.
dimensions_defined	8		An unsigned integer that indicates the number of dimensions defined for this region.
for (i=0; i<dimensions_defined; i++) {			Start of the dimensions defined "for" loop.
dimension_name_length	8		The length in bytes of the Multiple String Structure to follow.
dimension_name_text()	var.		A text string in Multiple String Structure format giving the name of this dimension.
reserved	3	111b	Reserved bits are set to 1.
graduated_scale	1		A flag that indicates whether or not this dimension is defined on a graduated scale.
values_defined	4		An unsigned integer that indicates the number of values defined for this rating dimension.
for (i=0; i<values_defined; i++) {			Start of the values defined "for" loop.
abbrev_rating_value_length	8		The length of the Multiple String Structure to follow.
abbrev_rating_value_text()	var.		A text string giving the abbreviated name of the rating value for the dimension being defined in this iteration of the dimensions-defined "for" loop.
rating_value_length	8		The length of the Multiple String Structure to follow.
rating_value_text()	var.		A text string giving the full name of the rating value for the dimension being defined in this iteration of the dimensions-defined "for" loop.
}			End of the values defined "for" loop.
}			End of the dimensions defined "for" loop.
reserved	6	111111b	Reserved bits are set to 1.
descriptors_length	10		This field indicates the length of any additional descriptors.

TABLE 10.16 RRT Syntax and Semantics (continued)

Field Name	Number of Bits	Field Value	Description
for (i=0; i<N; i++) {			Start of the additional descriptors "for" loop. The value of N is determined by descriptors_length. Zero or more descriptors are processed until the total number of descriptor bytes processed equals the value given in descriptors_length.
descriptor()	var.		Zero or more descriptors, formatted as type-length-data.
}			End of the additional descriptors "for" loop.
CRC_32	32		A 32-bit checksum designed to produce a zero output from the decoder defined in the MPEG-2 *Systems* Standard.
}			End of the rating_region_table_section().

region

This 8-bit field identifies the region to which the RRT data in this table instance is to apply. The field is in the position of table_id_extension, so that multiple instances of the RRT (one for each region) may be included in TS packets, using a common PID value.

rating_region_name_length

This field is an 8-bit integer giving the length in bytes of the multi-lingual rating region text field to follow.

rating_region_name_text()

Following the standard table section header, where rating region appears in the table ID extension portion of the header, is a multi-lingual text string in Multiple String Structure format that provides the name of the region. For the US table, the name has been defined in English to be "US (50 states + possessions)." The rating_region_name_text() field is a Multiple String Structure, so that it could be provided multi-lingually, if desired.

dimensions_defined

This is an integer that indicates the number of dimensions defined for this particular rating region. The data defined for each dimension includes its name, an attribute bit (graduated scale), and the definition of each of its levels.

dimension_name_length

A textual name is given for each dimension in this rating system. The dimension_name_length is an integer indicating the number of bytes in the dimension name text field to follow.

dimension_name_text()

The name of this dimension in the form of a Multiple String Structure.

graduated_scale

A flag that indicates, when set, that this dimension is a "graduated scale," meaning that higher values represent stronger and stronger content, where each level represents an increased level of the type of content defined in this rating dimension. We discussed this in Chapter 7 (see "Graduated Scales" on page 117).

values_defined

An integer value that indicates the number of values defined for this particular rating dimension.

abbrev_rating_value_length

Two text different strings may be defined for each level within each rating dimension. The abbreviated string is expected to be helpful when a receiver constructs text strings such as "This program is rated PG." The abbrev_rating_value_length field gives the number of bytes in the Multiple String Structure to follow.

abbrev_rating_value_text()

A text string in the form of a Multiple String Structure that represents the abbreviated string for this rating value.

rating_value_length

This field gives the length of the MSS structure to follow.

rating_value_text()

This is the full-length string that is the name of this particular rating value.

RRT descriptors

The RRT can have one or more descriptors if it is necessary to give any further information about this rating region. At present, no standard descriptors are defined for use with the RRT.

RRT transport rate and cycle time

A/65 states that the RRT describing a certain rating region, when sent, must be repeated at a rate of once per minute. If a typical RRT described six rating dimensions, each with an average of five levels, the RRT would be approximately 900 bytes in length. Such an RRT fits into five Transport Stream packets with 20 bytes to spare. Repeating five TS packets at a rate of once each 60 seconds yields a TS bitrate of 125 bps for this RRT.

Event Information Table (EIT)

Event Information Tables can provide program schedule information for any analog or digital channel listed in the Virtual Channel Table. A program schedule consists of the start and end times of each event, its name or title, linkage to a textual description of the event, and an optional list of descriptors that can give further information pertinent to the event such as its content advisory and the caption services available with the program.

In this section we look closely at the structure of the EIT and its detailed syntax and semantics.

Structure of the Event Information Table

Figure 10.6 illustrates the basic structure of the EIT section. EIT sections use the standard long-form MPEG-2 private section syntax, as usual. In the table header the source_id appears in the position of the table_id_extension. This allows multiple distinct EIT instances, each corresponding to a different virtual channel (linked by source_id), to appear in Transport Stream packets with common PID values. The source_id is what allows receivers to see the instances as separate from one another, and, by matching source_ids found in the table headers, to re-associate different parts of sectioned EITs.

Each EIT instance, then, gives a portion of the program schedule for the programming service identified by source_id. As the figure shows, the message body of the EIT consists of two "for" loops. The first contains event records and the second is composed of the additional descriptors loop found in most PSIP tables. The number of event records is given by the parameter just in front of the start of the loop, num_events_in_section.

Any given EIT describes a three-hour block of programming, where the three-hour blocks are aligned to UTC. Programs that run for any period of time within the three-hour time slot must be included. That means programs starting, ending, or continuing throughout the three-hour period are included. When a program is

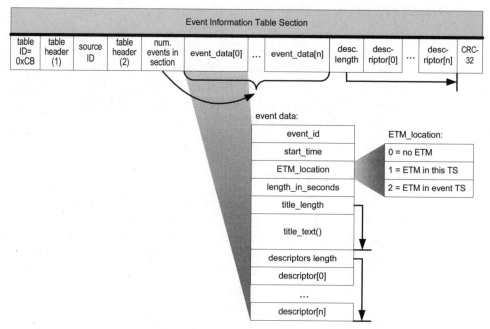

Figure 10.6 Event Information Table Section Structure

included in more than one EIT (EITs for different time slots), the same event_id must be used in each instance.

An event record consists of information pertaining to an event or program. Information given about each event includes:

- its starting time and duration.

- the program's title.

- flags indicating whether or not descriptive text exists, and if so, whether it is located in this same Transport Stream or in the TS carrying the actual event.

- the event_id field used to link the event to the textual descriptions carried in the Extended Text Tables.

- any descriptors pertinent to the event, such as Content Advisory Descriptors or Caption Services Descriptors.

EIT transport

Transport of the Event Information Table involves these considerations:

- EIT sections are carried in Transport Stream packets with PID values as identified in the MGT. All the of the EITs describing events in a given three-hour time slot must be transported in TS packets with a common PID. EITs for different time slots must be transported in packets with different PIDs.

- TS packets containing an EIT section cannot be scrambled (the transport_scrambling_control bits are set to zero).

- As with all PSIP tables, an adaptation field are not to be present in TS packets carrying an EIT section.

- Any EIT may be sectioned for delivery, and each EIT section can be as long as the MPEG-2 limit for private table sections, 4096 bytes.

- The total bitrate for transport packets with any single PID value used for EITs cannot exceed 250,000 bps. This means, for example, that the EITs representing the current three-hour period, EIT-0, cannot use more than 250 kilobits of the TS bandwidth. EIT-1 can, however, use another independent 250 kilobits.

Next we look in detail at the syntax and semantics of the EIT section.

EIT syntax and semantics

Table 10.17 shows the syntax of the fields making up an EIT section.

TABLE 10.17 Event Information Table Syntax and Semantics

Field Name	Number of Bits	Field Value	Description
event_information_table_section() {			Start of the event_information_table_section().
table_id	8	0xCB	Identifies the table section as being an event_information_table_section().
section_syntax_indicator	1	1b	The EIT uses the MPEG "long-form" syntax.
private_indicator	1	1b	Set to 1 in PSIP tables.
reserved	2	11b	Reserved bits are set to 1.
section_length	12		The EIT section can be 4096 total bytes in length, so section_length is limited to 4093.
source_id	16		Associates this EIT instance with the virtual channel record with a matching source_id field.
reserved	2	11b	Reserved bits are set to 1.
version_number	5		The version number reflects the version of a table section, and is incremented when anything in the table changes.
current_next_indicator	1	1b	Indicates that the EIT table section is currently applicable. "Next" EIT sections are not allowed to be sent.

TABLE 10.17 Event Information Table Syntax and Semantics (continued)

Field Name	Number of Bits	Field Value	Description
section_number	8		The EIT may be segmented into sections for delivery. When it is segmented, section_number indicates which part this section is. When the whole table fits into one table section, section_number is set to zero.
last_section_number	8		Indicates the section number of the last section in a segmented table. The last_section_number field is set to zero if the whole table fits into one section.
protocol_version	8	0	Indicates the protocol version of this table section. The only type of event_information_table_section() currently defined is for protocol_version zero.
num_events_in_section	8		Indicates the number of events being defined in this section of the EIT.
for (i=0; i<num_events_in_section; i++) {			Start of events "for" loop.
reserved	2	11b	Reserved bits are set to 1
event_id	14		Associates this event with a numeric tag, used to link it with a textual description in an Extended Text Table section.
start_time	32		Indicates the time this event starts; given as the number of GPS seconds since midnight January 6th, 1980.
reserved	2	11b	Reserved bits are set to 1
ETM_location	2		Indicates whether or not ETT text is available for this event, and if so whether it is in this Transport Stream or the one carrying the actual event.
length_in_seconds	20		Indicates the program or event's running time, in seconds.
title_length	8		Gives the size in bytes of the Multiple String Structure to follow.
title_text()	var.		This is a Multiple String Structure encoding of the title of the program or event.
reserved	4	1111b	Reserved bits are set to 1.
descriptors_length	12		Indicates the total number of bytes in the descriptors to follow.
for (i=0; i<N; i++) {			Start of event descriptors "for" loop. The value of N is given indirectly by descriptors_length. Zero or more descriptors should be processed until the total number of descriptor bytes processed equals descriptors_length.

TABLE 10.17 Event Information Table Syntax and Semantics (continued)

Field Name	Number of Bits	Field Value	Description
descriptor()	var.		Each of these descriptors pertains to the specific event being defined in this iteration of the event "for" loop.
}			End of event descriptors "for" loop.
}			End of events "for" loop.
CRC_32	32		A 32-bit checksum designed to produce a zero output from the decoder defined in the MPEG-2 *Systems* Standard.
}			End of the event_information_table_section().

source_id

All the events active in a three-hour time period are described in EITs transported in TS packets with the same PID. The Source ID is what distinguishes one EIT instance from another when collecting table sections from packets with one of these PID values. Source ID also ties each EIT instance to a virtual channel record with a matching Source ID.

As indicated in the discussion of the Virtual Channel Table, source_id values can be unique within the context of the Transport Stream or unique on a regional basis. A source_id value below 0x1000 is unique within the Transport Stream carrying the VCT and EIT, but that same value may be found within a different Transport Stream where it may refer to a different programming service. Source ID values 0x1000 and above are unique on a scale that spans across all of the Transport Streams that could be expected to be accessible by a single receiver. See Chapter 6 for a further discussion of the Source ID concept.

num_events_in_section

This integer value indicates the number of events being described in this EIT section. PSIP requires that events be placed in the EIT in order of their start times. Furthermore, all the events described in one EIT section must have their starting times earlier than those given in another, higher-numbered section. This means that all of the events will already be in time-ascending order after the sections are re-assembled into the full EIT, thus saving the receiver the trouble of sorting them.

event_id

Every event described in the EIT has an event_id tag, a 14-bit number used to link program data given in the EIT with descriptive text supplied in an Extended Text

Table. Event ID values must be unique for all of the events currently described in EIT/ETT data for a given virtual channel, for any given three-hour time slot.[*] Therefore, different events on two distinct virtual channels could use the same event_id value. Likewise, separate events in different time slots for the same source_id could use the same event_id value. It is the responsibility of the receiver to use the source_id and time slot in addition to event_id as an unambiguous identifier within an event's Transport Stream. Source_id and event_id values are used to link the event data given in the EIT with a piece of text supplied in an Extended Text Table.

If the duration of an event crosses a three-hour time slot boundary, it must appear in both EITs (the one it has come out of and the one it is going into). When this occurs, the same event_id must be used in both EITs.

start_time

One of the fundamental pieces of information about a program or event is its scheduled starting time. As with all times in the A/65 Standard, start_time is expressed as the number of GPS seconds since the beginning of the first second of the first full week of 1980, Coordinated Universal Time, or 00:00:00 January 6, 1980 UTC.

ETM_location

Each EIT record indicates with this field whether or not a textual description of the event is available in an accompanying Extended Text Table. If text is available, the flags indicate whether the ETT is carried in this same Transport Stream or in the TS carrying the event itself. ETM stands for Extended Text Message. ETM_location is coded according to the following table.

TABLE 10.18 ETM Location

ETM_location	Meaning
0x00	No ETM is available for this event.
0x01	An ETM is present for this event; it is located in this Transport Stream (the one carrying this EIT).
0x02	An ETM is present for this event; it is located in the Transport Stream carrying the event.
0x03	Reserved for future ATSC use.

---

[*] ATSC may extend the scope of uniqueness of event_id in a future update to form a better linkage between PSIP data and downloaded applications. Such a change, if it were to occur, would be backwards compatible with receivers built to the current specification.

length_in_seconds

This field simply gives the scheduled duration, in seconds, of the event being described in this iteration of the event's "for" loop.

title_length

This is the length field that precedes all Multiple String Structures. The title_length field indicates the number of bytes in the title_text() MSS structure to follow.

title_text()

Program providers should supply a title for every event in the program schedule. This is the same title that appears in printed TV guides, and when transmitted in an EIT it (or an abbreviation) can be displayed on-screen in Electronic Program Guide grids or listings. The first two Huffman encode/decode tables defined in A/65 are optimized for English-language program title text. Optimization takes the form of creating shorter Huffman codes for words starting with an initial capital letter, common for program titles. See "Huffman text compression" on page 139.

Event descriptors

Each event can include one or more descriptors pertinent to that program or event. Rules for terrestrial broadcast and cable differ from one another. For terrestrial broadcast some descriptors, when used, must be present in the EIT and can optionally be present in the PMT. For cable, these descriptors must be present in the PMT and also in the EIT, if the EIT itself is present.

This apparent disconnect arises from an inherent difference in philosophy between broadcasters and cable operators. Broadcasters felt that EITs should always be present, and in fact mandated that compliant terrestrial broadcast Transport Streams at least include EITs covering 9-12 hours into the future. Cable operators, on the other hand, have traditionally delivered program guide data only in proprietary ways to set-top boxes they owned and leased to customers.

Industry agreements regarding PSIP reached in February 2000 between the National Cable Television Association (NCTA) and the Consumer Electronics Association (CEA) stated a commitment on the part of the cable community to deliver PSIP data, including EIT/ETT data, in-band on Transport Streams modulated on cable, as long as this EPG data was made available to them by the content provider. These agreements are discussed in detail in Chapter 19 starting on page 388.

As a result, and given that cable operators did not have any direct control over whether content providers would supply PSIP data to their headends, operators wished to specify that a Transport Stream with an MGT, VCT, and STT, but no EITs, would still be compliant with SCTE standards. Proprietary set-top boxes they operate parse the PMT for all program-related descriptors and do not process the PSIP EIT at all.

Descriptors listed in Table 10.19 below are defined for use with the EIT.

TABLE 10.19 Event Descriptors in the EIT

Descriptor Tag	Descriptor Name	Where Defined	Description
0x80	Stuffing	A/65 PSIP	No function other than to make a placeholder of a given number of bytes. Stuffing bytes are disregarded by decoders.
0x81	AC-3 Audio	ATSC A/52	This descriptor can identify specific characteristics of audio services expected to be available when the associated event airs. Note that just one audio_stream_descriptor() can be present to describe any one event; if more than one audio track is offered, one descriptor won't tell the whole story.
0x87	Content Advisory	A/65 PSIP	Provides program content advisory information pertinent to the program. If a program is content-rated, for the terrestrial broadcast application, the Content Advisory Descriptor must appear in the transmitted EIT.
0x86	Caption Service	A/65 PSIP	Lists the caption services expected to be available when this event is aired. Information provided includes the language, format (EIA-608-B or advanced television closed captions defined in EIA-708-B) and type of each caption service. If a program is captioned, for the terrestrial broadcast application the Caption Service Descriptor must appear in the transmitted EIT.
0xAA	Redistribution Control	A/65 PSIP	Signals that the content owner has asserted rights with respect to redistribution of the associated program.

EIT transport rate and cycle time

A/65 states that for terrestrial broadcast the recommended maximum cycle time for EIT-0 is 500 milliseconds. If that recommendation is followed, on average, a receiver can pick up the current program schedule for channels delivered in a given Transport Stream within 250 milliseconds of acquiring that TS. The total maximum bitrate for EITs describing any given three-hour time slot is not allowed to exceed 250,000 bps.

Rates for cable have not been standardized, but content providers recognize it is in their best interest, if they wish to attract viewers, to provide EPG data in a timely manner.

Example EIT

Let's look at an actual example EIT table section in detail. Table 10.20 shows a set of event data for source_id 0x0001.

TABLE 10.20 Example Event Data

Event Name	Start Time	Duration	Event ID	Content Advisory	Has ETM?
"A Rash of Trouble"	December 16, 2001, 2:00:00 a.m. UTC	120 min.	0x0003	TV-PG	Yes, in TS
"Local News"	December 16, 2001, 4:00:00 a.m. UTC	30 min.	0x0084	-	Yes, in TS
"City Council Report"	December 16, 2001, 4:30:00 a.m. UTC	30 min.	0x0085	-	Yes, in TS

Table 10.21 shows the bit-stream representation of the EIT table section corresponding to this data set. Descriptors and Multiple String Structures are shown in the shaded cells.

TABLE 10.21 Event Information Table Example Bit-stream Representation

Field Name	Value	Description
table_id	0xCB	Identifies the table section as being an EIT.
section_syntax_indicator	1b	The EIT uses the MPEG "long-form" syntax.
private_indicator	1b	Set to 1 in PSIP tables.
reserved	11b	Reserved bits are set to 1.
section_length	0x0075	Length of the rest of this section is 117 bytes.
source_id	0x0001	Associates the data in this table section with a particular virtual channel.
reserved	11b	Reserved bits are set to 1.
version_number	10000b	Version number is 16.
current_next_indicator	1b	Section must be current EIT.
section_number	0x00	First section.
last_section_number	0x00	Last is first (only one section).
protocol_version	0x00	Zero for the present protocol.
num_events_in_section	0x03	This section has three events for Source ID 0x0001.
reserved	11b	Reserved bits are set to 1.
event_id	0x0003	Event ID for this event is 0x0003.

TABLE 10.21 Event Information Table Example Bit-stream Representation (continued)

Field Name	Value	Description
start_time	0x2946C2AD	December 16, 2001, 2:00:00 a.m. UTC.
reserved	11b	Reserved bits are set to 1.
ETM_location	01b	/ETM present in this Transport Stream.
length_in_seconds	0x01C20	7200 seconds (2 hours).
title_length	0x19	MSS length is 25 bytes.
number_strings	0x01	One string.
ISO_639_language_code	0x656E67	"eng" indicates English language.
number_segments	0x01	One segment.
compression_type	0x00	No compression.
mode	0x00	Selects ISO Latin-1 character encoding.
number_bytes	0x11	17 bytes in string to follow.
compressed_string_byte[k]	0x41 20 52 61 73 68 20 6F 66 20 54 72 6F 75 62 6C 65	"A Rash of Trouble" in ISO Latin-1.
reserved	1111b	Reserved bits are set to 1.
descriptors_length	0x015	21 bytes of event descriptors.
descriptor_tag	0x87	Descriptor is a Content Advisory Descriptor.
descriptor_length	0x13	Length is 19 bytes following length byte itself.
reserved	11b	Reserved bits are set to 1.
rating_region_count	0x01	One region.
rating_region	0x01	Region 1, US + possessions.
rated_dimensions	0x01	One dimension in the US system is rated.
rating_dimension_j	0x00	First one is dimension 0 (TV rating).
reserved	1111b	Reserved bits are set to 1.
rating_value	0x3	Value is 3, meaning "TV-PG".
rating_description_length	0x0D	Thirteen bytes of MSS text describing the rating.
number_strings	0x01	One string.
ISO_639_language_code	0x656E67	Language is English (code "eng").
number_segments	0x01	One segment.
compression_type	0x00	No compression.
mode	0x00	Mode zero selects ISO Latin-1 coding.
number_bytes	0x05	Length of the string to follow.
compressed_string_byte[k]	0x54562D5047	ISO Latin-1 encoding of "TV-PG".
reserved	11b	Reserved bits are set to 1.

TABLE 10.21 Event Information Table Example Bit-stream Representation (continued)

Field Name	Value	Description
event_id	0x0084	Event ID for this event is 0x0084.
start_time	0x2946DECD	December 16, 2001, 4:00:00 a.m. UTC.
reserved	11b	Reserved bits are set to 1.
ETM_location	01b	ETM present in this Transport Stream.
length_in_seconds	0x00708	1800 seconds (30 minutes).
title_length	0x12	MSS length is 18 bytes.
number_strings	0x01	One string.
ISO_639_language_code	0x656E67	"eng" indicates English language.
number_segments	0x01	One segment.
compression_type	0x00	No compression.
mode	0x00	Selects ISO Latin-1 character encoding.
number_bytes	0x0A	10 bytes in string to follow.
compressed_string_byte[k]	0x4C 6F 63 61 6C 20 4E 65 77 73	"Local News" in ISO Latin-1.
reserved	1111b	Reserved bits are set to 1.
descriptors_length	0x000	No event descriptors.
reserved	11b	Reserved bits are set to 1.
event_id	0x0085	Event ID for this event is 0x0085.
start_time	0x2946E5D5	December 16, 2001, 4:30:00 a.m. UTC.
reserved	11b	Reserved bits are set to 1.
ETM_location	01b	ETM present in this Transport Stream.
length_in_seconds	0x00708	1800 seconds (30 minutes).
title_length	0x1B	MSS length is 26 bytes.
number_strings	0x01	One string.
ISO_639_language_code	0x656E67	"eng" indicates English language.
number_segments	0x01	One segment.
compression_type	0x00	No compression.
mode	0x00	Selects ISO Latin-1 character encoding.
number_bytes	0x13	19 bytes in string to follow.

TABLE 10.21 Event Information Table Example Bit-stream Representation (continued)

Field Name	Value	Description
compressed_string_byte[k]	0x69 74 79 20 43 6F 75 6E 63 69 6C 20 52 65 70 6F 72 74	"City Council Report" in ISO Latin-1.
reserved	1111b	Reserved bits are set to 1.
descriptors_length	0x000	No event descriptors.
CRC_32	0x91A9EB2E	MPEG-2 32-bit table section CRC (example).

Extended Text Table (ETT)

PSIP defines a method to include multi-lingual text in the digital Transport Stream. The bulk of this text is event descriptions, giving a few words, a sentence, or a few sentences of information about the contents of a given program. Depending on the type of program, the text can include the title of an episode, a synopsis of the story line, the names of actors, the year of production, or anything the provider of the guide data wishes to include.

Extended Text Tables are linked to events announced in Event Information Tables by Source ID (to identify the virtual channel) and by event ID (to reference a specific event on that channel). ETTs may also be linked to data events announced in the Data Event Tables. In this case, data ID plays the same role as Source ID. Yet another way ETTs are used is to deliver textual information regarding a virtual channel; this kind of ETT is called a channel ETT.

Structure of the Extended Text Table

Figure 10.7 describes the Extended Text Table section. It is a very simple structure consisting of just two parts: the ETM ID, which identifies and links the text in the section to a specific television or data event, virtual channel, or other item; and a Multiple String Structure with the text itself. In the table header of the ETT is a table ID extension field. It is set as needed in each ETT instance to ensure that ETT sections occurring in Transport Stream packets with common PID values can be distinguished from one another at the level of the table header, per MPEG-2 rules.

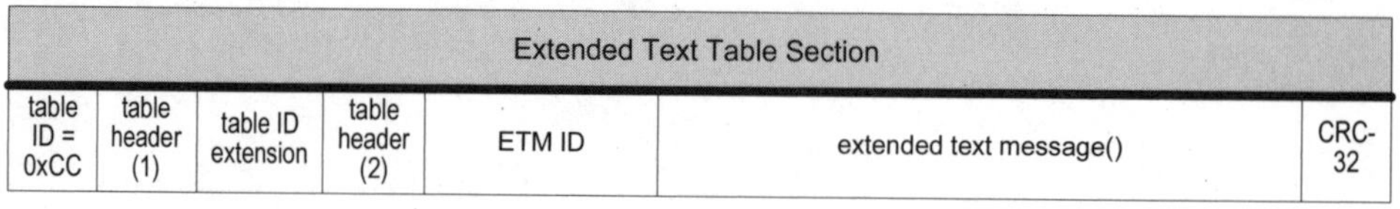

Extended Text Table Section						
table ID = 0xCC	table header (1)	table ID extension	table header (2)	ETM ID	extended text message()	CRC-32

Figure 10.7 Structure of the Extended Text Table Section

ETT transport constraints

Extended Text Table sections must be transported in accordance with certain rules:

- PID values in TS packets used to transport ETT sections must correspond to values indicated in the MGT for the associated table_type.

- As defined by the MGT's table_type structure, ETTs are either channel ETTs, or are ETTs corresponding to events in a given three-hour time slot. The PID value used to transport ETT sections must be unique for each MGT table_type. This means the PID used for channel ETTs cannot be used to transport any other type of table. Likewise, ETTs cannot be transported in the same PID as any EIT.

- Transport stream packets carrying ETT sections cannot be scrambled; the transport_scrambling_control bits must be set to zero.

- TS packets carrying ETT sections cannot have adaptation fields.

- An ETT cannot be segmented for delivery. The text corresponding to the data item identified in ETM ID must fit into a single table section, which can be a total of 4096 bytes in length. Given that each ETT instance corresponds to a single television or data event (or in the case of the channel ETT, virtual channel), and that English text can be compressed at an efficiency approaching 2:1, 4096 bytes represents a generous volume of text.

- The total bitrate for transport packets with any single PID value used for ETTs cannot exceed 250,000 bps. This means, for example, that the ETTs for events in the current three-hour period, ETT-0, cannot use more than 250 kilobits per second of the TS bandwidth. ETT-1 can, however, use another independent 250 kilobits per second.

ETT syntax and semantics

Table 10.22 gives the syntax of the Extended Text Table section.

table_id_extension

Earlier versions of ATSC A/65 set this field to zero, but MPEG purists pointed out that when parsing table sections that have arrived in Transport Stream packets with common PID values, one must be able to distinguish them as separate table instances by inspection of the MPEG-2 private section table header. The current A/65 states that table_id_extension has to be set to any value that will ensure that separate ETT instances have unique values when they are carried in TS packets with common PID values. The value of the table_id_extension field in the ETT has no intrinsic meaning.

TABLE 10.22 Extended Text Table Syntax and Semantics

Field Name	Number of Bits	Field Value	Description
extended_text_table_section() {			Start of the extended_text_table_section().
table_id	8	0xCC	Identifies the table section as being an extended_text_table_section().
section_syntax_indicator	1	1b	The ETT uses the MPEG "long-form" syntax.
private_indicator	1	1b	Set to 1 in PSIP tables.
reserved	2	11b	Reserved bits are set to 1.
section_length	12		The ETT section can be 4096 total bytes in length, so section_length is limited to 4093.
table_id_extension	16		Must be set such that all ETT instances of a given MGT table_type have unique values of table_id_extension. Has no intrinsic meaning.
reserved	2	11b	Reserved bits are set to 1
version_number	5		The version number reflects the version of a table section, and is incremented when anything in the table changes.
current_next_indicator	1	1b	Indicates that the ETT table section is currently applicable. It is not allowed to send "next" ETT sections.
section_number	8	0	The ETT cannot be segmented.
last_section_number	8	0	The ETT cannot be segmented.
protocol_version	8	0	Indicates the protocol version of this table section. The only type of extended_text_table_section() currently defined is for protocol_version zero.
ETM_id	32		Associates this ETT instance with a specific television or data event, virtual channel, or other entity. See text.
extended_text_message()	var.		A Multiple String Structure describing the entity given in ETM_id.
CRC_32	32		A 32-bit checksum designed to produce a zero output from the decoder defined in the MPEG-2 *Systems* Standard.
}			End of the extended_text_table_section().

ETM_id

This 32-bit field associates the Multiple String Structure contained within the table section with a television or data event, or in the case of the channel ETT, a virtual channel. Extension in a future version of the protocol of the ETM_id concept could allow the ETT to supply text describing yet another kind of SI data. Figure 10.8 defines the coding of the ETM_id field.

	MSB					
	31	16	15	2	1	0
channel ETM_ID	source_id		0x0000		0	0
event ETM_ID	source_id		event_id		1	0

Figure 10.8 ETM_id Coding

As shown in the Figure, the two least-significant bits of the 32-bit ETM_id field indicate the type of text object. Value 00b indicates a channel ETM_id, in which case the Source ID by itself is sufficient linkage to associate this ETM with one virtual channel. Value 10b indicates an event ETM, in which case the combination of Source ID and event ID forms an unambiguous reference to one event described in an EIT.

For channel ETMs, Source ID is sufficient to link this ETT instance to one virtual channel. For event ETMs, the Source ID in conjunction with the event ID is used.

Receivers could keep track of which type of ETM a given ETT instance delivered even without the two least-significant bits to distinguish the two: they could use the contextual reference given by the MGT table_type. However, given that the definition of ETM_id identifies the type of ETM (channel vs. event) , a receiver can collect and store text blocks labeled with ETM_id in memory without the need to add an additional accounting mechanism to distinguish between the types. In any case, it adds some robustness to the data structure.

extended_text_message()

The ETM is coded as a Multiple String Structure. There are no stated restrictions on the minimum or maximum sizes, except of course the maximum is limited by the ETT section size of 4096 bytes. Display of some text blocks, if sufficiently large, may be truncated by some receiver implementations. Receivers can be designed to show one page at a time, or support text scrolling to ensure that the user is able to see all available content.

ETT transport rate and cycle time

A/65 does not specify maximum cycle times for Extended Text Tables, so the rates used by any given broadcaster are totally discretionary. As with any PID associated with EITs, A/65 does restrict the maximum bitrate for Transport Stream packets carrying ETT sections on any given PID to 250,000 bps.

Since ETTs for different time slots are carried on different PIDs, the repetition rates for text describing programming further into the future can be lower than the rates used for current or near-term programming. In practice, ETT rates are likely be much lower than EIT rates, especially for those ETTs describing future programming. That's because the total amount of ETT data can be fairly large, and bandwidth used for delivery of text is then unavailable for compressed video (quality could suffer).

Example ETT

Consider an example situation where an ETT table section describes an event with event_id value 0x4003. Table 10.23 describes the bit-stream syntax for this event.

TABLE 10.23 Extended Text Table Example Bit-stream Representation

Field Name	Value	Description
table_id	0xCC	Identifies the table section as being an ETT.
section_syntax_indicator	1b	The ETT uses the MPEG "long-form" syntax.
private_indicator	1b	Set to 1 in PSIP tables.
reserved	11b	Reserved bits are set to 1.
section_length	0x005E	Length of the rest of this section is 94 bytes.
table_id_extension	0x0000	Must be set to a unique value for all ETT sections carried in TS packets using the same PID value.
reserved	11b	Reserved bits are set to 1.
version_number	01011b	Version number is 11.
current_next_indicator	1b	Section must be current ETT.
section_number	0x00	First section.
last_section_number	0x00	Last is first (only one section).
protocol_version	0x00	Zero for the present protocol.
ETM_id	0x0001000E	event_id 0x00004003 concatenated with 01b.
number_strings	0x01	MSS structure, one string.
ISO_639_language_code	0x65 6E 67	"eng" in ASCII, signifying English language.
number_segments	0x01	One segment.
compression_type	0x00	No compression.

TABLE 10.23 Extended Text Table Example Bit-stream Representation (continued)

Field Name	Value	Description
mode	0x00	Selects ISO Latin-1 code page in UTF-16.
number_bytes	0x48	72-byte string.
compressed_string_byte	4D 6F 76 69 65 20 28 31 39 38 35 29 2E 20 20 4D 61 72 79 20 4D 63 44 69 6C 6C 2C 20 47 61 72 79 20 4D 63 47 69 6C 6C 2E 20 20 41 20 64 6F 63 74 6F 72 20 69 73 20 61 6C 6C 65 72 67 69 63 20 74 6F 20 6C 61 74 65 78 2E	ISO Latin-1 encoded string: "Movie (1985). Mary McDill, Gary McGill. A doctor is allergic to latex."
CRC_32	0xF5555443	MPEG-2 32-bit table section CRC.

References

1. EIA/CEA-766-A, "U.S. and Canadian Rating Region Tables (RRT) and Content Advisory Descriptors for Transport of Content Advisory Information Using ATSC A/65A Program and System Information Protocol (PSIP)," Electronics Industries Alliance and Consumer Electronics Association, 2000.

2. ATSC Standard A/90, "Data Broadcast Standard," Advanced Television Systems Committee, 26 July 2000.

3. ANSI/SCTE 07 2000 (formerly DVS 031), "Digital Video Transmission Standard for Cable Television," Society of Cable Telecommunications Engineers.

4. ANSI/EIA/CEA-608-B, "Line 21 Data Services," Electronic Industries Association and Consumer Electron ics Association, 2001.

5. SCTE 54 2002A (formerly DVS 241), "Digital Video Service Multiplex and Transport System Standard for Cable Television," Society of Cable Telecommunications Engineers.

The PSIP Descriptors

ATSC A/65 defines several descriptors, some for use in the PSIP tables and others applicable to the MPEG-2 PSI tables (specifically, the PMT). In this chapter we look at each in detail. Even though it is not defined in A/65, we also take a look at the AC-3 Audio Descriptor specified in the Dolby AC-3 standard, ATSC A/52[1] and the ATSC Private Information Descriptor specified in ATSC A/53[*2]. At the end of the chapter, we review all of these descriptors and note the purpose and usage rules for each.

AC-3 Audio Descriptor

Dolby Laboratories created the multi-channel audio compression scheme ATSC chose for their digital television standard. ATSC A/52[1] specifies the AC-3 audio compression algorithm, and Annex B of the ATSC A/53[2] Digital Television Standard specifies certain constraints against the full AC-3 standard. A/52 defines the syntax and semantics of an audio_stream_descriptor() for use in the Program Map Table, and A/53 states that this descriptor shall be used for terrestrial broadcast to inform decoders of the characteristics of each transmitted AC-3 audio stream. The transport standard for cable, SCTE 54[3], has also asserted the requirement to transmit the audio_stream_descriptor().

One can learn the following information about an AC-3 audio stream from the AC-3 Audio Descriptor:

- the sample rate of the encoded audio.

- an indication of the type of AC-3 coding employed in the stream.

- the bitrate (or upper bound of the bitrate) of the audio stream.

---

*Note: as of this writing, the ATSC Private Information Descriptor is a proposed amendment to A/53B. It is expected to be included in A/53C, to be published in the second half of 2002.

- the number of channels encoded in the audio stream.

- whether or not the audio stream is Dolby Surround mode-encoded.

- an indication as to whether the audio stream is a "full service," suitable for presentation on its own, or is a partial service that must be combined with an associated service before it is ready for presentation.

- information pertaining to which other audio streams, if any, are associated with this one.

It may be helpful at this point to list some of the audio-related constraints against A/52 that the Digital Television Standard of A/53 imposes:

- The sample rate must be 48 kHz.

- Dolby 1+1 mode (also called "dual mono" mode) is disallowed. In 1+1 mode, a single stream can encode two separate, independent audio tracks.

- Complete audio services must be encoded at bitrates not to exceed 448 kbps.

- A single-channel associated service containing a single program element must be encoded at a bitrate not to exceed 128 kbps.

- A two-channel associated service containing only dialogue must be encoded at a bitrate less than or equal to 192 kbps.

- The combined bitrate of a main service and an associated service, where the two are intended to be decoded simultaneously, must not exceed 576 kbps.

AC-3 Audio Descriptor structure

Figure 11.1 diagrams the structure of the audio_stream_descriptor(), commonly known as the AC-3 Audio Descriptor, in the form in which it appears in an ATSC terrestrial broadcast TS. Inspection of the A/52 Standard shows a definition of the audio_stream_descriptor() with a data structure extended somewhat beyond what is shown here. That's because the version we have drawn is simplified by the constraints against A/52 that are imposed by the overriding A/53 document. In addition, our version reflects the "A" revision of A/52 in which the language code fields have been deprecated (the Standard states that they are being converted into reserved fields).

Note in Figure 11.1 that the last byte shown has a format that depends upon the value of the bsmod field. If bsmod is less than 2, the field is a three-byte main id followed by 8 reserved bits. If bsmod is greater than or equal to 2, the field is eight "associated service" flags.

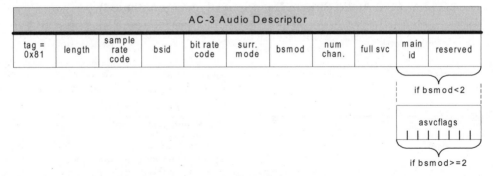

Figure 11.1 Structure of the AC-3 Audio Descriptor

Syntax and semantics

Table 11.1 gives the syntax of the AC-3 Audio Descriptor.

TABLE 11.1 AC-3 Audio Stream Descriptor Syntax

Field Name	Number of Bits	Field Value	Description
audio_stream_descriptor() {			Start of the audio_stream_descriptor().
descriptor_tag	8	0x81	Identifies the descriptor as the AC-3 Audio Stream Descriptor.
descriptor_length	8		Indicates the length, in bytes, of the data to follow.
sample_rate_code	3	0	A/53 Annex B constrains the sample rate to 48 kHz, so this field must be set to zero.
bsid	5	8	Bit stream identification. Set to 8 for the current version of this standard. See text.
bit_rate_code	6		This field indicates a nominal bitrate for the audio ES. See text for encoding. If the MSB of this field is set, the bitrate is an upper limit. If the MSB is clear, the bitrate is an exact rate.
surround_mode	2		Specifies whether the audio stream is Dolby Surround encoded or not, or that surround mode is not indicated.
bsmod	3		Bit stream mode. Indicates whether the audio stream is a Complete Main audio, or is one form or another of Associated Service. See text for coding.
num_channels	4		Indicates the number of audio channels present in this audio Elementary Stream. See text for coding.
full_svc	1		Indicates whether the audio service is a full service that is suitable for presentation as-is, or whether it is a partial service that must be combined with an associated service for presentation.

TABLE 11.1 AC-3 Audio Stream Descriptor Syntax (continued)

Field Name	Number of Bits	Field Value	Description
reserved	8	0xFF	Reserved fields always contain 1 (prior to mid-2001, this field was langcod but with the "A" version of A/52 it has been changed to a reserved field).
if (bsmod<2) {			The definition of the next 8 bits depends on bsmod.
mainid	3		If bsmod indicates a CM or ME channel, this provides a unique tag for the main audio for use with the asvcflags.
reserved	5	0x1F	Reserved bits are set to 1.
} else			
asvcflags	8		Associated service flags. If bsmod indicates an associated service, these flags associate this audio stream with a main audio. See text.
}			End of the audio_stream_descriptor().

sample_rate_code

A/52 defines several possible sample rates, including 32, 44.1, and 48 kHz. ATSC has chosen to limit audio to just one sample rate, 48 kHz. A sample_rate_code value of zero indicates 48 kHz sample rate.

bsid

This is the "bit stream identification" field. It is a 5-bit field set to the value 8 in this version of the standard. The intention is that current decoders will be capable of handling future versions of the Standard in which bsid is less than 8, so any bsid value in the 0 to 8 range should be acceptable to today's decoders. If a bsid value of greater than 8 is seen for a particular Elementary Stream, decoders built to today's standard are expected to mute audio if that Elementary Stream is selected for decoding (in other words, they will not attempt to create audio).

bit_rate_code

The 6-bit bit_rate_code field indicates the transmitted bitrate for the compressed audio in this stream. It is coded according to Table 11.2. As can be seen, depending upon the most-significant bit, the bitrate is either specified as an exact rate or as an upper limit.

TABLE 11.2 AC-3 Bitrate Code

bit_rate_code	Exact Bitrate (kbps)	bit_rate_code	Bitrate Upper Limit (kbps)
0x00	32	0x20	32
0x01	40	0x21	40
0x02	48	0x22	48
0x03	56	0x23	56
0x04	64	0x24	64
0x05	80	0x25	80
0x06	96	0x26	96
0x07	112	0x27	112
0x08	128	0x28	128
0x09	160	0x29	160
0x0A	192	0x2A	192
0x0B	224	0x2B	224
0x0C	256	0x2C	256
0x0D	320	0x2D	320
0x0E	384	0x2E	384
0x0F	448	0x2F	448
0x10	512	0x30	512
0x11	576	0x31	576
0x12	640	0x32	640

surround_mode

This is a two-bit field that tells whether or not the associated AC-3 Elementary Stream is encoded with one or more surround mode channels. Surround_mode is encoded as shown in the following Table.

TABLE 11.3 AC-3 Surround_mode Field Coding

surround_mode	Meaning
00b	Surround mode not indicated.
01b	Not Dolby surround encoded.
10b	Dolby surround encoded.
11b	Reserved.

bsmod

This 3-bit field is called bsmod, or bit stream mode, and it indicates the type of main or associated audio service. Note that the Emergency associated service, bsmod 6, is described in A/53B but not used in practice. Broadcasters use the Complete Main audio for all voice-overs and emergency alert announcements. A/53B states it like this:

> If a program contains an audio component, the primary audio shall be a complete main audio service (CM) as defined by ATSC Standard A/52 and shall contain the complete primary audio of the program including all required voice-overs and emergency messages.

TABLE 11.4 AC-3 bsmod Field Coding

bsmod	Type of Service
0	Main audio service: complete main (CM).
1	Main audio service: music and effects (ME).
2	Associated service: visually impaired. (VI).
3	Associated service: hearing impaired (HI).
4	Associated service: dialogue (D).
5	Associated service: commentary (C).
6	Associated service: emergency (E).
7	Associated service: voice over (VO).

num_channels

This 4-bit field indicates the number of audio channels in the AC-3 Elementary Stream. When the MSB is 0, the lower 3 bits indicate the number of full-bandwidth independent front and rear channels. In the "M/N" notation, M indicates the number of full-bandwidth front channels and N indicates the number of full-bandwidth rear channels. Whether or not a low-frequency effects (LFE) channel is available is not indicated (the LFE channel is the ".1" in the "5.1 channels" terminology).

When the most-significant bit of num_channels is set, the lower three bits indicate the maximum number of encoded audio channels (counting the LFE channel as one channel). Note that 1+1 (dual-mono) mode is not an allowed audio mode for ATSC terrestrial broadcast or in the cable standards specified by the SCTE Digital Video Subcommittee.

TABLE 11.5 AC-3 Num_channels Field Coding

num_channels	Audio Coding Mode	num_channels	Number of Encoded Channels
0000b	1+1*	1000b	1
0001b	1/0	1001b	≤ 2
0010b	2/0	1010b	≤ 3
0011b	3/0	1011b	≤ 4
0100b	2/1	1100b	≤ 5
0101b	3/1	1101b	≤ 6
0110b	2/2	1110b	Reserved
0111b	3/2	1111b	Reserved

* Note: 1+1 mode is disallowed for use in ATSC Terrestrial Broadcast

full_svc

This flag indicates whether or not the audio Elementary Stream is suitable for presentation to the listener as-is, or must be combined with another audio track prior to presentation. The bit is set for services that are complete in themselves and ready for presentation. For example, a visually impaired (VI) audio service could be indicated as being a full service if it included all of the elements of the program, including music, effects, dialogue, and the visual content description narrative. When an audio ES must be combined with another audio service before it is suitable for presentation, the full_svc bit is set to zero. A VI service would not be a full service if it contained the visual descriptive narrative but not the music and effects.

If bsmod is less than 2, the next three bits are mainid and the five bits following that are reserved. If bsmod is greater than or equal to 2, the next eight bits are associated service flags.

mainid

The purpose of the 3-bit mainid field is to assign a number in the range 0 to 7 to a main audio service so that other services can reference it with the associated service flags, asvcflags.

asvcflags

These flags tie this Associated Service audio track to one or more main audio services. If the most-significant bit (bit 7) is set it indicates that this Associated Service is linked with the main audio service identified with mainid value 7. Bit 6, if set, links this Associated Service with the main audio service identified with mainid value 6, etc. If any bit is zero, this Associated Service is not linked with the main service identified with a mainid value equal to that bit's bit position.

Usage rules

The cable transport standard SCTE 54[3] and the terrestrial broadcast transport specification in Annex C of ATSC A/53[2] state that an AC-3 Audio Descriptor must be present in the PMT section, in the inner (ES_info) descriptor loop describing each AC-3 audio program element. If the MPEG-2 program has n AC-3 audio tracks, there must be n AC-3 Audio Descriptors, one per track.

An AC-3 Audio Descriptor may also appear in the event descriptor loop in an EIT. When it appears in the EIT, the descriptor identifies the characteristics of the primary audio component for the given event.

ATSC Private Information Descriptor

One type of privately-defined data in the Transport Stream comes in the form of a descriptor that might appear in a PMT section or in some other place where descriptors are allowed. In 2002, ATSC clarified the rules for private descriptor-based data, and defined the ATSC Private Information Descriptor as the standard method to be used.

Prior to the development of this descriptor, some broadcasters included private descriptor-based data in the Transport Stream simply using descriptor_tag values in the "user private" range. This method is now discouraged because it offers no reliable way to identify the private entity that has supplied the descriptor. One might suggest that the MPEG-2 Registration Descriptor (MRD) could be used for this purpose, but MPEG did not prescribe such an interpretation. ATSC decided that the presence of an MRD in any given descriptor loop implies nothing about other descriptors that might appear in that same descriptor loop.

Receivers must be certain, when parsing data known to have originated from a private entity, of the identity of that private party. The ATSC Private Information Descriptor (APID) provides a robust method to identify the private organization and to tie a block of private data unambiguously to that entity by including the 32-bit company/organization identifier in the same descriptor as the private data. This identifier is the same one used in the MRD; users must specify a value registered with the SMPTE Registration Authority.

Figure 11.2 illustrates the simple structure of the ATSC Private Information Descriptor. As mentioned, the format_identifier is the same 32-bit field as is found in the MPEG-2 Registration Descriptor. Following the format_identifier field are zero or more data bytes. The number of bytes is given by the value of the descriptor length field minus four.

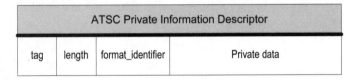

Figure 11.2 ATSC Private Information Descriptor Structure

Syntax and semantics

Table 11.6 gives the syntax and semantics of the ATSC Private Information Descriptor.

TABLE 11.6 ATSC Private Information Descriptor Syntax

Field Name	Number of Bits	Field Value	Description
ATSC_private_information_descriptor() {			Start of the ATSC_private_information_descriptor().
descriptor_tag	8	0xAD	Identifies the descriptor as the ATSC Private Information Descriptor.
descriptor_length	8		Indicates the length, in bytes, of the data to follow.
format_identifier	32		The registered identity of the entity that has supplied this descriptor.
for (i=0; i<descriptor_length-4; i++) {			The rest of the descriptor is private data bytes.
private_data_byte	8		Meaning of private data bytes is established by the private entity identified by format_identifier.
}			End of the private data "for" loop.
}			End of the ATSC_private_information_descriptor().

format_identifier

The format_identifier is the same 32-bit field as is defined in MPEG-2 *Systems*[7], section 2.6.8 and 2.6.9 for the registration_descriptor(). Only format_identifier values registered and recognized by SMPTE Registration Authority LLC, are allowed to be used (see http://www.smpte-ra.org/mpegreg.html). When it appears in the APID, the format_identifier establishes the entity that has supplied the private information contained in the descriptor.

private_data_byte

The syntax and semantics of this field are defined by the assignee of the format_identifier value. It is the responsibility of the assignee of the format_identifier to construct the private data bytes such that they are meaningful to the receiver.

Usage rules

An APID can appear in any PSI or SI descriptor loop, subject to any restrictions that might apply to a given type of table. An important point to note is that more than one APID can occur in the same descriptor loop.

Caption Service Descriptor

The function of the Caption Service Descriptor is to announce the presence of line-21 or DTV closed caption services carried with a given program. Figure 11.3 illustrates the structure of the Caption Service Descriptor.

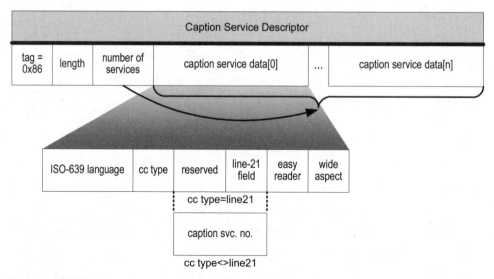

Figure 11.3 Structure of the Caption Service Descriptor

Following the tag and length fields, the Caption Service Descriptor has a 5-bit count of "number of services" in order to indicate the number of caption service descriptions to follow. Each block of caption service data is a fixed length and consists of an ISO 639 language identifier, a field indicating the type of caption service, six bits that identify the line-21 field for EIA-608-style captions or the caption service number for EIA-708-style captions, a flag indicating whether or not the caption service is geared to younger readers, and finally a flag that tells whether or not the caption service is formatted for widescreen video displays.

Syntax and semantics

Table 11.7 shows the syntax of the Caption Service Descriptor.

TABLE 11.7 Caption Service Descriptor Syntax

Field Name	Number of Bits	Field Value	Description
caption_service_descriptor() {			Start of the caption_service_descriptor().
descriptor_tag	8	0x86	Identifies the descriptor as the Caption Service Descriptor.
descriptor_length	8		Indicates the length, in bytes, of the data to follow.
reserved	3	111b	Reserved bits are set to 1.
number_of_services	5		An unsigned integer in the range 1 to 16 that indicates the number of caption services to be listed in the descriptor.
for (i=0; i<number_of_services; i++) {			Start of the caption services "for" loop.
language	24		Three characters indicating the language of the particular caption being described in this iteration of the "for" loop. The characters are coded according to ISO 8859-1 (Latin-1) and the 3-character language code is defined per ISO 639-2/B.
cc_type	1		This flag indicates the type of caption service, either an "advanced" caption service or a traditional line-21service. When the flag is set, the caption service being described in this iteration of the "for" loop is an "advanced" caption service per EIA-708-B[4]. When the flag is clear, a line-21 caption service per EIA/CEA-608-B[5] is indicated.
reserved	1	1b	Reserved bits are set to 1.
if (cc_type==line_21) {			The definition of the next 6 bits depends on the state of the cc_type flag. If cc_type is clear:
reserved	5	11111b	Reserved bits are set to 1.
line_21_field	1		Indicates whether the line-21 caption service is associated with field 1 or field 2 of the NTSC waveform. When the flag is set, field 2 is indicated. When the flag is clear, the service is found in field 1.
} else {			if cc_type is set...
caption_service_ number	6		An unsigned integer that ties this iteration of the "for" loop with a particular caption service carried in video user data. The Service Number referenced here is described in the EIA-708-B[4] Standard.
}			End of the "if" statement.

TABLE 11.7 Caption Service Descriptor Syntax (continued)

Field Name	Number of Bits	Field Value	Description
easy_reader	1		When this flag is set, the caption service is geared towards beginning readers. "Easy reader" caption services are discussed in EIA-708-B[4].
wide_aspect_ratio	1		A flag that indicates this particular caption service is formatted for wide-screen (16:9) displays. When the flag is clear, the service is formatted for 4:3 displays but caption text can be displayed centered on a 16:9 display.
reserved	14	0x3FFF	Reserved bits are set to 1.
}			End of the caption services "for" loop.
}			End of the caption_service_descriptor().

number_of_services

This is an unsigned integer in the range 1 to 16 that indicates the number of caption services present.

language

Like all the other language tags in PSI and PSIP data structures, this one is three lower-case ISO Latin-1 characters coded in accordance with ISO 639-2/B. Its purpose here is to identify the language of the associated caption service.

cc_type

Two types of captions can be described by the Caption Service Descriptor. The first type corresponds to the standard line-21 captioning defined in EIA-608-B[5] for analog NTSC services and is indicated when the cc_type flag is clear. For analog NTSC captions, the line_21_field indicates whether the indicated caption service is to be found in field 1 or field 2 of the vertical blanking interval of the NTSC waveform. The second type is an advanced television closed caption service as defined in EIA-708-B[4], and that type is indicated when the cc_type flag is set. For advanced captioning, the caption_service_number field links the service being described in this iteration of the caption services "for" loop with the like-numbered service in the transmitted advanced captioning data stream.

Note that the line_21_field and the five bits preceding it are interpreted one way or the other depending on the state of the cc_type field. For line-21 caption services (when this flag is set), the bits are five reserved bits followed by the line_21_field field. When the flag is clear, the bits are the six-bit caption_service_number of an advanced caption service.

line_21_field

For line-21 analog NTSC caption services, the line_21_field indicates which field carries the caption service. When line_21_field is set, field 2 is indicated. When the line_21_field is clear, the captions are carried in field 1.

caption_service_number

For advanced captioning, this 6-bit field associates the caption service being described in this iteration of the "for" loop with the like-numbered caption service in the advanced captioning stream. Refer to EIA-708-B[4] for details.

easy_reader

Analog captions were limited to at most two simultaneous services, one carried in field 1 and the other in field 2. Digital captions, on the other hand, can use ten times the transmitted bandwidth and hence can deliver more than two services at the same time. Captions in several languages are possible. Another possibility is that a caption service might be produced to meet the needs of young or beginning readers or those just learning the language. Such caption services are called "easy reader" services. When set, the easy_reader flag in the Caption Service Descriptor signals that this particular caption service is an easy-reader service. When clear, the caption service is not tailored in this way.

wide_aspect_ratio

This flag, when it is set, indicates the caption service is formatted for displays with wide-screen (16:9) aspect ratios. When the flag is clear, the service is formatted for 4:3 displays, but can of course be displayed centered within 16:9 displays.

Usage rules

Caption Service Descriptors may appear in two types of tables, the PMT and the EIT. If a program is captioned, for terrestrial broadcast Transport Streams the descriptor must be present in the EIT and may optionally also be present in the PMT. For Transport Streams delivered on cable, if the current program is captioned, the Caption Service Descriptor must be present in the PMT section for that program and if an EIT is sent, must be present in the EIT as well. When it is present in the PMT section, the Caption Service Descriptor is placed in the Elementary Stream information descriptor loop of the video program element. Caption Service Descriptors may also appear in any EIT describing a future program.

Component Name Descriptor

MPEG-2 *Systems*[7] does not specify any standard methods for encoding text strings for use in PSI or SI tables. Standards such as ATSC A/65 extend MPEG-2 *Systems* in that direction by defining the Multiple String Structure. While MPEG gives us the Program Map Table, which groups Elementary Stream components into MPEG programs, it doesn't define any way to associate a textual label with an Elementary Stream component of a program. That's the function of the Component Name Descriptor.

Figure 11.4 shows the structure of the Component Name Descriptor. It is in standard tag-length-data format, where the data portion is a Multiple String Structure. The Component Name Descriptor is specified for use in the PMT section, and in certain cases its use is mandatory.

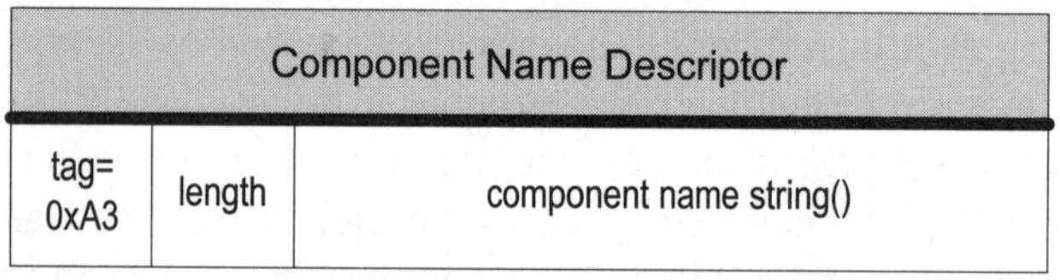

	Component Name Descriptor	
tag= 0xA3	length	component name string()

Figure 11.4 Structure of the Component Name Descriptor

Syntax and semantics

Syntax and semantic definition of the component_name_descriptor() data structure are given in Table 11.8.

TABLE 11.8 Component Name Descriptor Syntax

Field Name	Number of Bits	Field Value	Description
component_name_descriptor() {			Start of the component_name_descriptor().
descriptor_tag	8	0xA3	Identifies the descriptor as the Component Name Descriptor.
descriptor_length	8		Indicates the length, in bytes, of the data to follow.
component_name_string()	var.		A Multiple String Structure giving the name of the associated ES component.
}			End of the component_name_descriptor().

Usage rules

Amendments introduced in 2002 to A/65 and Annex C of ATSC A/53 established the following rules regarding usage of the Component Name Descriptor:

1. A Component Name Descriptor must be used whenever a service includes two or more audio components of the same type and language. The type of an audio program element is given by the bit stream mode (bsmod) field in the AC-3 Audio Descriptor. Different types of audio components include Complete Main audio (regular program audio) and audio tracks designed for the visually or hearing-impaired. When it is used, the Component Name Descriptor must be included in the ES info descriptor loop of the audio Elementary Streams.

2. A Component Name Descriptor should be included whenever an audio component without an associated language is present (a "music and effects" track). Again, the descriptor is included in the ES info descriptor loop (inner loop) for that ES.

One might ask, how could an audio component not have a language? Some examples of audio tracks without language are such things as crowd noise at a sporting event (without any announcer commentary), and ambient sound from a live camera location. Such applications have been rare with analog television services, but given the greater bandwidths and compression efficiencies offered by digital technology, they may now be realized.

The only other usage requirement is that no two Component Name Descriptors in a given PMT section are allowed to include identical text. This restriction comes from the desire that the text in the descriptor is to be used to help the user distinguish between and choose ES components. If the names of two components were the same, such differentiation would not be possible.

Content Advisory Descriptor

Figure 11.5 diagrams the Content Advisory Descriptor. As with all descriptors, the format is tag-length-data. The descriptor is structured with a header followed by one or more data blocks containing data specific to a particular rating region. These are called Region data in the Figure. The number of sets of Region data is given by the "Rating Region count" field, shown as "rating region count" in the figure.

Each set of Region data is associated with a particular rating region by the "rating region" field and contains one or more sets of Dimension data. The number of sets is given by the "dimension count" parameter. Each set of Dimension data simply identifies the rating value for the rating dimension given by "dimension index."

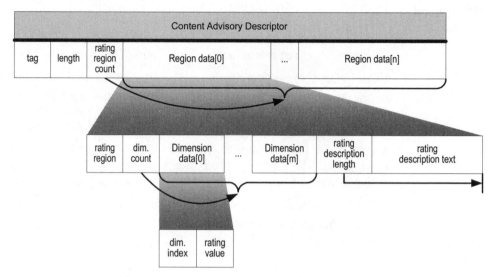

Figure 11.5 Structure of the Content Advisory Descriptor

Syntax and semantics

Table 11.9 defines the syntax of the Content Advisory Descriptor.

TABLE 11.9 Content Advisory Descriptor Syntax

Field Name	Number of Bits	Field Value	Description
content_advisory_descriptor() {			Start of the content_advisory_descriptor().
descriptor_tag	8	0x87	Identifies the descriptor as the Content Advisory Descriptor.
descriptor_length	8		Indicates the length of the data to follow.
reserved	2	11b	Reserved bits are set to 1.
rating_region_count	6		Unsigned integer in the range 1 to 8. Indicates the number of regions for which content advisory data is to be supplied in this descriptor.
for (i=0; i<rating_region_count; i++) {			Start of the rating region "for" loop.
rating_region	8		An unsigned integer that specifies the rating region for which rating data in this iteration of the rating region "for" loop applies.
rated_dimensions	8		An unsigned integer that indicates the number of rating dimensions to be given in the "for" loop to follow.

TABLE 11.9 Content Advisory Descriptor Syntax (continued)

Field Name	Number of Bits	Field Value	Description
for (j=0; j<rated_dimensions; j++) {			Start of the dimension data "for" loop.
rating_dimension_j	8		An unsigned integer that points to a particular rating dimension in the instance of the RRT corresponding to rating_region. The value given is used as an index, so that value zero points to the first defined dimension, value one points to the second, etc. Values for rating_dimension_j may range from zero to the value of the dimensions_defined parameter minus one, in that RRT. Dimension indices must be listed in numerical order.
reserved	4	111b	Reserved bits are set to 1.
rating_value	4		This 4-bit field indicates the rating level for the dimension given by rating_dimension_j, for the RRT corresponding to the region given by rating_region.
}			End of the dimension data "for" loop.
rating_description_length	8		Indicates the length, in bytes, of the rating_description_text() field to follow.
rating_description_text()	var.		This field can provide a textual description of the program's rating, if desired. If present, the string is limited to a length of 16 displayable characters.
}			End of the rating region data "for" loop.
}			End of the Content Advisory Descriptor.

Let's take a closer look at some of the fields in the Content Advisory Descriptor.

rating_region

The rating_region field ties the content advisory information in this iteration of the rating region "for" loop to the RRT for the corresponding rating region. It is possible for one content_advisory_descriptor() to include content advisory information pertaining to as many as eight rating regions.

rated_dimensions

This is the count of the number of dimensions, for the rating system given in rating_region, for which content advisory information is given for this program.

The number of rated dimensions can be as few as one or as many as the total number of dimensions defined in the RRT for this rating_region.

rating_dimension_j

Now that the rating system has been identified (by rating_region), the descriptor provides (dimension, level) data pairs. The rating_dimension_j field identifies a dimension for which a rating value is supplied. It is required that the values of rating_dimension_j be given in ascending order, starting with lower rating dimension indices first.

rating_value

The rating_value field gives the content rating for rating_dimension_j. Let's say the RRT for rating_region R defines dimension 0 as an age-based rating. And say level 3 for dimension 0 is a rating of "14 and above." If a certain program has content advisory data for rating region R, and is age-rated at 14+, the program could include a content_advisory_descriptor() with rating_region of R, rating_dimension_j of 0, and rating value of 3.

rating_description_length

This is simply the number of bytes in the Multiple String Structure to follow.

rating_description_text()

It is possible for the content_advisory_descriptor() to include a text string that is related to the program's content. If it is included, it must correspond to no more than 16 displayed characters for any language. There is no requirement for receivers to process or display the rating description text, and it does not appear to be commonly used. In fact, for US and Canadian ratings, EIA/CEA-766-A[6] indicates that no rating description text shall be sent.

Usage rules

There is no requirement that every program carry a content advisory. When a program is rated, however, for terrestrial broadcast a Content Advisory Descriptor must appear in the EIT describing the program and a copy may also be placed into the PMT section. For cable, when a program is rated a Content Advisory Descriptor must appear in the PMT transmitted during the duration of the program, and if an EIT-0 is present, in that EIT as well. Content Advisory Descriptors can appear in EITs describing future programs as well.

Extended Channel Name Descriptor

Channel names of at most seven characters are specified in the Virtual Channel Table in the short_name field. If a broadcaster like WXYZ wishes to multi-cast several SD channels, it can be tough to retain the four-letter call sign and also differentiate the different channels when constrained to just seven characters. There are two ways to provide more information about a programming service. Including an Extended Channel Name Descriptor is one. The other way is to include an Extended Text Table that is tied to the virtual channel. Typically, the ETT method is used to give a sentence or two of descriptive information about the channel, while the Extended Channel Name Descriptor is just a full-length channel name.

Structurally, the Extended Channel Name Descriptor is so simple we won't bother to diagram it. Following the tag and length bytes is simply a Multiple String Structure. Because the MSS format is used, it is possible to provide the extended channel name multi-lingually, if desired.

The Extended Channel Name Descriptor is defined for use in the channel information descriptors loop in either the Terrestrial or Cable VCT.

Table 11.10 shows the syntax of the Extended Channel Name Descriptor.

TABLE 11.10 Extended Channel Name Descriptor Syntax

Field Name	Number of Bits	Field Value	Description
extended channel_name_descriptor(){			Start of extended_channel_name_descriptor().
descriptor_tag	8	0xA0	Identifies the descriptor as the Extended Channel Name Descriptor.
descriptor_length	8		Indicates the length, in bytes, of the data to follow. In this case, it's the length of the MSS.
long_channel_name_text()	var.		The name of the virtual channel in the form of a Multiple String Structure.
}			End of extended_channel_name_descriptor().

long_channel_name()

A Multiple String Structure that gives the name of the virtual channel. There are no formal rules or recommendations regarding the length or format of the long channel name. The intention was that the extended channel name would allow a programming content provider like Discovery Networks or Home & Garden Network to spell out the name of each of their programming services. It can be challenging, if not virtually impossible, to compress some of these service names into the seven-character short version.

Because of the large number of cable services that have to be handled, the long channel name may find more appeal in the cable application than for terrestrial broadcasting. But consider a group of CSPAN channels. The short form channel name might label them as CSPAN-1, CSPAN-2, etc., while the long form could call them "CSPAN-1 National," "CSPAN-2 State" or whatever descriptive name best fits the theme of the particular minor channel.

Usage rules

The Extended Channel Name Descriptor is defined only for use in one position within PSIP tables: the inner (channel information) loop of the VCT. Its use is optional. Any given virtual channel in one VCT section may or may not include an Extended Channel Name Descriptor (some may have it while others do not).

Redistribution Control Descriptor

The Redistribution Control Descriptor has two claims to fame at this point in time. First, as of this writing it is the newest descriptor to be defined in ATSC. Its second claim to fame is that, at two bytes in length, the Redistribution Control Descriptor is the shortest descriptor, and probably the shortest separately named data structure in the ATSC Standard. Curiously, it doesn't carry any data bytes—the presence of the descriptor is what imparts meaning, not the values of any data payload or flags it carries.

After many spirited debates, the descriptor was boiled down to its core purpose for being: to signal that the content owner is asserting his or her rights with regard to re-distribution of the program. Exactly what rights the content owner actually has (or will have) are not enumerated. It is left to appropriate future legislation or regulation to specify what one can legally do with an in-the-clear broadcast that is marked with the Redistribution Control Descriptor, and what treatments are not legal.

As more lawyers than engineers were involved in its development, a considerable number of billable hours were undoubtedly charged during the course of the derivation of the words that sum up the purpose of the Redistribution Control Descriptor: "The descriptor's existence within the ATSC stream shall mean: 'technological control of consumer redistribution is signaled.'"

The stated purpose of the Redistribution Control Descriptor is to address situations where audio, video and data are sent unscrambled via terrestrial broadcast; that is, not using the ATSC A/70 conditional access system specification. A particular instance of a Redistribution Control Descriptor applies to a television (audio/video) or data event, not to a virtual channel. Hence, the descriptor appears in the Event Information Table or (for data events) the Data Event Table.

For currently broadcast events, when rights related to redistribution are being asserted, the Redistribution Control Descriptor must appear in the Program Map Table section as well. For use in the PMT, the descriptor must appear in the outer (program information) descriptor loop.

Syntax and semantics

Table 11.11 describes the (very simple) syntax of the Redistribution Control Descriptor.

TABLE 11.11 Redistribution Control Descriptor Syntax

Field Name	Number of Bits	Field Value	Description
redistribution_control_descriptor() {			Start of the redistribution_control_descriptor().
descriptor_tag	8	0xAA	Identifies the descriptor as the Redistribution Control Descriptor.
descriptor_length	8	0	Indicates the length, in bytes, of the data to follow—in this case zero. Note: additional bytes may be defined in the future, so non-zero values for descriptor_length must be properly handled.
}			End of the redistribution_control_descriptor().

Note that, as with all descriptors, a future revision of the standard could define new fields that would extend the descriptor's length. Receiving devices are expected to always process the descriptor_length field, and make no assumptions about its value. To reach the beginning of the data structure just following any descriptor, one simply skips ahead the number of bytes given by descriptor_length. Designers are strongly encouraged to test to make sure an implementation can handle non-zero values for the descriptor_length field in the Redistribution Control Descriptor.

Service Location Descriptor

Decreasing the time between the moment a viewer enters a new channel number and the time decoded audio and video from the new channel can be presented to that user's ears and eyes has always been a goal of digital television systems engineering. Terrestrial broadcasters have acknowledged that the Terrestrial Virtual Channel table will always be sent, and sent at a high repetition rate. They felt it could give receiving equipment a chance to acquire a new channel faster if PID information were to be included in the TVCT, and would potentially avoid the need for terrestrial receivers to look at the PMT. Also, PID assignments for a given virtual channel were expected to be set to relatively fixed values. Out of this logic the Service Location Descriptor was born.

Figure 11.6 shows the Service Location Descriptor's structure. The first data field is the PCR PID, needed by the decoder to establish the time base used for decoding and presentation of the Elementary Streams. Next is a count of the number of program elements the descriptor carries, followed by that many data blocks, each describing one Elementary Stream. Data provided for each program element includes its stream_type (audio or video), the TS packet PID used to transport it, and a language code.

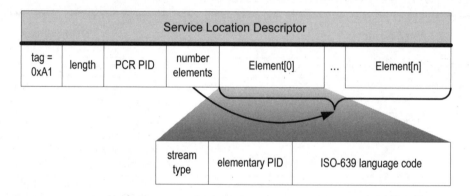

Figure 11.6 Structure of the Service Location Descriptor

Service Location Descriptors are mandated for terrestrial broadcast, but they are optional for cable. The cable community, as we have noted, felt that Transport Streams compliant with the cable transport standard may or may not include EIT data. Therefore, they designed cable set-top boxes to always collect and process the PMT upon acquisition of a service. They felt it was not helpful to repeat PID data in the Cable VCT that they already transmitted in the PMT.

A/65 specifies that the Elementary Stream data given in the Service Location Descriptor must exactly match the corresponding portion of the Program Map Table. The video ES, if one is present, must be described, as must all audio stream components. A/65 does not, however, require that all of the other Elementary Streams that may be defined in the PMT for a given service be included in the Service Location Descriptor.

Data in the descriptor for each program element includes an ISO 639-2 language code. For stream types such as the video, it is allowable to fill in null bytes (0x000000) for the language code to indicate "no language" is specified.

Syntax and semantics

Table 11.12 gives the syntax and semantics of the Service Location Descriptor.

TABLE 11.12 Service Location Descriptor Syntax

Field Name	Number of Bits	Field Value	Description
service_location_descriptor(){			Start of the service_location_descriptor().
descriptor_tag	8	0xA1	Identifies the descriptor as the Service Location Descriptor.
descriptor_length	8		Indicates the length, in bytes, of the data to follow.
reserved	3	111b	Reserved bits are set to 1.
PCR_PID	13		The Program Clock Reference PID for this service.
number_elements	8		Indicates how many ES components are described here.
for (i=0; i<number_elements; i++) {			Start of the services "for" loop.
stream_type	8		Gives the type of stream, for example MPEG-2 video, AC-3 audio, A/90 data service, etc.
reserved	3	111b	Reserved bits are set to 1.
elementary_PID	13		The Packet Identifier for this elementary stream.
ISO_639_language_code	24		The language associated with this program element. If no language is specified, the value 0x000000 is used.
}			End of the services "for" loop.
}			End of the service_location_descriptor().

PCR_PID

This is the 13-bit Packet Identifier of the Transport Stream packets that carry the Program Clock Reference in their adaptation fields for the service being described by this Service Location Descriptor. It must reflect the same PCR_PID value that appears in the Program Map Table, and it represents an alternate location in the Transport Stream where this value can be found.

number_elements

This is simply an integer number that gives the number of program elements that are described in this Service Location Descriptor. Because descriptors are always limited by the 8-bit field size of their length fields to 256 total bytes, number_elements cannot exceed 41. Typically, of course, a service consists of a video stream, one or perhaps a few audio components, and possibly one or a few data components.

stream_type

Indicates the type of stream being defined in this iteration of the services "for" loop. Values for stream type are aligned with those defined for use in the stream_type field in the Program Map Table. A/65 specifies, in fact, that the Service Location Descriptor must quote the same values given in the PMT section.

elementary_PID

Indicates, just as it does in the PMT, the PID value associated with TS packets carrying this ES component.

ISO_639_language_code

These three lowercase ISO-Latin-1 coded characters indicate the language of this component. If the component has no specified language, these 24 bits may be set to zero. Language codes are as specified by ISO 639-2/B, the "bibliographic" encoding.

Usage rules

For terrestrial broadcast, the Service Location Descriptor must always be present in the TVCT, and as noted its contents must always match the Program Map Table currently being transmitted. For cable, the Service Location Descriptor can be placed into the CVCT. The rules state that it must be present in the CVCT *when used*. That isn't much different than saying its use in the CVCT is optional.

Time-Shifted Service Descriptor

In some applications, a certain programming service may be identical to another service offering except that it is time-shifted with respect to the other service by a fixed amount of time. When that occurs, it is possible to avoid the need to repeat Event Information Table and Extended Text Table data for the time-shifted channel. Instead, a time_shifted_service_descriptor() can be included in the VCT to list channels that have the exact same program schedule as this channel except with a given time offset.

Near Video On Demand

One possible use for the Time-Shifted Service Descriptor is for so-called Near Video On Demand (NVOD) applications. Anyone interested in watching a certain program or movie offering will find that new showings are scheduled to start at regular intervals, such as every 15 minutes or every half-hour. A two-hour movie could, for example, be set up such that a new showing starts every 30 minutes. In this case, four virtual channels would be used for this movie.

One virtual channel carries a Time-Shifted Service Descriptor that lists each of the other virtual channels that have identical but time-shifted program schedules, and the amount of time shift involved for each. The virtual channel carrying the Time-Shifted Service Descriptor is called the "base" channel of the set of NVOD channels.

An NVOD base channel's EITs are the same as a normal channel except that EIT-0, in addition to its normal entries, may list some events that have *expired* in the base channel's schedule. Events must be carried until they have expired in all of the time-shifted channels. The reader is encouraged to study the NVOD examples in Annex Section D7 of the A/65 Standard.

Time-Shifted Service Descriptor structure

The structure of the time_shifted_service_descriptor() is shown in Figure 11.7 It is a rather simple structure, formatted in the standard way as tag, length, and data. The data is a count of the number of channel references followed by that number of

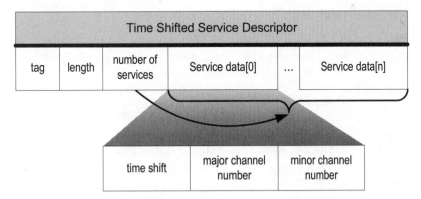

Figure 11.7 Structure of the Time-Shifted Service Descriptor

service data blocks. Each service data block consists of a time-shift value (given in minutes) and the major and minor channel numbers of a virtual channel that is time-shifted from the channel bearing the descriptor.

Syntax and semantics

Table 11.13 defines the syntax of the Time-Shifted Service Descriptor.

TABLE 11.13 Time-Shifted Service Descriptor Syntax

Field Name	Number of Bits	Field Value	Description
time_shifted_service_descriptor() {			Start of the time_shifted_service_descriptor().
descriptor_tag	8	0xA2	Identifies the type of this descriptor.
descriptor_length	8		Indicates the length, in bytes, of the data to follow.
reserved	3	111b	Reserved bits are set to 1.
number_of_services	5		Gives the number of time-shifted services described in this descriptor.
for (i=0; i<number_of_services; i++) {			Start of the number of services "for" loop.
reserved	6	111111b	Reserved bits are set to 1.
time_shift	10		Indicates the amount of the time shift.
reserved	4	1111b	Reserved bits are set to 1.
major_channel_number	10		The major channel number of a time-shifted service.
minor_channel_number	10		The minor channel number of a time-shifted service.
}			End of the services "for" loop.
}			End of the time_shifted_services_descriptor().

number_of_services

Unsigned integer in the range 1 to 20. Indicates the number of programming services for which time-shift data is to be supplied in this descriptor.

time_shift

An unsigned integer in the range 1 to 720 that specifies the number of minutes the service indicated by major and minor channel numbers is time-shifted with respect to the virtual channel associated with this descriptor.

major_channel_number

An unsigned integer in the range 1 to 999 that represents the major channel number of a time-shifted service.

minor_channel_number

An unsigned integer in the range 1 to 999 that represents the minor channel number of a time-shifted service.

Usage rules

A time_shifted_service_descriptor() is allowed to be used only in the case that the indicated time shift is constant and consistent across all of the programs covered in the time period spanned by the included Event Information Tables. Let's say we have two programming services called Service A and Service B. Service B's schedule of programming is currently identical to Service A's except it is delayed by 15 minutes. Let's also say that, beginning next Tuesday at noon Service B will switch to a different and independent program schedule. As long as no transmitted EIT extends as far out into the future as noon next Tuesday, a time_shifted_service_descriptor() may be used. But as soon as a service wishes to announce the program schedule for Tuesday noon, the descriptor may no longer be used. At that point, Services A and B must have their own independent EITs and ETTs.

Summary of Usage Rules for Descriptors

Table 11.14 lists each type of terrestrial broadcast table that can carry descriptors and the type of descriptor applicable to that table. Note that Stuffing Descriptors (descriptor_tag value 0x80) can appear anywhere, in any table. Table 11.15 lists descriptors applicable to PSIP and PSI tables for the cable application.

TABLE 11.14 Descriptor Usage for Terrestrial Broadcast Tables

MPEG or PSIP Table	Applicable Descriptor	Rule
Program Map Table	AC-3 Audio Stream Descriptor	Required if service has audio ES.
	Caption Service Descriptor	Optional if service is captioned.
	Content Advisory Descriptor	Optional if event is rated for content.
	Component Name Descriptor	Required if multiple same-language, same-type audio tracks; should be used to label music & effects tracks.
Virtual Channel Table	Extended Channel Name Descriptor	Optional, used as needed or desired.
	Service Location Descriptor	Required unconditionally.
	Time-shifted Service Descriptor	Required if service meets specified conditions (has same schedule as another service except for fixed, uniform time shift).

TABLE 11.14 Descriptor Usage for Terrestrial Broadcast Tables (continued)

MPEG or PSIP Table	Applicable Descriptor	Rule
Event Information Table	AC-3 Audio Stream Descriptor	Required if event has audio.
	Caption Service Descriptor	Required if event is captioned.
	Content Advisory Descriptor	Required if event is rated for content.
Directed Channel Change Table	DCC Arriving Request Descriptor	Optional.
	DCC Departing Request Descriptor	Optional.

TABLE 11.15 Descriptor Usage for Tables on Cable

MPEG or PSIP Table	Applicable Descriptor	Rule
Program Map Table	AC-3 Audio Stream Descriptor	Required if service has audio ES.
	Caption Service Descriptor	Required if service is captioned.
	Content Advisory Descriptor	Required if event is rated for content.
	Component Name Descriptor	Required if multiple same-language audio tracks; should be used to label music & effects tracks.
Virtual Channel Table	Extended Channel Name Descriptor	Optional, used as needed or desired.
	Service Location Descriptor	Optional, not typically used except for services derived from terrestrial broadcast.
	Time-shifted Service Descriptor	Used if service has same schedule as another service except for fixed, uniform time shift).
Event Information Table	AC-3 Audio Stream Descriptor	Optional, applies if service has audio.
	Caption Service Descriptor	Required if event is captioned.
	Content Advisory Descriptor	Required if event is rated for content.
Directed Channel Change Table	DCC Arriving Request Descriptor	Optional.
	DCC Departing Request Descriptor	Optional.

References

1. ATSC Standard A/52A, "Digital Audio Compression Standard (AC-3)," 20 August 2001.

2. ATSC Standard A/53B, "ATSC Digital Television Standard," 7 August 2001.

3. SCTE 54 2002A (formerly DVS 241), "Digital Video Service Multiplex and Transport System Standard for Cable Television," Society of Cable Telecommunications Engineers.

4. EIA-708-B, "Digital Television (DTV) Closed Captioning," Electronic Industries Alliance, December 1999.

5. ANSI/EIA/CEA-608-B, "Line 21 Data Services," Electronic Industries Alliance and Consumer Electronics Association, 2001.

6. EIA/CEA-766-A, "U.S. and Canadian Rating Region Tables (RRT) and Content Advisory Descriptors for Transport of Content Advisory Information Using ATSC A/65A Program and System Information Protocol (PSIP)," Electronic Industries Alliance and Consumer Electronics Association, 2000.

7. ISO/IEC 13818-1, 2000, "Information Technology—Generic coding of moving pictures and associated audio information—Part 1: Systems."

The Electronic Program Guide

As we have seen, PSIP allows the broadcaster or cable content or service provider to include program schedule information along with the audio/video service itself. A receiver can use this information to offer an Electronic Program Guide (EPG) function in a terrestrial broadcast or cable-compatible receiver.

Certain aspects of the operation of PSIP Event Information and Extended Text Tables and the references to EIT and ETT table instances given in the Master Guide Table are best shown by detailed examples. In the case studies that follow here we illustrate principles including:

- ways to manage the passage of time through the three-hour time slot boundary.

- how to manage and deal with changes to EIT and ETT data, both at the three-hour time slot boundary and at other times.

- the proper interpretation of the table-type version number given in the MGT.

We start our discussion of Electronic Program Guides with a review of general principles of EITs.

General EIT Principles

ATSC Standard A/65 describes the Event Information Table as a table containing information about programming (events) to be carried on defined virtual channels. In this context, the term "event" refers to a television program (talk show, news broadcast, situation comedy, movie, football game, or whatever), typically scheduled for an integral number of half-hour time increments.

Information provided in the EIT includes event titles, start times and durations, and may include other data such as content advisories and information regarding caption services that are scheduled to be offered when the given program is aired. Reference pointers to textual descriptions of programs in Extended Text Tables may also be given.

Generally speaking, EIT data is organized in the same way a typical on-screen Electronic Program Guide is organized: by virtual channel and by time. Figure 12.1 shows an example EPG on-screen display showing programming for three digital channels listed on the left. The time axis moves left to right. Along the top row, the

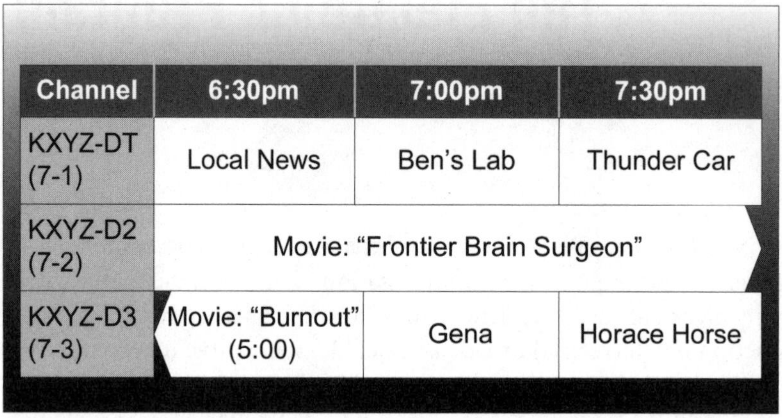

Figure 12.1 Example Electronic Program Guide On-screen Display

local time of day in half-hour time segments is shown. In this example, the on-screen real estate covers a 90-minute time period horizontally, and three channels vertically, but more or less time or channels may be displayed if different choices are made regarding the font size and line spacing.

Given the program schedule and event description data provided by PSIP, many display options are possible. An implementation may choose to display event information in the form of lists; for example, as a list of channels on the left and a list of scheduled events for a selected channel on the right. In the example traditional EPG grid of Figure 12.1, programs whose end times fall beyond the end of the third displayed half-hour are shown with a triangular right-side edge. Programs that started before the first half-hour are shown with a triangular left-side edge. These pointer-like edges are meant to indicate "program continues" into or beyond the indicated time period. Of course, many other graphical treatments are possible, limited only by the imagination of the human interface designer.

In accordance with MPEG-2 *Systems*, tables have a table_id_extension field that can be used to distinguish different instances of a table. In the case of the EIT, the table_id_extension field contains the source_id of a virtual channel, so that one instance of an EIT corresponds to all of the data describing events associated with one virtual channel for a particular three-hour time slot. Each such EIT instance can be delivered in a single table section if it fits in 4095 bytes, or (if not) delivered in two or more sections. If it is segmented for delivery, each part is labeled (according to MPEG convention) "part M of N" using the section_number and last_section_number fields in the table section header.

From the point of view of the EPG grid, all of the EIT instances describing events scheduled for a particular three-hour time slot, taken together, can be thought of as representing a particular three-hour wide vertical slice of the grid. For convenience in discussions, we sometimes number these three-hour-wide slices as "EIT-k," where k indicates the particular time slot. Value zero for k always indicates the time slot corresponding to the current time. Therefore, "EIT-0" encompasses the set of EIT instances representing EIT data for a set of virtual channels including all of the events starting in or continuing into or through the current three-hour time slot. "EIT-1" includes the set of EIT instances describing events starting in or continuing into or through the time slot just following the current one, and so on.

Figure 12.2 depicts a wider perspective of the same EPG grid example of the prior figure and illustrates several aspects of our discussion. The three-channel by 90-minute segment of the grid we saw in the previous Figure appears here within the area encompassed by the thick-bordered rectangle. We can now look at the previous view as being a window into a larger space that extends vertically to other channels and horizontally to past and future times.

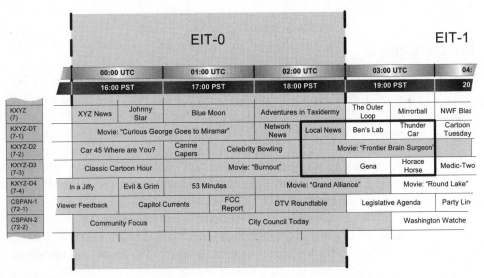

Figure 12.2 EPG Grid—Wider Perspective

As we discussed in Chapter 6, EIT data and Virtual Channel Table data are linked together using Source IDs (see "Source IDs" on page 109). For convenience, in Figure 12.2 we have performed the same lookup/association the EPG display application uses to display the channel names alongside the program schedule data.

Figure 12.1 displayed local time, since that of course is most relevant to the viewer. Here in Figure 12.2 we have added the equivalent UTC time-line. Pacific Standard Time is eight hours behind UTC, so for example 4 p.m. PST is midnight UTC. For this example we assume the current time of day is 6:45 p.m. PST, or 02:45 UTC. As we have noted, the three-hour time-span of any given EIT is aligned to UTC. Therefore, at 6:45 p.m. PST, EIT-0 covers the time period from midnight to 03:00 UTC (4:00 to 7:00 p.m. PST).

EIT-0 must include all of the programs that start within its three-hour window and also those that started prior to that window but continue on into (or through) its window. In Figure 12.2, programs described in EIT-0 are shaded. The programs "In a Jiffy" and "Viewer Feedback" are included in EIT-0 because they started prior to midnight UTC and continue into the 00:00-03:00 UTC time period.

In Figure 12.2 we can also see some of the programs included in EIT-1. The un-shaded ones on the right are all included as are three that started in EIT-0 and continue on into the EIT-1 time slot.

If we look back again at Figure 12.1 and notice in Figure 12.2 where the three half-hours fall with respect to the UTC time line, we see that to build the display for the 6:30 to 8:00 p.m. PST time period the receiver must use data from both EIT-0 and EIT-1.

EIT Transport

We now take the example to the next level of detail and see how EIT table sections deliver the EPG data shown in Figure 12.2. The first thing to note is that the EIT data itself does not give any information about the virtual channel that is to carry the programming. Each EIT instance provides only a schedule of program start times, durations, titles, and other information about the events associated with one programming service. But, as we saw in Chapter 6, each EIT instance is tagged with a Source ID that links the schedule to a virtual channel identified with the same value of Source ID.

Let's assume that one broadcaster manages the seven channels shown in Figure 12.2. KXYZ (7) is an analog NTSC channel. KXYZ-DT, KXYZ-D2, KXYZ-D3, and KXYZ-D4 are related digital services. This broadcaster provides, as a community service, bandwidth for two CSPAN channels as well. These are called CSPAN-1 and CSPAN-2 and are accessible to viewers on virtual channels 72-1 and 72-2. All of the digital services are delivered in a single Transport Stream, which also delivers the PSIP data describing them plus analog channel KXYZ (7).

The corresponding Virtual Channel Table for this example is shown in Table 12.1 below.

TABLE 12.1 Example Virtual Channel Table

Channel Name	Major Chan. No.	Minor Chan. No.	Source ID	Service Type	TSID	program_number
KXYZ	7	0	0x0800	Analog TV	0x0974	0xFFFF
KXYZ-DT	7	1	0x0801	Digital TV	0x0975	0x0200
KXYZ-D2	7	2	0x0802	Digital TV	0x0975	0x0201
KXYZ-D3	7	3	0x0803	Digital TV	0x0975	0x0202
KXYZ-D4	7	4	0x0804	Digital TV	0x0975	0x0203
CSPAN-1	72	1	0x0805	Digital TV	0x0975	0xE100
CSPAN-2	72	2	0x0806	Digital TV	0x0975	0xE101

EIT-0

In this example, the EIT-0 delivered in this TS is composed of seven EIT instances, one for each of the seven virtual channels. Table 12.2 shows each EIT instance and its contents. Note that here we show the source_id values that tie each schedule to the virtual channel with the corresponding source_id from Table 12.1.

TABLE 12.2 Example EIT Instances for EIT-0

Source ID	Version No.	No. Events	Event Title	Start Time	Duration
0x0800	1	4	XYZ News	00:00	30 min.
			Johnny Star	00:30	30 min.
			Blue Moon	01:00	60 min.
			Adventures in Taxidermy	02:00	60 min.
0x0801	1	3	Movie "Curious George Goes to Miramar"	00:00	120 min.
			Network News	02:00	30 min.
			Local News	02:30	30 min.
0x0802	1	4	Car 45 Where Are You?	00:00	60 min.
			Canine Capers	01:00	30 min.
			Celebrity Bowling	01:30	60 min.
			Movie: "Frontier Brain Surgeon"	02:30	120 min.
0x0803	1	2	Classic Cartoon Hour	00:00	60 min.
			Movie "Burnout"	01:00	120 min.
0x0804	1	4	In a Jiffy	23:00	90 min.
			Evil & Grim	00:30	30 min.
			53 Minutes	01:00	60 min.
			Movie: "Grand Alliance"	02:00	90 min.

TABLE 12.2 Example EIT Instances for EIT-0 (continued)

Source ID	Version No.	No. Events	Event Title	Start Time	Duration
0x0805	1	4	Viewer Feedback	23:30	60 min.
			Capitol Currents	00:30	60 min.
			FCC Report	01:30	30 min.
			DTV Roundtable	02:00	60 min.
0x0806	1	2	Community Focus	00:00	60 min.
			City Council Today	01:00	150 min.

Important note: each of the seven EIT table instances that make up EIT-0 must have the same value for version_number. This restriction arises due to the way the Master Guide Table references EITs. Since the MGT gives the set of information relevant to all the EIT instances that make up EIT-k for a given value of k, that information must be consistent for all instances.

Table 12.3 shows EIT-1. Note that all of the EIT-1 instances are at version_number zero.

TABLE 12.3 Example EIT Instances for EIT-1

Source ID	Version No.	No. Events	Event Title	Start Time	Duration
0x0800	0	5	The Outer Loop	03:00	30 min.
			Mirrorball	03:30	30 min.
			NWF Blast	04:00	30 min.
			Medic-One	04:30	30 min.
			Biography	05:00	60 min.
0x0801	0	4	Ben's Lab	03:00	30 min.
			Thunder Car	03:30	30 min.
			Cartoon Tuesday	04:00	60 min.
			Cartoon Monday (R)	05:00	90 min.
0x0802	0	2	Movie: "Frontier Brain Surgeon"	02:30	120 min.
			Movie: "The Spectrum Wars"	04:30	120 min.
0x0803	0	4	Gena	03:00	30 min.
			Horace Horse	03:30	30 min.
			Medic Two	04:00	30 min.
			Movie: "Sleepless in Kirkland"	04:30	150 min.
0x0804	0	3	Movie: "Grand Alliance"	02:00	90 min.
			Movie: "Round Lake"	03:30	120 min.
			NWF Blast	05:30	30 min.

TABLE 12.3 Example EIT Instances for EIT-1 (continued)

Source ID	Version No.	No. Events	Event Title	Start Time	Duration
0x0805	0	4	Legislative Agenda	03:00	60 min.
			Party Line	04:00	30 min.
			Weekend in Review	04:30	60 min.
			Capitol Commentary	05:30	60 min.
0x0806	0	3	City Council Today	01:00	150 min.
			Washington Watcher	03:30	60 min.
			Synchronicity	04:30	90 min.

MGT References

The Master Guide Table references EIT-k for each k from zero up to a maximum of 127. Table 12.4 shows example MGT entries for the EIT-0 through EIT-3 tables, valid at 6:45 p.m. PST.

TABLE 12.4 MGT Valid at 02:45 UTC (6:45 p.m. PST)

Table Type	Table Type PID	Table Type Version Number	Number Bytes	Time Slot
0x0100 (EIT-0)	0x7100	1	1,388	00:00 to 03:00 UTC
0x0101 (EIT-1)	0x7101	0	1,340	03:00 to 06:00 UTC
0x0101 (EIT-2)	0x7102	4	1,361	06:00 to 09:00 UTC
0x0101 (EIT-3)	0x7103	5	1,404	09:00 to 12:00 UTC

The table shows that EIT-0 is delivered in Transport Stream packets with PID value 0x7100, EIT-1 is delivered in PID 0x7101, and so on. All the EIT-0 instances have their version_number fields set to 1 and all the EIT-1 instances have their version_number fields set to 0.

Passage of Time

Now let's watch what happens as time moves forward to the three-hour EIT boundary coming up at 03:00 UTC (7 p.m. PST). Those building PSIP data generators can choose to design their products so that the tables are moved into TS packets with different PID values at the three-hour boundary (we call this the "PID-shift" method), or they can choose to transport the tables in TS packets with the same

PIDs and only move the MGT PID pointers (this is called the "pointer-shift" method). We take a look at both scenarios here.

PID-shift vs. pointer-shift

Figure 12.3 illustrates the difference between the PID-shift and the pointer-shift methods for a simple case in which only EIT-0 and EIT-1 are transmitted. At the

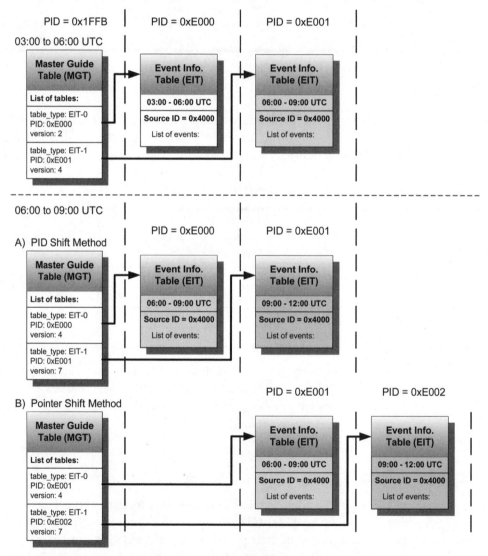

Figure 12.3 PID Shift vs. Pointer Shift at the Three-Hour Boundary

top, the time is in the range 03:00 to 06:00 UTC. EIT-0 (at version 2) is being sent in TS packets of PID value 0xE000 and EIT-1 (at version 4) is being sent in TS packets of PID value 0xE001.

Case A illustrates the PID-shift method, so called because the PID values used to transport the EIT tables for a given time slot are shifted at the boundary. At 6:00 UTC, the PID value used to transport the EIT describing the 6:00 to 9:00 time slot shifts from 0xE001 to 0xE000. At the time slot boundary, the PID values indicated in the MGT for EIT-0 and EIT-1 stay constant. But in this example, at the time slot boundary (06:00 UTC), the MGT changes to indicate EIT-0 is at version 4 (it is actually now referencing a completely different set of EIT section instances, the ones that were EIT-1 just a moment before).

Case B illustrates the pointer-shift method. Here, the PID values used to transport the EIT tables do not change. Instead, the MGT PID reference is shifted (hence the name "pointer-shift.") The tables that were EIT-1 a moment ago continue to be transported in TS packets of PID value 0xE001, and just after the time slot boundary the MGT indicates that EIT-0 uses this PID value. EIT-1 uses PID value 0xE002 at the boundary.

The first thing to notice about the passage of time across the three-hour boundary is that most of the events defined in the set of tables that was EIT-0 (in our example, they covered 00:00 to 03:00 UTC) have now expired (the events are over). The set of tables that had been EIT-1 form the basis for the new EIT-0.

Table 12.5 shows the contents of the new EIT-0. Comparison of Table 12.5 with Table 12.3 shows that they are exactly the same.

TABLE 12.5 The New EIT-0

Source ID	Version No.	No. Events	Event Title	Start Time	Duration
0x0800	1	5	The Outer Loop	03:00	30 min.
			Mirrorball	03:30	30 min.
			NWF Blast	04:00	30 min.
			Medic-One	04:30	30 min.
			Biography	05:00	60 min.
0x0801	1	4	Ben's Lab	03:00	30 min.
			Thunder Car	03:30	30 min.
			Cartoon Tuesday	04:00	60 min.
			Cartoon Monday (R)	05:00	90 min.
0x0802	1	2	Movie: "Frontier Brain Surgeon"	02:30	120 min.
			Movie: "The Spectrum Wars"	04:30	120 min.

TABLE 12.5 The New EIT-0 (continued)

Source ID	Version No.	No. Events	Event Title	Start Time	Duration
0x0803	1	4	Gena	03:00	30 min.
			Horace Horse	03:30	30 min.
			Medic Two	04:00	30 min.
			Movie: "Sleepless in Kirkland"	04:30	150 min.
0x0804	1	3	Movie "Grand Alliance"	02:00	90 min.
			Movie: "Round Lake"	03:30	120 min.
			NWF Blast	05:30	30 min.
0x0805	1	4	Legislative Agenda	03:00	60 min.
			Party Line	04:00	30 min.
			Weekend in Review	04:30	60 min.
			Capitol Commentary	05:30	60 min.
0x0806	1	3	City Council Today	01:00	150 min.
			Washington Watcher	03:30	60 min.
			Synchronicity	04:30	90 min.

Figure 12.4 shows an example EPG grid view of the world at the 03:00 UTC boundary.

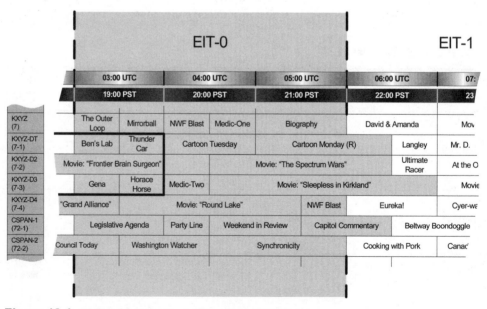

Figure 12.4 EPG Grid at the Next Three-hour Boundary

Let's say the MGT must be updated now because one of the tables it references has been updated. We look at two cases. In the first case, the PSIP generator has shifted the PID values associated with each EIT-k but left the tables themselves in the same TS packets as before. We call this the "pointer shift" approach.

Pointer shift approach

Table 12.6 shows the pointer-shift method of dealing with the three-hour boundary.

TABLE 12.6 MGT Valid at 03:00 UTC (7 p.m. PST)—Pointer Shift Method

Table Type	Table Type PID	Table Type Version Number	Number Bytes	Time Slot
0x0100 (EIT-0)	0x7101	1	1,387	03:00 to 06:00 UTC
0x0101 (EIT-1)	0x7102	4	1,361	06:00 to 09:00 UTC
0x0101 (EIT-2)	0x7103	5	1,404	09:00 to 12:00 UTC
0x0101 (EIT-3)	0x7104	9	1,334	12:00 to 15:00 UTC

There is a very important point illustrated here (hopefully you've followed along this far!). Compare the table type 0x0100 (EIT-0) row in Table 12.4 and the same row in Table 12.6. Processing an updated MGT should tell the receiver exactly which tables have been updated and therefore which tables to refresh. In this example, there is a right way and a wrong way to interpret this MGT update. Here are some incorrect and some correct inferences:

- **Wrong interpretation:** EIT-0 has not changed, since the old version number and the new version number are the same (both are 1). EIT-1 changed, since the old version number was 0 and now it is 4.

- **Correct interpretation:** The version numbers of EIT table instances carried in TS packets with PID value 0x7101 have changed from 0 to 1 and must be reloaded. The version numbers of EIT table instances carried in TS packets with PID value 0x0702 have not changed (the version number remained at 4). Where before the tables associated with EIT-0 were carried in TS packets with PID value 0x07100, the EIT-0 tables are now carried in TS packets with PID value 0x07101.

As we have noted elsewhere in this book, in accordance with MPEG-2 *Systems* philosophy and rules, the MGT indicates the version numbers of tables carried in TS packets identified with certain PID values. In the receiver, table versions must be organized by table_id and PID value, not by MGT table_type.

Table shift approach

Another way the three-hour boundary can be handled by the PSIP generator is to change the PID values associated with the table sections delivering EIT data for a particular time slot. For example, at the boundary the tables that were EIT-1 can be moved over to use the same PID value that the old EIT-0 once used.

Table 12.7 shows an MGT that might result if the table shift approach were to be used.

TABLE 12.7 MGT Valid at 03:00 UTC (7 p.m. PST)—Table Shift Method

Table Type	Table Type PID	Table Type Version Number	Number Bytes	Time Slot
0x0100 (EIT-0)	0x7100	2	1,387	03:00 to 06:00 UTC
0x0101 (EIT-1)	0x7101	5	1,361	06:00 to 09:00 UTC
0x0101 (EIT-2)	0x7103	6	1,404	09:00 to 12:00 UTC
0x0101 (EIT-3)	0x7104	10	1,334	12:00 to 15:00 UTC

In this method, the tables that had been sent in TS packets with PID value 0x7100 (the old EIT-0) are no longer sent. Instead, the tables that had been EIT-1 are sent in their place, using the same TS packet PID values. Because these are different tables than the ones that were previously in TS packets with PID value 0x7100, the version numbers *must* be incremented. As shown in Table 12.7 the version number for tables appearing in TS packets with PID value 0x7100 has changed from 1 to 2. Any interpretation of the MGT by the receiver yields a result indicating that these tables have changed and must be reloaded.

The pointer-shift method is generally preferred over the table-shift technique as it typically results in less work on the part of receivers.

Other Considerations

Here are a couple of other miscellaneous EPG/EIT-related considerations.

Short events and odd starting times

The A/65 Standard does not in any way constrain the length of events beyond the fact that the length_in_seconds field is a 20-bit number. With twenty bits an event lasting as long as a little over 12 days could be indicated.

While an event lasting a short amount of time (say 10 minutes) could be described in the EIT, typical receivers would probably have difficulty adequately representing such an event in an on-screen program guide, so events shorter than 30 minutes are rarely seen in practice.

A/65 also does not constrain starting times to fall on even half-hours. Receiver designers should accommodate events whose starting times do not fall on half-hour boundaries and should gracefully handle odd-sized events should they appear in EIT data. A common case is a program that starts at 35 minutes after the hour. Such a program could be displayed as if it started on the half-hour, but could include "(:35)" somewhere to indicate the true starting time.

Inactive channels

As of this writing, A/65 does not define a method whereby a broadcaster can indicate the scheduled status of a virtual channel with regard to its being either on-the-air (broadcasting) or inactive (not currently broadcasting). ATSC committees are currently contemplating whether or not to add a new flag to the EIT, or to specify that any periods of time for which no EIT data is specified shall indicate the channel is not broadcasting during those periods.

Directed Channel Change

As we have seen, when a broadcaster transmits audio and video programming in digital format, more than one program can be included in the transport multiplex delivered in that broadcaster's RF channel. Broadcasters recognized several scenarios in which it would be advantageous for them to direct certain receivers at given times to different virtual channels in the transmitted signal. When a change in virtual channel originates from a broadcaster command, it is called a "directed" channel change, or DCC. Directed Channel Change is the subject of this chapter.

Broadcasters can use DCC in a variety of ways to enhance the viewing experience. For a consumer to enjoy its benefits, a DTV receiver capable of supporting DCC features is required.

DCC was originally published as Amendment 1 to A/65A in May, 2000. In late 2001, ATSC began preparing a proposed revision to DCC that made significant changes to the original specification. The DCC specification discussed here is reflected in ATSC A/65B, issued in 2002.

Our discussion begins with an overview of the concept and philosophy of Directed Channel Change. We then look at detailed syntax and semantics of the two DCC-related tables and the two DCC-related descriptors and finally finish by exploring some examples of DCC applications.

Overview of Directed Channel Change

The Directed Channel Change mechanism has the potential to add a new dimension of richness to a viewer's experience of digital television. Fundamentally, the DCC protocol offers ways broadcasters could indicate, to compatible receiving devices, situations in which a change in channel is suggested or requested. Receiving devices designed to comply with the DCC protocol are called "DCC-capable Reference Receivers" (DCCRRs).

Before most DCC functions can work, the viewer must set up his or her DCC-capable receiver, supplying it with information, interest categories and preferences specific to the viewer and other TV watchers in the household. A particular receiver may prompt the user to supply this information at initial setup. It must be possible to interact with the receiver to add or modify the data at a later time. The specific items of information relevant to the DCC function are discussed below.

In many (perhaps most) cases, the broadcaster may wish to target only a subset of the population of DCCRRs for a given channel change opportunity. The protocol offers a number of "selection" criteria for this purpose. Examples of possible target test sets for a given DCC event include:

- receivers located in a certain geographic location (by zip code or by state/county).

- receivers whose user or users have indicated affiliation with certain demographic categories (categories include gender- and age-based categories and employment status).

- receivers whose user(s) have indicated an affinity to certain programming genres or interest categories.

- those receivers that are unable to view the program because they are not authorized (due to the enforcement of conditional access provisions) to do so.

- receivers that are unable to view the current program because it has been blocked due to objectionable content (V-chip, or content advisory).

Many of these tests have logical variants as well. For example, the set could be specified to include those receivers whose users have *not* indicated an interest in a certain programming genre, or who are not in a certain age range. Furthermore, the test can specify that the DCC is targeted only to receivers whose users have expressed interest in one or more of a set of programming genres.

It is also possible to further refine the set of targeted receivers by specifying combinations of these conditions. Two or more tests can be specified in one DCC event definition such that the event targets only those receivers who meet the conditions in all of the tests. With this mechanism, for example, one can target receivers in a certain geographic location whose users have specified an interest in a certain subject or genre (in this case we have a combination of the geographic location and the genre category tests).

Examples of the kinds of sophisticated selection criteria that a DCC can specify include directed channel change events targeting:

- viewers residing in zip code 98034 who are interested in Football or Basketball but not Hockey or Tennis.

- viewers who live in households in which both males and females are present, but no one is younger than 50.

- households in which no one works, some family members are under the age of 35, and no one is interested in comedy or musicals.

- viewers who are unable to watch the current program because of content advisory data, unless no one is under the age of 18.

Of course, it is also possible for a DCC event to target all DCC-capable receivers.

Viewer direct-select

Another feature offered by the DCC protocol is called "viewer direct-select." The viewer direct-select feature makes it possible for the broadcaster to associate a DCC event that involves different programming choices with as many as four buttons available on the DCCRR remote-control unit (or through on-screen graphics). Any given direct-select opportunity can involve one, two, three, or all four of the buttons. The buttons are labeled "A" through "D" to allow the broadcaster to refer to the choices verbally or with on-screen text.

Let's say a broadcaster wishes to support the viewer direct-select function on a news broadcast. A four-way branch may be offered, with each different branch tied to a different viewer direct-select button. Through on-screen prompting, the viewer learns that pushing "A" will take him or her to a local weather segment, "B" branches to a report of local weather in an adjoining region, "C" is for local news, and "D" is for local news for a nearby metropolitan area.

Each time a set of viewer direct-select branches is specified, the broadcaster can associate a unique tag to each of the branches. The DCCRR can use the tags and take note of the choice made by the viewer when a certain set of branches is offered. Then later (perhaps the next day), if another viewer direct-select branch is offered and one of the branches has a tag value matching a choice made at a previous opportunity, the DCCRR can take as the default path the one the viewer chose last time. In this way, the receiver can learn the user's preference and use it when the same (or a similar) choice is presented again. This function is called viewer direct-select "persistence."

For consistency of implementation, all DCC-capable receiving devices must support the viewer direct-select feature and must include four buttons (no more, no less) dedicated to its operation. The buttons may be physically present on the remote control unit, or they can be implemented graphically. With a graphical method, the buttons are depicted on-screen, and the viewer uses arrow keys on the RCU to choose a button and a "Select" or "OK" key to activate the chosen button.

Specification mask

A given user may or may not specify all (or perhaps any) of the various pieces of information needed to perform a given DCC selection. In general, when one of the DCC tests involves a data item or category for which the user has not provided input, that part of the DCC selection criterion has no effect. When we say "has no effect," we mean it neither causes that particular receiver to be included in nor to be excluded from that DCC event. The DCC specification is very clear, for each test, how that test should deal with the case in which the required information is unavailable.

As an example, one of the selection tests involving demographic category selection is called the "one or more non-members" test. For this type of test to evaluate true, one or more of the indicated categories must correspond to "non-member" (logic value zero) in the set of data representing the viewer's indicated demographic category membership, *and* one or more of those non-member categories must be one in which the viewer made a definitive assertion with regard to membership or non-membership in that category. If, when setting up the DCCRR, the viewer gave no answer at all to the question "Are you working?", then references to the "Working" demographic category do not influence DCC selection one way or the other for that receiver.

In order to describe required behavior in the DCCRR in this regard, a concept called the "specification mask" (S) is used. S is a bit-vector where each bit corresponds to a demographic group or genre category. For demographic groups, a zero in a given bit position of S means the user *has not* specified whether or not he or she considers himself or herself a member of that group. A one (1) in a given position means that the user *has* made an assertion one way or the other. For the genre categories, a zero means nothing has been specified for that particular genre category while a one (1) means that either interest or lack of interest in the associated category has been declared.

Continuing with the "one or more non-member" demographic category test example, the DCCRR uses the following logic. The bit-vector representing the different demographic tests of interest (given in the DCC message) is bit-wise ANDed with the one's complement of the bit-vector representing the viewer's membership (or non-membership) in the demographic categories. That result is ANDed with the specification mask. If the result is non-zero, the selection test evaluates True and it evaluates False otherwise. If the viewer indicates non-membership in a given category (for example, he or she answers "no" to the question "Are you working?"), then the specification mask has a "1" in the bit position corresponding to the Working category. A "one or more non-member" type of test involving the Working category selects this particular viewer. If no answer is given to the "Are you working?" question, the selection mask has a "0" in that bit position and any "one or more non-member" test entirely disregards the Working category.

Overview of DCC logic

A given DCC table section defines one or more selection tests. Each selection test targets only those receivers tuned to specific virtual channels and involves one or more "terms." For the selection to be valid for a given DCCRR, all of the terms must evaluate True.

Figure 13.1 gives a simplified logic flow diagram for processing the DCC table section. In this example implementation, the AND function involved in processing the various terms is achieved by exiting the loop if any term evaluates False.

The first decision box, "Any further tests?" reflects the fact that any given DCC table section can specify more than one distinct DCC selection test. At the second decision point, receivers not tuned to one of the virtual channels targeted for this particular test skip past it to the next one. The third decision point involves a time check: if the start time of this DCC event has not been reached, processing continues on to other tests (perhaps to return at a later time).

In the lower portion of Figure 13.1, the terms are evaluated for the particular test. If any test evaluates False, processing continues at the next test (if any). If all of the terms evaluate True, the DCC event is acted upon by the DCCRR.

DCC context

Two distinctly different types of Directed Channel Change events are defined. Within the context of channel navigation, one type operates exactly as if the user had changed channels while the other switches programming tracks in a (potentially) seamless and non-obvious way. The type that is a straightforward command from the broadcaster to the DCCRR to change channels is called the Channel Redirect type of DCC. The type that takes the viewer to an alternate programming track without making that fact obvious is called the Temporary Retune type.

Channel Redirect DCC

When the DCC-capable receiver responds to a Channel Redirect DCC event, it simply changes channels to the virtual channel indicated in the DCC message. If the DCCRR wishes to display a channel number, the number displayed is the newly acquired channel, just as if the viewer had initiated the change.

Processing of a Channel Redirect DCC event involves no state memory. Once the event is processed and the channel has been changed, this type of DCC event has no further effect on subsequent DCC-related processing.

A typical use for the Channel Redirect DCC event might be to handle the case in which a standard-definition minor channel is going off the air temporarily (moving to "inactive" state) because the full transmitted bandwidth is to be dedicated to High-Definition programming for the next portion of the broadcast schedule. At the appropriate time, all DCC-capable receivers tuned to the standard-definition chan-

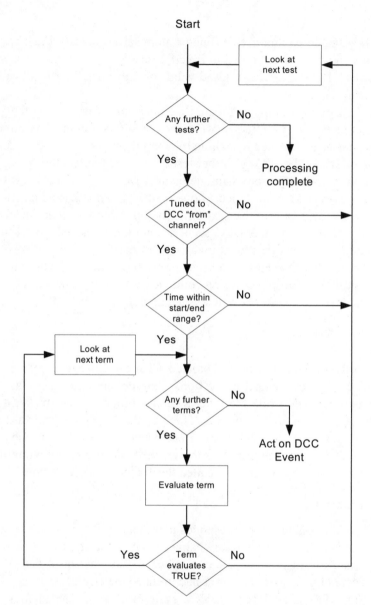

Figure 13.1 DCC Logic Flow

nel may be commanded by a Channel Redirect DCC to change minor channels to the one carrying the HD programming.

Temporary Retune DCC

With the Temporary Retune DCC function, different subsets of the viewing population may be presented with different audio/video content in the context of the same virtual channel. A viewer will typically not be aware that the content being viewed may be different from the content being seen by that viewer's neighbor, even though both DTV receivers would, if asked to do so, display the same virtual channel number.

This DCC function is called "temporary" because it involves a well-defined and limited time duration. Unlike the Channel Redirect type of DCC event, which is state-free, processing the Temporary Retune DCC event does change the DCCRR to a state in which certain behaviors are modified from normal behavior. Once the DCCRR has accepted and acted on a Temporary Retune DCC event:

1. no DCC events other than one to cancel the current Temporary Retune may be accepted.

2. when displaying the channel number, the number that must be shown is the original channel number (the one that had been acquired prior to re-tuning), not the true current virtual channel number.

3. if the DCC end time is reached, a return to the original channel must be made.

Any time the viewer initiates a channel change, the state is reset back to normal viewing and all memory of prior DCC events is lost.

A broadcaster could use the Temporary Retune DCC function to direct a subset of the viewing population to a programming track suitable to the interests or demographics of that subset. For example, it could be used for targeted advertising. Let's say a tire dealership has two locations within the station's broadcast service area. A DCC could be designed such that viewers closer to the first location would see that location's address and telephone number, while those closer to the other site would see that site's information instead.

DCC example time line

The principles of Channel Redirect and Temporary Retune DCC events may be illustrated with an example time line. Figure 13.2 maps time on the horizontal axis and programming choices on the vertical axis. The Figure shows three ordinary virtual channels, 13-1, 13-2 and 13-3, plus one hidden channel, 13-81.

The heavy dotted line traces the viewing path of one DCC-compatible receiver as it starts on the left tuned to channel 13-2. At 5:25 p.m., a Temporary Retune DCC event is processed that takes this DCCRR to hidden channel 13-81. At 5:40, a DCC message is processed that cancels the Temporary Retune and commands the DCCRR to return to the original channel, 13-2. For the 15-minute period 5:25 to

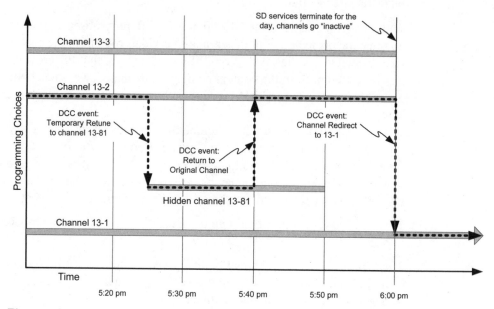

Figure 13.2 Example DCC Time Line

5:40, the DCCRR is in a state in which the displayed channel is 13-2 even though it is actually tuned to 13-81. In this state, it disregards any DCC events other than one to cancel the currently active Temporary Retune.

As a result of processing the DCC that canceled the Temporary Retune at 5:40, the DCCRR returns to normal viewing mode. At 6:00 p.m., channels 13-2 and 13-3 go inactive, as the broadcast multiplex switches format to one High Definition programming service on channel 13-1. At that time, a Channel Redirect is processed that causes an immediate (stateless) change to channel 13-1.

Avoiding a tuning glitch

In some instances, a broadcaster may wish to schedule back-to-back DCC events. Figure 13.3 shows an example with two back-to-back events. At 5:25 p.m., as with the prior example, a Temporary Retune DCC event is processed that takes the DCCRR to hidden channel 13-81. Two DCC-related events are scheduled at 5:40 p.m.: the end of this first Temporary Retune, and the start of a new Temporary Retune DCC event targeted at our DCCRR.

Conceptually, when the DCC end time is reached, the DCCRR must return to the original channel (13-2) before it can process the next DCC command. However, if the DCCRR actually tuned and acquired the original channel before processing the next Temporary Retune, a tuning glitch would be visible to the viewer. The heavy

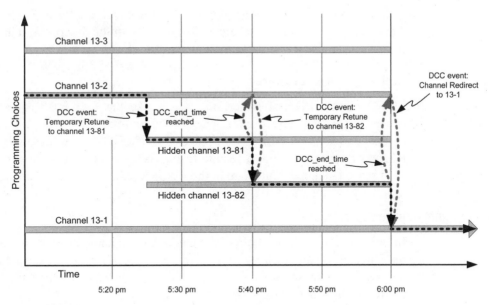

Figure 13.3 Back-to-back DCC Events

black dotted line path in Figure 13.3 shows that the DCCRR can skip the step of physically retuning to the original channel before taking action on the next Temporary Retune. In the example, it can go directly from virtual channel 13-81 to channel 13-82 without going back to 13-2 in between steps.

A second situation where a tuning glitch can be avoided is shown at the right side of Figure 13.3. Here, at 6:00, the end point of the second Temporary Retune is reached. Since the DCCRR takes action on a Channel Redirect at that same instant, it can skip the step of physically re-acquiring the original channel (13-2) prior to tuning and acquiring channel 13-1.

Additional functions

The DCC mechanism includes some additional functions and features.

Arriving Request Descriptor

A broadcaster may wish to cause a certain text string to be displayed when DCC-capable receivers tune to a given virtual channel (either manually via the remote control or as a result of processing a DCC event). The Arriving Request Descriptor can include a text string to be displayed on-screen in a centered window. Two forms of the descriptor are described, one in which the display is for ten seconds (or until the user takes action to cancel the display) and one for an indefinite time (or until the user takes action to cancel it).

It is important to note that support in a given receiver for the Arriving Request Descriptor is optional. For receivers that do support the descriptor, it is important to offer the user a way to disable it if is his or her choice to do so.

Departing Request Descriptor

The Departing Request Descriptor is just like the Arriving Request Descriptor except that it takes effect upon detection of an imminent channel change (either a manual one or one resulting from processing of a DCC event). As with the Arriving Request Descriptor, support for the feature is optional. If it is supported, the standard states that the default state in the DCCRR should be "not enabled."

Minimum requirements

In order to comply with the ATSC Standard, the manufacturer who wishes to offer the Directed Channel Change feature on their receiving device must support a certain minimum subset of DCC-related functions. These functions include support of the following types of DCC operations:

- unconditional channel change.
- numeric postal code test (all forms).
- return to original channel.
- the viewer direct-select feature (including the persistence function).
- both the Channel Redirect and Temporary Retune forms of the DCC.

All the other types of DCC tests (including genre interests, demographic category, geographic location based on FIPS codes, the rating-blocked and not-authorizable tests, and the alphanumeric postal code test) may or may not be implemented at the discretion of the product designer.

Directed Channel Change Table (DCCT)

A broadcaster can place a Directed Channel Change table section into the Transport Stream at any time to indicate to DCC-enabled receivers that alternate programming is available.

Structure of the DCCT

Figure 13.4 illustrates the structure of the Directed Channel Change section as it is formatted within transmitted PSIP data.

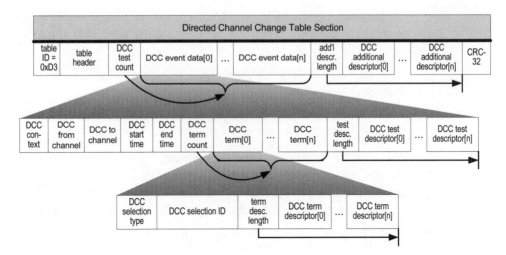

Figure 13.4 Structure of the Directed Channel Change Table Section

At the outer level, the Directed Channel Change table section consists of a number of "DCC event data" blocks followed by a number of descriptors pertaining to the entire table section. Each DCC event data block defines one situation that is an opportunity for some number of DCC-enabled receivers to make a channel change. Each event is described in terms of the following pieces of information:

- the virtual channel number to which the DCC-enabled receiver must be currently tuned to participate in this channel change event.

- the virtual channel number that will be switched to if the conditions required for this DCC event are satisfied.

- the date and time of day that is the starting point for this event.

- the date and time of day that is the end point for this event.

- one or more selection criteria that have to be met if the channel change is to occur.

- zero or more descriptors pertinent to this DCC event.

Each selection criterion consists of a type field and a 64-bit data field, followed by zero or more descriptors related to this selection.

DCCT transport

Transport of the DCCT involves some constraints against MPEG-2 *Systems*:

- A Directed Channel Change Table section, if carried in the Transport Stream, must be placed into Transport Stream packets with PID value 0x1FFB, the SI base_PID.

- TS packets containing the DCCT section cannot be scrambled (the transport_scrambling_control bits must be set to zero).

- As with all PSIP tables, an adaptation field is not to be present in TS packets carrying the DCCT section.

- Each DCCT instance may not be segmented for delivery; it must fit into a single section with a maximum length of 4096 bytes. Practically, this is really not a limitation because multiple DCCT sections may be present in the Transport Stream at any given time, differentiated from one another by the dcc_id field (see below).

DCCT syntax and semantics

Table 13.1 describes the table section syntax for the directed_channel_change _section().

TABLE 13.1 Directed Channel Change Table Section Syntax

Field Name	Number of Bits	Field Value	Description
directed_channel_change_section() {			Start of the DCCT section.
table_id	8	0xD3	Identifies the table section as part of the Directed Channel Change Table.
section_syntax_indicator	1	1b	Indicates that the section is formatted in MPEG "long-form" syntax.
private_indicator	1	1b	This flag is not defined in MPEG and is set to 1 in PSIP tables.
reserved	2	11b	Reserved bits are set to 1.
section_length	12		An unsigned integer that specifies the length, in bytes, of data following the section length field itself to the end of this table section.
dcc_subtype	8	0x00	Set to zero for the current protocol. Could be used in future standards for a different type of DCC.

TABLE 13.1 Directed Channel Change Table Section Syntax (continued)

Field Name	Number of Bits	Field Value	Description
dcc_id	8		Used to differentiate multiple instances of DCCs. More than one different DCC table section may be transmitted together in the same Transport Stream (in TS packets with PID value 0x1FFB) as long as each different instance is identified with a unique value of dcc_id.
version_number	5		The version number reflects the version of a table section, and is incremented when anything in the table section changes.
current_next_indicator	1	1b	Indicates, in general, whether the table section is currently applicable (value 1) or is the next one to be applicable (value 0). For the DCCT, only the current table is sent.
section_number	8	0	Indicates which part this section is for tables segmented into sections. For the DCCT, this parameter is set to zero, as the DCCT is only one section long.
last_section_number	8	0	Indicates the section number of the last section in a segmented table. For the DCCT, this parameter is set to zero.
protocol_version	8	0	Indicates the version of this DCCT. At present, only protocol version zero is defined.
dcc_test_count	8		Specifies the number of tests this DCC table section specifies. It can be zero if the DCCT table section is being sent only to define Departing or Arriving Request Descriptors.
for (i=0; i<dcc_test_count; i++) {			Start of the test "for" loop
dcc_context	1		Indicates whether this DCC event is a Temporary Retune (value 0) or a Channel Redirect (value 1) (see "DCC context" on page 265).
reserved	3	111b	Reserved bits are set to 1.
dcc_from_major_channel_number	10		Indicates the major channel number of a virtual channel targeted for this DCC test.
dcc_from_minor_channel_number	10		Indicates the minor channel number of a virtual channel targeted for this DCC test.
reserved	4	1111b	Reserved bits are set to 1.

TABLE 13.1 Directed Channel Change Table Section Syntax (continued)

Field Name	Number of Bits	Field Value	Description
dcc_to_major_channel_number	10		Indicates the major channel number of the destination virtual channel used if the DCCRR acts on this channel change event.
dcc_to_minor_channel_number	10		Indicates the minor channel number of the destination virtual channel used if the DCCRR acts on this channel-change event.
dcc_start_time	32		The time of day, given in GPS seconds, when this DCC event is to take place.
dcc_end_time	32		The time of day, given in GPS seconds, when this DCC event is over.
dcc_term_count	8		Indicates the number of terms present in the "for" loop that follows.
for (i=0; i<dcc_term_count; i++) {			Start of the term "for" loop
dcc_selection_type	8		Indicates the type of test associated with this term. See text.
dcc_selection_id	64		A field that provides a parameter or parameters used in the logic of this test. See text.
reserved	6	111111b	Reserved bits are set to 1.
dcct_term_descriptors_length	10		Indicates the number of bytes of descriptors to follow.
for (i=0; i<N; i++) {			Start of the term descriptors "for" loop. The value of N is determined indirectly from dcc_term_descriptors_length
dcc_term_descriptor()	var.		Zero or more term descriptors, each formatted as type-length-data.
}			End of the term descriptors "for" loop.
}			End of the term count "for" loop.
reserved	6	111111b	Reserved bits are set to 1.
dcc_test_descriptors_length	10		Indicates the number of bytes of descriptors to follow.
for (i=0; i<N; i++) {			Start of the test descriptors "for" loop. The value of N is determined indirectly from dcc_test_descriptors_length.
dcc_test_descriptor()	var.		Zero or more test descriptors, each formatted as type-length-data.
}			End of the test descriptors "for" loop.

TABLE 13.1 Directed Channel Change Table Section Syntax (continued)

Field Name	Number of Bits	Field Value	Description
}			End of the term "for" loop.
reserved	6	111111b	Reserved bits are set to 1.
dcc_additional_descriptors_length	10		Indicates the number of bytes of descriptors to follow.
for (i=0; i<N; i++) {			Start of the additional descriptors "for" loop. The value of N is determined indirectly from dcc_additional_descriptors_length.
dcc_additional_descriptor()	var.		Zero or more additional descriptors, each formatted as type-length-data.
}			End of the additional descriptors "for" loop
CRC_32	32		A 32-bit checksum designed to produce a zero output from the decoder defined in the MPEG-2 *Systems* standard.
}			End of the DCCT section.

DCC selection type

Table 13.2 lists the selection tests usable as DCC terms. Please refer to the latest version of A/65 for details. Shaded rows highlight selection types that must be supported in every DCC-capable receiver implementation.

TABLE 13.2 Defined DCC Selection Types

dcc_selection_type value	Name	Function
0x00	Unconditional Channel Change	Term always evaluates True. Can be used to make a DCC in which all DCCRRs accept the DCC request.
0x01	Numeric Postal Code Inclusion	Term evaluates True if the zip code supplied by the viewer matches the one supplied in the dcc_selection_id field. A "?" character in any digit position matches any digit.
0x02	Alphanumeric Postal Code Inclusion	Same as the Numeric Postal Code Inclusion test, except the postal code (and selection characters) can include letters A-Z.
0x05	Demographic Category—One or More	Test evaluates True if the viewer has indicated membership in any of the demographic categories specified in the dcc_selection_id field.
0x06	Demographic Category—All	Test evaluates True if the viewer has indicated membership in all of the demographic categories specified in the dcc_selection_id field.

TABLE 13.2 Defined DCC Selection Types (continued)

dcc_selection_ type value	Name	Function
0x07	Genre Category—One or More	Test evaluates True if the viewer has indicated interest in any of the genre categories specified in the dcc_selection_id field.
0x08	Genre Category—All	Test evaluates True if the viewer has indicated interest in all of the genre categories specified in the dcc_selection_id field.
0x09	Cannot Be Authorized	Test evaluates True if the DCCRR cannot view services on this channel due to conditional access limitations (no subscription, etc.)
0x0C	Geographic Location Inclusion	Test evaluates True if the viewer has specified a geographic location in the form of State and County FIPS codes, and that location matches the location specified in the dcc_selection_id field.
0x0D	Rating Blocked	This test evaluates True if the current program cannot be viewed because of a Content Advisory limitation applied at the receiver
0x0F	Return to Original Channel	If this term is present, and the other DCC terms (if present) are all True, a DCCRR that is processing a Temporary Retune is requested to return to the original channel.
0x11	Numeric Postal Code Exclusion	Just like test 0x01, except the term evaluates True if the DCCRR's postal code does not match the one specified in the dcc_selection_id field.
0x12	Alphanumeric Postal Code Exclusion	Just like test 0x02, except the term evaluates True if the DCCRR's postal code does not match the one specified in the dcc_selection_id field.
0x15	Demographic Category—One or More Non-member	This test evaluates True if the viewer has specified non-membership in some or all of the demographic categories specified in the dcc_selection_id field.
0x16	Demographic Category—All Non-member	The test evaluates True if the viewer has specified non-membership in all of the demographic categories specified in the dcc_selection_id field.
0x17	Genre Category—One or More Non-member	This test evaluates True if the viewer has specified non-interest in some or all of the genres specified in the dcc_selection_id field.
0x18	Genre Category—All Non-member	The test evaluates True if the viewer has specified non-interest in all of the genres specified in the dcc_selection_id field.
0x1C	Geographic Location Exclusion	A test involving geographic location just like type 0x0C, but this version evaluates True if the DCCRR's geographic location does *not* match the location specified in the dcc_selection_id field.
0x20	Viewer Direct-Select— Button A	Term evaluates True at the instant the viewer presses button A.
0x21	Viewer Direct-Select— Button B	Term evaluates True at the instant the viewer presses button B.

TABLE 13.2 Defined DCC Selection Types (continued)

dcc_selection_ type value	Name	Function
0x22	Viewer Direct-Select— Button C	Term evaluates True at the instant the viewer presses button C.
0x23	Viewer Direct-Select— Button D	Term evaluates True at the instant the viewer presses button D.

DCC selection ID coding

The 64-bit DCC selection ID field is encoded in different ways, depending upon the type of selection:

- for the Demographic Category tests (dcc_selection_id 0x05, 0x06, 0x15, or 0x16), dcc_selection_id is a bit field, where each bit corresponds to one demographic category (see below for details).

- for the Genre Category tests (dcc_selection_id 0x07, 0x08, 0x17, 0x18), dcc_selection_id is organized as eight bytes, where each byte represents a category in the Categorical Genre Code Assignment Table (see below).

- for the Postal Code tests (dcc_selection_id 0x01, 0x02, 0x11, 0x12) dcc_selection_id is a text string of eight numeric or alphanumeric ASCII-encoded characters.

- for the Viewer Direct-Select tests (dcc_selection_id 0x20 through 0x23), dcc_selection_id is a 64-bit identifier that can be used by the DCCRR to repeat the viewer's previous choice when the same set of branch options appears again.

- for the Geographic Location tests (dcc_selection_id 0x0C, 0x1C), the dcc_selection_id includes an 8-bit state code, a 4-bit "county subdivision" code, and a 10-bit county code. The state and county codes are defined in Annex H of the A/65 Standard. The values listed were derived from the Federal Information Processing Standards (FIPS) Pub 6-4. Information on FIPS publications may be found on the web at http://www.itl.nist.gov/fipspubs/.

Demographic categories

Table 13.3 shows the demographic categories defined for the Directed Channel Change function and the bit position within the dcc_selection_id field that corresponds to each.

A dcc_selection_id mask that specifies two or more demographic categories may be made by adding together the bit mask for each of the categories. For example, to

TABLE 13.3　Demographic Categories

Bit position	Bit Mask	Demographic Category
0	0x0000000000000001	Males
1	0x0000000000000002	Females
2	0x0000000000000004	Ages 2-5
3	0x0000000000000008	Ages 6-11
4	0x0000000000000010	Ages 12-17
5	0x0000000000000020	Ages 18-34
6	0x0000000000000040	Ages 35-49
7	0x0000000000000080	Ages 50-54
8	0x0000000000000100	Ages 55-64
9	0x0000000000000200	Ages 65+
10	0x0000000000000400	Working
11-63	-	Reserved

specify Males + Ages 55-64 + Working, the bit masks for these three categories are added together to form 0x0000000000000501.

Categorical Genre Code Assignment Table (CGCAT)

The Directed Channel Change specification defines an 8-bit encoding for approximately 150 genre codes. The Categorical Genre Code Assignment Table (CGCAT) can be considered to be a two-part dictionary. One part specifies "basic" categories such as Movie, Education, Entertainment, Sports, and Religious. The other part lists a large number of subcategories that may be applicable to one or more of the basic categories. Nearly all types of sports, for example, are listed in the subcategory portion of the table.

To support DCC events based on genre-based interests, the DCCRR implementation should try to organize the genre codes defined in the Categorical Genre Code Assignment Table in some intelligent way. For example, in the user interface dialogue that allows the viewer to express interest in genre categories, the subcategories related to Sports can be collected together so the viewer can focus on sports-related interests before moving on to some other interest category.

The Categorical Genre Code Assignment Table is expandable by the Directed Channel Change Selection Code Table (described below) or by a revision to the ATSC Standard. Any given DCCRR implementation may choose to support the genre codes defined in the CGCAT and not to support table expansion via the DCCSCT.

Please refer to the A/65 specification for the definition of the CGCAT. A quick look at the table reveals that the subcategories corresponding to codes in the range 0x27 through 0xAD are organized as two alphabetically-sorted groups. The first group, in the 0x27 through 0x7F range, corresponds directly with genre codes defined in Sec. 9.5.1.4 of ANSI/EIA/CEA-608-B[1]. Those in the range 0x80 to 0xAD are additional subcategory genres beyond those defined in 608-B.

DCCT transport rate and cycle time

While a DCC request is in progress, it is recommended that the DCCT be repeated at a cycle time not to exceed 150 milliseconds. This high rate is specified because a receiver just tuning to the Transport Stream needs to determine very quickly (as part of the service acquisition process) whether or not a DCC request is applicable, and to take action if it is.

Several seconds just prior to the DCC start time, the DCCT should be repeated at a cycle time not to exceed 400 milliseconds.

DCC Selection Code Table (DCCSCT)

As previously mentioned, one of the functions of the DCC Selection Code Table is to expand the Categorical Genre Code Assignment Table, thus defining new genre categories or subcategories that could be used for DCC event selection in the future. Another function of the DCCSCT is to establish codes for states and counties that are not in the A/65 definition. If a county changes its name, typically a new county code number is assigned rather than re-defining the existing code, so the DCCSCT supports name changes as well as newly-created counties and states.

The structure of the DCCSCT is general enough that it could be updated in the future to reference and expand any number of new table types as needed.

Structure of the DCCSCT

Figure 13.5 illustrates the structure of the DCCSCT section. In the position of table_id_extension in the DCCSCT section, is a field called dccsct_type. In the current DCCSCT definition, the value of dccsct_type is zero, but in the future a different table structure with the same DCCSCT table_id value could be standardized by using a different value for dccsct_type.

Structurally, the DCCSCT section consists of a count of "update" blocks followed by that number of data blocks, then an "additional" descriptors length field followed by a number of descriptors covering that length. Each data block begins with an update type field that establishes the kind of update being supplied. Three types are defined in the current protocol: an update to the genre category table, an

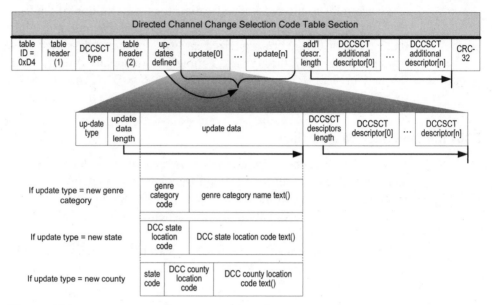

Figure 13.5 Structure of the Directed Channel Change Selection Code Table Section

update to the state code table, and an update to the county code table. The format of the genre category update consists of the genre code byte followed by the textual definition to be associated with that code. The format of the state code update is similar, with the state code followed by the name of the new state. The county code update includes the state within which the new county is located, the new county code, and finally the textual name of the new county. Zero or more descriptors may be associated with any given update. None are defined for current use.

DCCSCT transport

Transport of the DCCSCT involves a few constraints on MPEG-2 *Systems*:

- A Directed Channel Change Table section, if carried in the Transport Stream, must be placed into Transport Stream packets with PID value 0x1FFB, the SI base_PID.

- TS packets containing the DCCSCT section cannot be scrambled (the transport_scrambling_control bits are required to be set to zero).

- As with all PSIP tables, an adaptation field must not be present in TS packets carrying the DCCSCT.

- The DCCSCT cannot be segmented for delivery; it must fit into a single section with a maximum length of 4096 bytes.

DCCSCT syntax and semantics

Table 13.4 gives the syntax of the dcc_selection_code_table_section().

TABLE 13.4 Directed Channel Change Selection Code Table Section Syntax

Field Name	Number of Bits	Field Value	Description
dcc_selection_code_table_section() {			Start of the DCCSCT section.
table_id	8	0xD4	Identifies the table section as part of the DCC Selection Code Table.
section_syntax_indicator	1	1b	Indicates the section is formatted in MPEG "long-form" syntax.
private_indicator	1	1b	This flag is not defined in MPEG and is set to 1 in PSIP tables.
reserved	2	11b	Reserved bits are set to 1.
section_length	12		An unsigned integer that specifies the length, in bytes, of data following the section length field itself to the end of this table section.
dccsct_type	16	0x0000	Set to zero for the current protocol. Could be used in future standards to indicate a different type of DCCSCT.
version_number	5		The version number reflects the version of a table section, and is incremented when anything in the table changes.
current_next_indicator	1	1b	Indicates, in general, whether the table section is currently applicable (value 1) or is the next one to be applicable (value 0). For the DCCSCT, only the current table is sent.
section_number	8	0	Indicates which part this section is for tables segmented into sections. For the DCCSCT, this parameter is set to zero, as the DCCSCT is only one section long.
last_section_number	8	0	Indicates the section number of the last section in a segmented table. For the DCCSCT, this parameter is set to zero.
protocol_version	8	0	Indicates the version of this DCCSCT. At present, only protocol version zero is defined.
updates_defined	8		Specifies the number of update items to be provided in this table section.
for (i=0; i<updates_defined; i++) {			Start of the updates_defined "for" loop.

TABLE 13.4 Directed Channel Change Selection Code Table Section Syntax (continued)

Field Name	Number of Bits	Field Value	Description
update_type	8		Indicates what type of table is being extended. Three types are currently defined: an update to the genre code table, an additional state code, and an additional county code.
update_data_length	8		Indicates the side of the update data to follow.
if (update_type==new_genre_category) {			When the update_type is a genre category update.
genre_category_code	8		Gives the code value for the new genre category to be defined.
genre_category_name_text()	var.		Gives the textual name of the new genre category.
}			End of "new_genre_category" "if" statement.
if (update_type==new_state) {			When the update_type is a state update.
state_location_code	8		Gives the code value for the new state to be defined.
dcc_state_location_code_text()	var.		Gives the textual name of the new state.
}			End of "new_state" "if" statement.
if (update_type==new_county) {			When the update_type is a county update.
state_code	8		Indicates which state the new county belongs to.
reserved	6	111111b	Reserved bits are set to 1.
dcc_county_location_code	10		Gives the code value for the new county to be defined.
dcc_county_location_code_text()	var.		Gives the textual name of the new county.
}			End of "new_county" "if" statement.
}			End of "updates_defined" "for" loop.
reserved	6	111111b	Reserved bits are set to 1.
dccsct_additional_descriptors_length	10		Indicates the number of bytes of descriptors to follow.
for (i=0; i<N; i++) {			Start of the additional descriptors "for" loop. The value of N is determined indirectly from dcc_additional_descriptors_length.
dccsct_additional_descriptor()	var.		Zero or more additional descriptors, each formatted as type-length-data.
}			End of the additional descriptors "for" loop.
CRC_32	32		A 32-bit checksum designed to produce a zero output from the MPEG-2 *Systems* CRC-32 decoder.
}			End of the DCCSCT section.

DCCSCT transport rate and cycle time

The DCCSCT may be repeated at an extremely slow rate, although it is recommended that the rate not be longer than one repetition per hour.

DCC Descriptors

Two descriptors are defined for use with the DCC function: the DCC Arriving Request Descriptor and the DCC Departing Request Descriptor. The Arriving Request Descriptor may be used by a DCC-capable receiver when arriving at a new channel. The Departing Request Descriptor may be used when the DCCRR exits the channel tuned by the DCC event either because the event is now over or because the user has manually changed the channel.

Each of these descriptors is formatted in the same way: the data portion of the descriptor consists of a type byte, a text length byte, and finally a text string formatted as a multiple-string structure.

DCC Arriving Request Descriptor

Table 13.5 gives the syntax of the DCC Arriving Request Descriptor.

TABLE 13.5 DCC Arriving Request Descriptor Syntax

Field Name	Number of Bits	Field Value	Description
dcc_arriving_request_descriptor() {			Start of the descriptor.
descriptor_tag	8	0xA9	Identifies the descriptor as the DCC Arriving Request Descriptor.
descriptor_length	8		Indicates the length, in bytes, of the data to follow.
dcc_arriving_request_type	8		Gives the type of this DCC Arriving Request Descriptor. Two types are defined: see text below.
dcc_arriving_request_text_length	8		Indicates the length in bytes of the MSS to follow.
dcc_arriving_request_text()	var.		The text to be associated with the DCC Arriving Request, in the form of a Multiple-String Structure.
}			End of the dcc_arriving_request_descriptor().

dcc_arriving_request_type

The dcc_arriving_request_type is an unsigned integer in the range 0 to 0xFF that indicates the type of the arriving request. Two types are currently defined. Table 13.6 shows the coding for dcc_arriving_request_type.

TABLE 13.6 DCC Arriving Request Type Coding

dcc_arriving_request_type	Meaning
0x00	Reserved for future ATSC use.
0x01	Display arriving request text in a centered window for a minimum of 10 seconds after performing the channel change requested by the viewer, or for a smaller amount of time if the viewer issues a "continue", "OK", "proceed", or equivalent command.
0x02	Display arriving request text in a centered window indefinitely after performing a channel change request requested by the viewer, until the viewer issues a "continue", "OK", "proceed", or equivalent command.
0x03-0xFF	Reserved for future ATSC use.

DCC Departing Request Descriptor

Table 13.7 gives the syntax of the DCC Departing Request Descriptor.

TABLE 13.7 DCC Departing Request Descriptor Syntax

Field Name	Number of Bits	Field Value	Description
dcc_departing_request_descriptor() {			Start of the dcc_departing_request_descriptor().
descriptor_tag	8	0xA8	Identifies the type of the descriptor.
descriptor_length	8		Indicates the length, in bytes, of the data to follow.
dcc_departing_request_type	8		Gives the type of this DCC Departing Request Descriptor. Three types are defined, see text.
dcc_departing_request_text_length	8		Indicates the length in bytes of the MSS to follow.
dcc_departing_request_text()	var.		The text to be associated with the DCC Departing Request, in the form of a Multiple-String Structure.
}			End of the dcc_departing_request_descriptor().

dcc_departing_request_type

This is an unsigned integer in the range 0 to 0xFF that indicates the type of the departing request. Three types are currently defined. Table 13.8 shows the coding for dcc_departing_request_type.

The A/65 standard states that a given DCCRR implementation may offer the user the choice to disable response to DCC Departing Request Descriptors with values 0x02 or 0x03 for dcc_departing_request_type.

TABLE 13.8 DCC Departing Request Type Coding

dcc_departing_request_type	Meaning
0x00	Reserved for future ATSC use.
0x01	Cancel any outstanding departing request type and immediately perform a channel change upon request by the viewer.
0x02	Display departing request text in a centered window for a minimum of 10 seconds prior to performing the channel change requested by the viewer, or for a smaller amount of time if the viewer issues another channel change request or a "continue", "OK", "proceed", or equivalent command.
0x03	Display departing request text in a centered window indefinitely until viewer issues another channel change request or a "continue", "OK", "proceed", or equivalent command.
0x04-0xFF	Reserved for future ATSC use.

DCC Examples

Table 13.9 illustrates nine examples of DCC-capable receivers, each having been programmed with a different set of demographic category membership data by the viewers in the household. In the top part of the Table, a "✓" indicates the viewer specified membership in that particular demographic category (by one or more viewers), "-" indicates the viewer specified non-membership in that category, and "?" indicates that particular piece of data was not specified.

As the table shows, in case a) viewers did not specify anything about genre categories at all. In case b), the DCCRR registered the fact that one or more male viewers were present, no female viewers were present, that the males were in the age range 35 to 49, and that one or more were employed.

The bottom part of the table shows seven different DCC tests, each designed to specify a specific subset of the DCCRR population. Arrows in the table cell indicate that this particular DCCRR would accept the DCC request, while "x" indicates that particular receiver would disregard it.

Let's look at one example in detail. The DCCRR in case b) has been set up for DCC operation by the viewers in the household to indicate that male but not female viewers were present in the house and that the age of the viewers was in the 35-49 range. They also indicated that one or more viewers was employed. With that setup, a DCC test for "anyone working or over 65" would result in a channel change request because someone in this house is employed. A test for "not working or over 65" would not result in a channel change. A test for "working males" would pass because both "working" and "males" had been specified in this DCCRR.

Inspection of the table shows just how not specifying a particular demographic category affects DCC processing. In case a), no information at all has been given by

TABLE 13.9 DCC Example with Demographic Categories

Demographic Category	Case								
	a)	b)	c)	d)	e)	f)	g)	h)	i)
Males	?	✓	✓	✓	-	✓	✓	✓	✓
Females	?	-	-	✓	✓	?	✓	✓	✓
Ages 2-5	?	-	-	-	-	-	?	-	-
Ages 6-11	?	-	-	✓	-	✓	?	-	-
Ages 12-17	?	-	-	-	-	-	?	-	-
Ages 18-34	?	-	-	✓	-	✓	?	✓	-
Ages 35-49	?	✓	✓	✓	✓	✓	?	-	-
Ages 50-54	?	-	-	-	✓	-	?	-	-
Ages 55-64	?	-	-	-	-	-	?	-	-
Ages 65+	?	-	✓	-	-	-	?	✓	✓
Working	?	✓	-	✓	✓	?	-	-	-
1 Anyone working or over 65	x	→	→	→	→	x	x	→	→
2 Anyone not working or over 65	x	x	→	x	x	x	→	→	→
3 Working males	x	→	x	→	x	x	x	x	x
4 Non-working males	x	x	→	x	x	x	→	→	→
5 Female-only household, no one younger than 35	x	x	x	x	→	x	x	x	x
6 Elderly couple (>65)	x	x	x	x	x	x	x	→	→
7 Elderly couple, no kids in home	x	x	x	x	x	x	x	x	→

the viewers in the household; the result is that none of the seven tests result in a channel change. For case f), the user indicated that one or more viewers was male, but no answer was given to the question "Are you working?". As a result, neither the test for "working males" nor "non-working males" results in a channel change for that unit.

References

1. ANSI/EIA/CEA-608-B, "Line 21 Data Services," 2001.

14

PSIP Expandability

At the time the A/65 Standard was first finalized and released in December of 1997, the designers were certain of one thing: while the initial specification was complete, the work would continue. It was clear that PSIP would be expanded in the years to come, and would become an important part of developments such as data broadcasting, interactive television, downloadable applications, and advanced formats for electronic program guide data. In this chapter we describe the concepts and data structures that enable the A/65 Standard to grow and change to accommodate a wide variety of future developments.

It is always possible for a standards-developing organization like ATSC to simply revise and re-issue a standard. The challenge is to design a protocol such that enhancements, additions, updates, and changes do not cause existing equipment to malfunction. We say that "backwards compatibility" must be maintained whenever an update to the protocol is issued.

An example of such an update is the Directed Channel Change functionality added to A/65 Revision A during 2000 (and refined in early 2002). DCC involved the definition of table types not previously specified in A/65. Table ID values 0xD3 and 0xD4 had been identified as "reserved for future ATSC use." Digital Television receivers had been designed and fielded to the original A/65 specification at the time DCC was defined. The new tables are backwards compatible with the previous version of PSIP because receivers are able to (in fact, they are required to) disregard table sections with unrecognized or unsupported values of table ID.

Whenever standards makers look at expanding or updating a protocol like PSIP, they always ask themselves, "How would existing equipment respond to this change?" If the answer is anything other than "no change to current behavior," they must take a hard look at whether or not the change is feasible to make.

There may be times when a manufacturer fields a consumer product that is not adequately tested for requirements related to protocol extensibility. There are many ways that shortcuts or simplifications in a product's software or firmware design can lead to problems. Understanding the concepts in this chapter can aid in the development of a design and test regimen that can help avoid difficulties that could otherwise occur as protocols evolve after a product is shipped.

There is a famous example of a television product that had trouble when broadcasters began to employ a new feature of the closed captioning protocol, EIA-608[1]. This product had difficulty synchronizing the caption extraction process when both field-1 and field-2 captions began to be transmitted. The manufacturer complained to the broadcast community that they should quit transmitting the new VBI data because it was causing problems with their receivers. Ultimately, the broadcasters refused to stop using the feature, and affected consumers had to learn to live with the problem or take their televisions in for an upgrade/repair.

PSIP, like other related digital television protocols, can be expanded and updated in a variety of ways. We take a look at each area in turn.

Reserved Fields

A quick look at the PSIP tables and descriptors shows a generous sprinkling of reserved fields of various bit lengths. Prior to the time a reserved field is given a meaning (and is therefore no longer "reserved"), the specification requires generating equipment to set all of the bits in the field to 1. One might ask, why not use zeros instead? Actually, using 1's instead of zeros helps designers uncover flaws in software. Consider the simple data structure shown here in Table 14.1.

TABLE 14.1 Reserved Fields Example

Field Name	Number of Bits	Field Value	Description
some_data_structure() {			Start of a data structure.
reserved	3	111b	Reserved fields always contain 1.
foo	5		Some 5-bit variable.
...			

In this example, the variable named "foo" is positioned in the least-significant 5 bits of the first byte of the data structure. A 3-bit reserved field occupies the upper (most-significant) three bits of that first byte. When parsing this data structure, a programmer will typically need to extract foo into a separate variable. Typically, a bit mask is used to separate the bits in the byte that are part of the variable of interest from those that are not part of it. In this example, a mask value of 0x1F would be appropriate. The first byte of the data structure ANDed with the 0x1F mask would yield just the bits of foo.

Let's say the programmer made a mistake with the value used as the mask, and instead of five bits (0x1F), used six instead (0x3F). If the reserved bits were zeros, testing would not reveal a problem! But if ones are used instead, the value of foo is out of range and is easily detected.

Although less likely, it is possible to design flawed software that works *unless* a reserved bit is set. It is probably a good idea for software testers to make sure all values of reserved bits are dutifully ignored. PSIP generating equipment or test sequences can be designed to set all reserved bits to ones, set them all to zeros, or to randomize them. The goal in each case is the same: to verify that a given design properly disregards reserved fields.

Reserved Table IDs

Perhaps the most common method for expanding a protocol such as PSIP is to introduce table sections identified with a previously-reserved value of table ID. The assignment of new Table ID values is a function managed by the ATSC Code Points Registrar. ATSC maintains a "code points" registry containing not only a list of currently assigned table ID values, but lists of other items as well.

The requirement on receiving devices with regard to table_id is simple: the table_id value must be checked as part of the table section parsing process. If the value of table_id represents an unrecognized or unsupported table type, the section must be discarded.

You might ask: How could a software design appear to function if the table_id field was *not* validated? Consider that a firmware designer might use the following flawed logic when writing code to process the PMT: the Program Association Table in PID 0 gives me the PID value for each PMT in the Transport Stream. Therefore, if I look at transport stream packets with any of the PID values given by the PAT, any table section I find there must be a PMT. So I don't need to look at the table_id field; I simply assume it must be a PMT.

The flaw lies in the assumption that only PMT sections can appear in transport packets identified with PID values given in the PAT. In fact, in cable applications today, table types other than the PMT are sent in the same PID as the PMT. Furthermore, ATSC could define new tables at some future time that are designed to be transported in TS packets with the same PID value as the PMT. For a fielded product that did not verify the value of table_id, these new tables would cause serious trouble. Receivers are therefore expected to use this rule: "never assume, by context or any other means, that a received table section is a certain type of table without inspecting the table_id field."

Reserved Table ID Extensions

The 16-bit field occupying the table_id_extension portion of the MPEG-2 private section header is fully specified for the table types defined in the PSIP Standard. For some types of tables, any value in the 16-bit range is valid, while for other types only certain values are allowed or defined.

Following the design approach and philosophy of MPEG-2 *Systems*, the function of table_id_extension is to allow different instances of tables with common table ID values to be transported in TS packets with common PID values. Saying it another way, table_id_extension allows one to distinguish between two tables, both with the same table ID but delivering different data, when both are carried in the same PID.

If a given type of table is currently specified with the table_id_extension field set to zero, what should a receiver do if it sees a table with that same table ID, in the PID where that type of table is expected, but with a non-zero table_id_extension? It should discard it as being an instance of this type of table that is not supported by this version of the firmware.

Strictly speaking, it should be possible then for ATSC to standardize new tables with values of table_id_extension that were previously not defined. Whether or not this is practical goes back to the issue we raised at the introduction to this chapter: Would such a change break existing receivers? In this particular example, the present standards do not warn implementers to validate the table_id_extension fields for table types like the MGT, where it is specified to be zero. So it is likely a second MGT instance with non-zero table_id_extension could never be carried in the SI base_PID, 0x1FFB. Proposed changes are looked at on a case-by-case basis.

Let's look closely at the one type of PSIP table that has a reserved field within the 16 bits of table_id_extension: the Rating Region Table. In the case of the RRT, those 16 bits are defined as 8 reserved bits followed by the 8-bit rating_region field. Since reserved bits must be ignored by receivers, that 8-bit reserved field is not used by current receivers to distinguish between different RRT instances. Only the 8-bit rating region field can do that. Those reserved bits cannot be used to create new RRT instances, because such use would not be backwards-compatible with existing receivers.

In conclusion, receiver designers should always check the non-reserved parts of the table_id_extension field (or whatever those 16 bits are named for the particular type of table) and discard the table section if they do not correspond to a valid value in the version of the Standard to which one is building.

Reserved Descriptor Tags

Descriptors may appear in a variety of locations within PSIP table sections. There are various rules governing the use of descriptors, the conditions under which they must appear, and the types of tables where they may appear. Designers must accommodate descriptors that may be present in any location in a table section where descriptors are allowed.

An example of flawed logic that can cause trouble is to say "The only descriptor defined for use in this location is X, therefore any descriptor we find in this location must be X." If this logic is followed, the descriptor could be processed without looking at the descriptor_tag and problems would occur later if ATSC extends the protocol to allow other descriptors to appear in that location.

Another example is: "The only descriptor we support in this location is X, therefore if a descriptor is found, it must be X." This logic leads to similar problems where one type of descriptor is interpreted as another.

It may be that a receiver design supports one type of descriptor in a certain location. It may also be that a certain version of the Standard only allows that one type of descriptor in that location. The receiver design may inspect the descriptor_tag, as it should, but not allow for the fact that a new descriptor could appear in the same location in the table *in front of* the supported one. It could, for example, simply discard all descriptors if the first one wasn't a recognized type. Software designs should not assume anything about the order in which descriptors appear or how many descriptors might be present in a given location.

The rule of thumb is, therefore, to always process descriptor_tag bytes (skipping unrecognized or unsupported types), and to always loop through all of the descriptors that might be present to find those that are supported.

Reserved Stream Types

Stream type codes primarily appear in the Program Map Table, but they also appear in the PSIP Service Location Descriptor. Stream type codes defined for terrestrial broadcast or cable use are derived from those defined in MPEG-2 *Systems*, with standards such as ATSC and DVB extending the MPEG definition with new values. We tabulated currently defined stream type code values in Table 3.13 (see "Stream Type Codes" on page 71).

In general, the stream type code tells a receiver what kind of decoder is needed to decode the content to be found in the TS packets with the PID value associated with that program element. By checking the stream_type code associated with a certain program element, a receiver can know that one of the following two statements is true for this component:

1. no decoder for this stream_type is available, hence this program element must be discarded; or

2. a decoder is available that may be able to handle content identified with this stream_type.

As an example of the latter case, one might have an MPEG-2 video decoder capable of handling Main Profile at Main Level content (standard definition television). Such content is identified with stream_type code 0x02, ISO/IEC 13818-2 Video. But this decoder cannot handle Main Profile at High Level content (HD television), which also uses the stream_type code value 0x02. To find out whether the decoder can handle a certain stream_type 0x02 content, a receiver can look for a descriptor that might give a clue (in this case it could be a Video Stream Descriptor), or it can begin parsing the stream contents itself to see how far it can get. For this video example, the video syntax itself includes the profile_and_level_indication field.

OK, having said all that, what about stream_type codes specified in the current standards as "reserved"? As we said, a receiver must disregard any stream_type code seen in the PMT for which no decoder is available. In this way, ATSC can define new types of streams in the future (and the decoders needed to decode and present content formatted for that stream type) and identify the new types of streams with previously-reserved values for stream_type. Pre-existing receivers are expected to dutifully ignore them.

Reserved PIDs

When we looked at the MPEG-2 Transport Stream in Chapter 3, we listed the allocations and the reserved values and ranges of the 13-bit Transport Stream Packet Identifier field. We saw, for example, that MPEG has reserved PID values in the 0x0000 to 0x000F range and that DVB has defined SI tables in the 0x0010 to 0x001F range. We indicated that PID values 0x0050 through 0x1FEF are available for general use. In terms of reserved values, ATSC has set aside PID values 0x1FF0 through 0x1FFA and 0x1FFC through 0x1FFE for possible future use.

What do reserved PID values mean in terms of protocol expandability? It means that in the future, a Transport Stream could come into a receiver with TS packets identified with one of these reserved PID values. What does a receiver designer need to do now to ensure that their design does not encounter problems if that happens? A receiver built to the current standards has no reason to extract TS packets from the multiplex identified with those reserved PID values, so it's hard to see how they would not be disregarded, as they should be. In any case, designers should be aware that the Standard defines these PID values as reserved for future use and should make no assumptions that no TS packets will ever be encountered that are identified with those values.

Length Extensions

In many cases, the lengths of data structures defined in the current standards could be extended in future releases of the standards by adding new fields at their ends. Here, we take a look at some examples of how this works.

Extensions to the lengths of table sections

Many of the PSIP tables are designed so that new fields could be added at a future time to the end of the table section structure. An example is the MGT, where the last fields preceding the CRC are descriptors_length followed by a number of descriptors. In parsing the MGT section, one knows when the end of the last of the descriptors has been reached because exactly the number of bytes indicated in descriptors_length will have been traversed. Therefore, an update to the Standard could add some new fields to the MGT just after that descriptor loop. In addition to the MGT, the types of tables in which length extension is possible include the TVCT, CVCT, RRT, EIT, DCCT and DCCSCT.

Such an extension to any of these types of tables should not break backwards compatibility. Note that receivers are expected to use the section_length field to determine the section length (no surprise here). How else could the section length be determined? Or, asking the question another way, is there some method besides counting the section_length number of bytes beyond the section_length field to find the end of the section? The answer must be no.

For a table like the MGT section, then, one must not assume that the CRC is the field following the last descriptor. Instead, one must assume that no other known data fields are present beyond that point. As mentioned, the location of the CRC must be determined by the knowledge of which byte is the last in the section, which in turn must be determined by section_length.

To test to make sure a receiver can handle such table section length extensions, receiver designers can create table sections that include dummy data in front of the CRC. They then can test to make sure that their design dutifully ignores it, and that its presence does not disrupt regular processing.

Extensions to descriptor length

The definition of all currently-defined descriptors can be expanded in the future to add one or more new fields. To ensure compatibility with such a protocol update, receiver designers must always process a descriptor's descriptor_length field. After parsing the last supported field, one can discard any additional bytes in a descriptor by resuming the parsing operation at a point descriptor_length bytes beyond the byte just following the descriptor_length field itself.

Appearance of Descriptors

Another way the ATSC SI standard can be extended in the future is to define a meaning for the presence of a certain descriptor in a syntactic position for which no prior definition existed. For example, the current standards do not allow a Component Name Descriptor to appear in an EIT section. Receivers built to the current standards are expected to ignore descriptors appearing in any location disallowed by these standards.

References

1. ANSI/EIA/CEA-608-B, "Line 21 Data Services," American National Standards Institute, Electronics Industries Alliance and Consumer Electronics Association, 2001.

Private Data

As we have seen, "private" is a relative term. From the point of view of MPEG-2 *Systems*, "user private" refers to a semantic structure not defined within the scope of the MPEG standards. In addition to not being defined in the current MPEG standards, when a range of values is designated "user private," MPEG is indicating that these values will not be assigned in future revisions or extensions to the MPEG standards. Regional standards such as DVB and the family of ATSC standards have been built by extending MPEG into the MPEG user private ranges for table ID values, stream type assignments, descriptor tag values, and other semantic structures.

Within the ATSC standards, the term "user private" is used as well in an analogous way. While ATSC standardizes values in the MPEG user private ranges, ATSC also sets aside portions of these ranges as "ATSC user private." ATSC has pledged not to assign values in the ATSC user private ranges in future versions of the Standard, so a private party is free to make assignments in these ranges without the worry that a future extension will assign any of these values.

Collision avoidance

Data provided by a private company or consortium may be present in a Transport Stream that is fully compliant with the ATSC or SCTE standards. In fact several pieces of private data, each provided by a different private entity, may be present in the same Transport Stream. Each private party is free to choose among the values in the user private range appropriate for the type of data being sent. It is not possible to coordinate the values that might be chosen, so it is entirely possible that Entity A and Entity B might use the same value for a user private stream type or table ID. If Entity A and Entity B both decide to use stream_type value 0xC4 for example, it might be said a "collision" of stream type values has occurred. What's to prevent such collisions? How can a receiving device make sense of user private data and, for example, find Entity A's data while rejecting that of Entity B?

For reliable operation, when processing any given private data element it must be possible for the receiving device to determine the identity of the private entity that supplied it. The digital television standards provide several mechanisms for such determination. We look at these mechanisms in this chapter.

Multi-standard receivers

From the point of view of MPEG user-private, ATSC might be thought of as Entity A and DVB as Entity B. For the MPEG user private data that has been defined and standardized by ATSC, the receiver must determine that the Transport Stream is ATSC compliant and that all of the ATSC values will be recognized. Receiving devices built for an ATSC country such as the United States or Korea can be built to simply assume that a Transport Stream is ATSC-compliant and to act accordingly. Or, a more cautious design could look for PSIP data in Transport Stream packets identified with PID value 0x1FFB. If a System Time Table and a Master Guide Table were to be found, chances are very good the TS is ATSC and/or SCTE compliant.

If one wishes to build a multi-standard receiver, such as one that can process either DVB or ATSC SI, one can implement an algorithm to determine whether any received Transport Stream is ATSC or DVB compliant. One helpful test is to check for SI tables on the well-known PID values associated with each of the standards. For example, one could look for the ATSC Master Guide Table in PID 0x1FFB or for the DVB Network Information Table in PID 0x0010.

For data that is user private from the point of view of the regional standard, such as ATSC user private data, how does one determine the entity that has defined that data? One useful mechanism is defined in MPEG-2 *Systems*—the MPEG-2 Registration Descriptor. We looked at the structure and semantics of the MRD in Chapter 3 (see "MPEG-2 Registration Descriptor (MRD)" on page 81). Let's look at how the MRD is used to identify the supplier of user private tables, stream types, and other syntactic elements.

User Private Tables

From the MPEG-2 point of view, table ID values 0x00 through 0x3F are either already defined in the MPEG standards or are reserved for future assignment by MPEG. MPEG-2 *Systems* indicates that table ID values in the 0x40 through 0xFE range are user private. Starting with the MPEG ranges, ATSC has mapped out the 8-bit range for the table_id value as well, stating that table ID values in the 0x40 through 0xBF range are user private to ATSC. ATSC assigns or reserves for its future use table ID values in the 0xC0 through 0xFE range.

Figure 15.1 shows the ranges of table_id values as defined in MPEG-2 *Systems*, ATSC, and DVB.

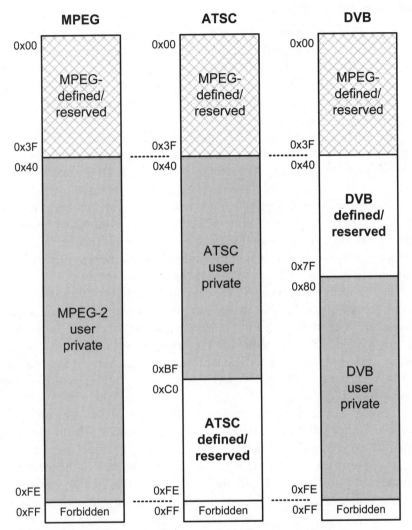

Figure 15.1 MPEG, ATSC, and DVB User Private Table ID Ranges

Note that ATSC standardizes table ID values in the 0xC0 through 0xFE range and leaves the range 0x40 thorough 0xBF as user private. DVB standardizes table ID values 0x40 through 0x7F and leaves 0x80 through 0xFE user private.

How would a private party include table sections specific to its private application in an ATSC-compliant Transport Stream? There are two ways, depending upon the application. In one case the table sections are linked into PSIP data while in another case, the table sections could be considered to be an element of a program.

MGT references to private tables

If the tables were related to system or service information or were applicable network-wide or to the full Transport Stream rather than to a specific service, the private entity defining them might want to use the features of the PSIP Master Guide Table to announce the size and version of the tables and to indicate the PID values used to transport them. A/65 allows the MGT to point to ATSC user private table sections to address just this need.

Recall that the way the MGT indicates the types of the tables it references is through the 16-bit table_type field. One of the ranges defined for table_type is an ATSC user private range, 0x0400 through 0x0FFF. To include references to ATSC user private tables in the MGT, the entity defining the private table uses a table_type value in this private range and includes an appropriate MPEG-2 Registration Descriptor in the iteration of the "for" loop in the MGT describing that table. In the MPEG-2 Registration Descriptor, the format_identifier field uniquely identifies the private entity. This is entirely analogous to the way private stream types are identified with the MRD in a PMT section.

Figure 15.2 illustrates an MGT referencing an Event Information Table, a Terrestrial Virtual Channel Table, and a user private table. As shown, for the user private table, an MRD is included in the iteration of the "for" loop in the MGT in which table_type indicates a value in the user_private range. In this example, the table_type is 0x0401. The MRD indicates the private table has been defined by an entity registered with the MPEG Registration Authority with format_identifier value 0x00442201.

A receiving device may or may not be aware of the company or organization whose format_identifier is 0x00442201. If the receiving device recognizes this format ID as being one that may be supported, it may collect the table sections found in transport packets with the PID value given (0xE000 in the example), and try to parse the table sections. If the table ID value 0x8E associated with this format_identifier code is also recognized, the table can be processed. If not, it is discarded.

Private tables as program elements

The second way user private table sections can be included in a Transport Stream is when the table sections are associated with a program as a program element. In this method, the Program Map Table indicates the PID value used to transport the private table sections. Again, an MPEG-2 Registration Descriptor is used to identify the entity that has defined the user private table ID. Typically in this case the stream_type code in the Program Map Table indicates a type 0x05 stream, meaning that it is formatted as long- or short-form private sections defined in accordance with MPEG-2 *Systems*.

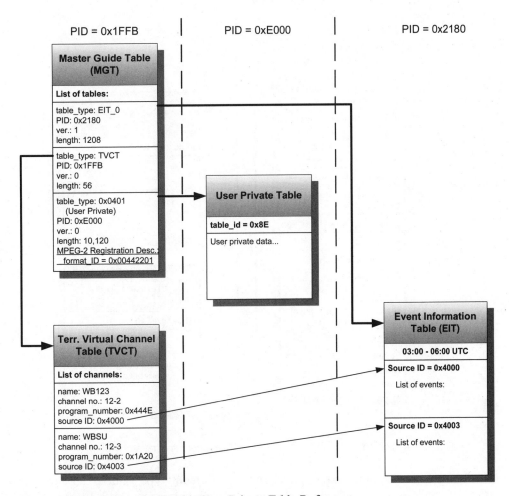

Figure 15.2 Master Guide Table User Private Table Reference

Figure 15.3 illustrates an example Program Map Table section in which the program consists of three elements: an MPEG-2 video Elementary Stream, a Dolby AC-3 audio stream, and a private program element transported in the form of MPEG-2 private table sections.

This PMT includes three MPEG-2 Registration Descriptors. The program-level descriptor has a format_identifier of "GA94" to identify the program as complying with the ATSC Digital Television Standard. Inside the description of the AC-3 audio stream component may be found an MRD indicating that the type 0x81 stream conforms to the ATSC A/52 standard. This MRD carries a format_identifier code value of "AC-3."

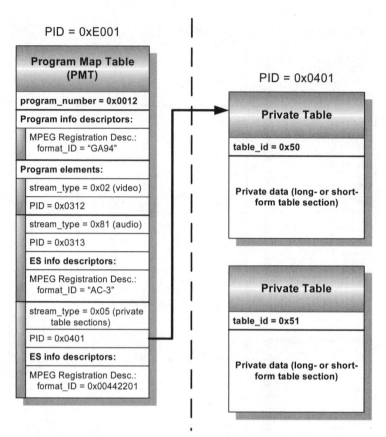

Figure 15.3 Program Map Table References to Private Tables

The third program element has a stream_type value 0x05 and indicates that the private table sections are transported in TS packets with PID value 0x0401. To use stream_type value 0x05, an MPEG-2 Registration Descriptor must be present in the Elementary Stream information loop for this program element. In the example, the format_identifier code in that MRD is 0x00442201. Within the TS packets identified with PID value 0x0401 may be found two private tables, one with table_id value 0x50 and the other with table_id 0x51.

It is important to note that the only receiver implementations that are able to parse and thereby make use of the table sections found in TS packets with PID 0x0401 are those that recognize the format_identifier code 0x00442201. That value may be registered to a private company or organization or it could be assigned to a consortium comprising a cooperative group of companies.

In some instances, for a manufacturer to be able to build a receiving device that can use the private tables, a technology license must be signed. The terms of that license would include the detailed information necessary to implement the technology, and that information includes the syntax and semantics of the private tables (in the example, the tables identified with table ID values 0x50 and 0x51).

The example in Figure 15.3 also provides an opportunity to say something about the values of the table_id field that could be assigned to the private table types. In the example, table_id values 0x50 and 0x51 are used. These values correspond to tables defined in the European DVB Service Information Standard, yet in this application, the private entity identified with format_identifier 0x00442201 can associate the values with an entirely different syntax and semantics at the discretion of the registered private entity. This points out the fact that the table_id field is scoped at the PID level, because references to the PID value of the TS packets carrying the sections of a table establish a clearly identified context. By the time we get to the table sections carried in PID 0x0401, we already know they are the ones defined by the entity registered with the ATSC, and identified with format_identifier 0x00442201 and no others.

In spite of the fact that one could distinguish by context between instances of tables in the multiplex that had identical values of table_id, standards organizations world-wide usually try to avoid re-using each other's values. Such a practice helps keep clear the global view of System/Service Information and makes it a bit more straightforward to build a multi-standard receiver, should anyone wish to do so.

User Private Stream Types

As mentioned earlier, a Program Map Table section can include a privately defined program element that consists of long- or short-form MPEG-2 table sections. It is also possible to use the PMT to define a program that includes a privately defined element that doesn't consist of table sections, but some other structure. If that structure conforms to the Packetized Elementary Stream structure defined in MPEG-2 *Systems*, stream_type value 0x06 is used. If the structure does not conform to any of the types of streams defined in MPEG-2 *Systems*, a stream_type value in the ATSC user private range (0xC4 to 0xFF) may be used. In any case, an MPEG-2 Registration Descriptor must be present in the Elementary Stream information loop in the PMT section to identify the private entity. Receiving devices must disregard any privately defined program elements found in the structure of a program unless they recognize and support the private entity identified in the MPEG-2 Registration Descriptor accompanying that private element.

Let's say the National Association of Broadcasters (NAB) defined a stream type with a stream_type code 0xD0 for a type of data used by network affiliates for some kind of automated control. Let's also say that NAB registered a format_identifier code of 0x00000123 with SMPTE-RA (the ISO/IEC-designated Registration Authority for MPEG). If a consumer receiver were to encounter a PMT section including a stream_type value of 0xD0, it would likely disregard that program element because it was not prepared to process a stream of that type. A piece of commercial receiving equipment, on the other hand, could be designed to recognize and process the NAB control data stream. It would find the 0xD0 stream_type code and then look for and process the MRD. If the format_identifier code of 0x00000123 were to be seen, it would be assured that this stream could be properly decoded and processed, otherwise it would discard the stream.

Other Syntactic Elements

Private data can appear in various places in the Transport Stream. For example, the syntax of the TS packet adaptation field can accommodate one or more private data bytes. In this way, private data can be present within standard stream types such as MPEG-2 video or AC-3 audio. As with private-range stream types, an MRD in the PMT inner loop indicates the private entity to be associated with private data bytes in the associated stream.

Syntactic structures defined in ATSC standards such as the A/90 Data Broadcast Standard also accommodate private data. Typically, the MRD is again used to identify the entity responsible for the private data portions of a given program element.

User Private Descriptors

ATSC and SCTE have declared that the range of descriptor_tags 0xF0 through 0xFE are to be considered user-private from the point of view of ATSC/SCTE standards. That means these standards-defining organizations have vowed not to define new descriptors using tag values in that range.

In 2002 ATSC, however, reviewed the question of private tag-range descriptors to address some identified concerns. As mentioned, it must be possible for receiving equipment to unambiguously identify the private entity that has supplied a given private data element. The expectation had been that the MPEG-2 Registration Descriptor (MRD) would be used for this purpose for user-private descriptors, but the detailed rules had not been written down. After lengthy analysis, ATSC decided that all descriptor-based private information must be placed into a new type of descriptor called the ATSC Private Information Descriptor (APID) (see "ATSC Private Information Descriptor" on page 224). The structure of the APID includes the same 32-bit format_identifier as is found in the MRD.

More than one APID can appear in a given descriptor loop, making it possible for private data from multiple private sources to be associated with the same item (program, program element, table, etc.).

PSIP and the Digital Cable-Ready Receiver

In response to the Telecommunications Act of 1996, the FCC stated that it would work with standards-making organizations to enable any manufacturer to design and build devices that could connect to cable systems and that would be able to receive video programming "and other services" from cable Multiple System Operators. Section 629[1] is titled "Competitive Availability of Navigation Devices." The introductory portion is quoted here:

> The Commission shall, in consultation with appropriate industry standard-setting organizations, adopt regulations to assure the commercial availability, to consumers of multichannel video programming and other services offered over multichannel video programming systems, of converter boxes, interactive communications equipment, and other equipment used by consumers to access multichannel video programming and other services offered over multichannel video programming systems, from manufacturers, retailers, and other vendors not affiliated with any multichannel video programming distributor.

Types of devices covered by Section 629, also called the "Navigation Devices Order," include televisions, VCRs, cable set-top boxes, personal computers and cable modems. An important aspect of the order is related to the conditional access security employed by cable operators to protect their services from theft: it states that manufacturers can build navigation devices for retail sale that do not include built-in conditional access (CA) functions. Instead, these devices would support cable operator-supplied Point of Deployment (POD) modules that will descramble any digitally scrambled service on the cable network. FCC rules required cable Multiple System Operators (MSOs) to be ready to supply POD modules by July 1, 2000. It also requires them to stop, after January 1, 2005, providing new set-top boxes with integrated conditional access functions. In other words, all cable boxes built after 2005 must use the POD module as the CA element, even those sold to the cable MSO.

One of the provisions of Section 629 called for the nation's cable MSOs to provide the information needed to allow anyone to manufacture and sell at retail devices that would work on their systems. The rules recognized the MSOs need to create requirements and standards that would protect them from potential harm that could be done to their network by devices consumers might buy and attach to it. In many ways, these provisions for opening up cable to retail devices mirrored the provisions that opened up the telephone network, allowing retail devices to connect directly to the telephone line.

So the 1996 Act, especially Section 629, set in motion a flurry of activity in the Digital Video Subcommittee (DVS) of the Society of Cable Telecommunications Engineers (SCTE), in the Consumer Electronics Association (CEA) (part of the Electronic Industries Alliance), and in CableLabs, the cable MSO's research consortium. Cable MSOs, represented and supported by CableLabs, were in a tricky position: they needed to see to it that industry standards came together to fulfill the FCC's navigation devices rulings, yet they could not afford to give Consumer Electronics (CE) companies free rein to define them. On the other hand, it was the CE side that would be actually trying to build and market the retail navigation devices, so the standards had to be fundamentally acceptable to them.

In this chapter we review the standards designed to enable any manufacturer to build a device that can be directly connected to cable to access the digital audio/video services offered by the cable service provider. We start with a discussion of OpenCable™, an initiative championed by CableLabs, and then look at the some of the applicable standards that have come from SCTE and CEA. In summary, we focus on those aspects of the current cable standards involving PSIP and MPEG Program Specific Information to get ready for the discussion of the cable out-of-band channel to follow in Chapter 17.

OpenCable™

CableLabs' approach was to develop a set of specifications for interoperability of cable-ready devices by coordinating the work with engineers from the cable MSOs and from the two dominant cable equipment vendors, General Instrument (GI, now Motorola Broadband) and Scientific-Atlanta (S-A). This effort was called "Open-Cable™." The term "open" had a nice marketing ring to it, but not everyone agreed that the process was truly open in the sense of open standards-making. It was "open" in the sense that S-A and GI were required to put some of their proprietary technologies on the table, and the net effect was intended to open the market and give MSOs the ability to purchase equipment on a more competitive basis. For more information on OpenCable, visit their website at http://www.opencable.com. An excellent overview of OpenCable can be found in Michael Adams' book *OpenCable Architecture (Fundamentals)*[2].

Figure 16.1 illustrates how the OpenCable architectural model labels the various interfaces in the cable system for the purposes of defining their specifications.

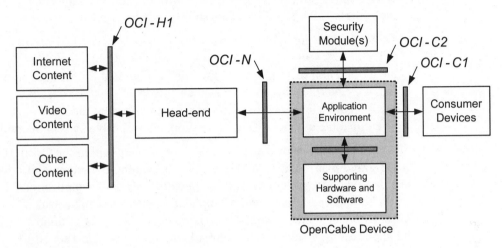

Figure 16.1 OpenCable Interfaces

The four major OpenCable interfaces indicated in the Figure are:

- **OCI-H1**: this set of interfaces is designed to specify the way content is supplied to any OpenCable-compliant headend. Types of data that can be provided to the cable headend include audio/video content, Internet content, and other content including downloadable applications and data for interactive television. As of this writing, CableLabs has not published any specifications in the HCI-H1 family. It may be that the de-facto standards already in use are sufficient.

- **OCI-N**: the OpenCable Network interface defines the way digital signals are modulated on cable, including what frequency ranges and plans are to be used, and it gives detailed characteristics of the RF parameters describing signal amplitude and purity. It specifies that audio/video services are to be transported in MPEG-2-compliant Transport Streams, and states requirements for transport, Service Information and MPEG Program Specific Information, audio and video compression coding, and closed captioning. OCI-N also describes characteristics of the upstream and downstream out-of-band channel, citing the use of one of two SCTE standards—one supplied by GI/Motorola and the other by Scientific Atlanta.

- **OCI-C1**: CableLabs recognized the fact that digital cable set-top boxes must interoperate with consumer devices including digital television sets and recording devices such as VCRs and hard-disk video recorders. The OCI-C1 interface covers the interface between an OpenCable-compliant set-top box and such consumer devices. CableLabs specified a digital interface based on IEEE 1394 *Standard for a High Performance Serial Bus*[3], aligning its approach with standards work coming from CEA committees working on DTV receiver interfaces. We look at this 1394-based interface shortly.

- **OCI-C2**: This interface specifies the protocols needed to separate the conditional access security functions in a retail OpenCable-compliant device. When the term "Point of Deployment module" was coined for the security module, the OCI-C2 interface became known as the POD-Host Interface. A cable-ready retail digital television or set-top box must implement the POD-Host Interface to be compatible with a POD module supplied by a cable system operator.

Figure 16.1 also shows another interface, located within the OpenCable Device. In the grand plan outlined by CableLabs, all OpenCable-compliant devices would be able to support cable- operator-supplied downloadable applications. Each device would be required to provide an "application environment" consisting of hardware and software elements, coupled with a transport mechanism suitable for delivery of the applications. CableLabs has embarked upon a specification effort called Open-Cable Application Platform, or OCAP, bringing in a variety of companies to supply technology and expertise. Prominent companies in the OCAP effort include Microsoft, Sun Microsystems, and Liberate with other contributors including Canal Plus, OpenTV, and PowerTV. Draft specifications were circulated for general comment during 2001 and in December, OCAP version 1.0 was published. Version 2.0 followed in April 2002.

To gain the acceptance of other stakeholders in the cable arena (the CE companies, content providers, broadcasters, and the non-traditional equipment vendors—new players wishing to get into the game), CableLabs followed an approach in which most of the key OpenCable specifications were submitted to the SCTE Digital Video Subcommittee upon their completion. SCTE DVS is an ANSI-accredited standards-defining organization, and documents adopted there can be submitted to the International Telecommunications Union (ITU) as candidates for international standardization.

Figure 16.2 shows the three interfaces surrounding the OpenCable Device (OCI-N, OCI-C1, and OCI-C2), and indicates the primary SCTE DVS standards that reflect the respective OpenCable specification documents. EIA/CEA standards relevant to the OpenCable interfaces are listed in the Figure as well.

OCI-N and the Cable Network Interface

Of all the OpenCable interfaces, the one that initially generated the most controversy and garnered the most attention was OCI-N. It is possible to specify the cable network interface such that lower-cost terminal devices function properly, but that

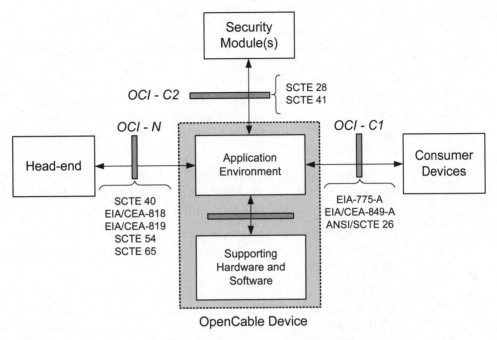

Figure 16.2 Relationship of OpenCable Interfaces to SCTE DVS and EIA Standards

approach requires cable operators to maintain all of the cable plants in such a way that all of the signals presented to those terminal devices are within strict tolerances for signal strength and purity. Practically speaking, such an approach can be prohibitively expensive to the cable operators. Another approach is to define the interface such that a wide range of cable signal amplitudes and purities are acceptable. But that method imposes a costly burden on the terminal devices that must produce high quality audio and video from weaker and noisier carriers, some of which may be polluted with interference from neighboring channels or with ingress from RF sources outside the cable plant. Clearly, a trade-off is needed between these two extremes.

In the early months of 2000, representatives from the cable and the Consumer Electronics manufacturers communities engaged in a series of intense negotiations aimed at reaching agreement on the network interface parameters described in OCI-N. Some CE companies felt that the CableLabs specification was asking for higher performance from a retail cable-ready device than from the set-top boxes owned and operated by the cable MSOs themselves. The tug-of-war involved experts from both sides, who argued the technical merits of their respective cases. On February 23, 2000, the National Cable Television Association (now called the National Cable Telecommunications Association) issued a joint announcement with the Consumer

Electronics Association to say that an agreement had been reached that would allow digital television sets and digital cable systems to work together.

CableLabs updated OCI-N to reflect the newly-formed agreements and submitted it to SCTE as DVS 313[4] in March, 2000. DVS 313 was approved by the Engineering Committee of SCTE and is now known as SCTE 40[4]. We won't go into the details of the RF parameters or of all the cable standards encompassed in SCTE 40, but we give here an overview that provides the background we need in order to discuss how PSIP fits into the equation. SCTE 40[4] includes these PSIP-related standards in a normative fashion:

- **SCTE 65**[5]—defines requirements for out-of-band System/Service Information, including formats for electronic program guide data suitable for out-of-band delivery. We review out-of-band SI in Chapter 17.

- **EIA 814/SCTE 18**[6]—defines requirements related to the Emergency Alert System for cable systems and connected compatible devices. We discuss the EAS system for cable in Chapter 18.

- **SCTE 54**[7]—specifies transport-specific requirements for cable. Although the basic transport method is based on the Transport Stream defined in ISO/IEC 13818-1 MPEG-2 *Systems*, the SCTE 54 document specifies certain constraints against MPEG and defines some "private" extensions to MPEG. We review the aspects of SCTE 54 that pertain to SI and PSI data below.

- **ATSC A/65**—SCTE 40[4] specifies PSIP itself in a normative fashion. We look at the details below in our discussion of the SCTE 54 cable transport specification.

Figure 16.2 lists the following two EIA standards alongside SCTE 40[4]:

- **EIA-818** Cable Compatibility Requirements[8]. Part I of this document was originally prepared by a group of consumer electronics and cable industry technical representatives. Later it was transferred to the newly-formed EIA R8 committee in the Consumer Electronics Association. R8 was created to deal with cable compatibility issues. Currently it is made up only of consumer electronics manufacturers. Companies representing cable were invited to participate, but they have chosen instead to support the SCTE and CableLabs OpenCable efforts. EIA-818 defines "Minimum requirements for receiver-compatible digital cable TV systems" in Part I, and "Minimum requirements for cable-compatible digital TV receivers" in Part II. Part I echoes requirements found in SCTE 40. Part II does not correspond to any SCTE standards, because none of the OpenCable specifications for receiving devices have been brought to SCTE. With Part II of EIA-819, CEA made a policy statement that CE companies do not accept the notion that the cable MSOs have the right, on their own, to dictate functional and performance requirements for cable-ready devices available for retail sale.

- **EIA-819** Cable Compatibility Requirements for Two-Way Digital Cable TV Systems[9]. EIA R8 has prepared EIA-819 as a companion to EIA-818. It adds requirements related to upstream communications on the out-of-band channel that are needed to support impulse pay-per-view (IPPV) functions in a retail device. Without the return channel, IPPV is not possible, as the DCRR must be able to report purchases back to the cable headend for billing purposes.

OCI-C1 and the Home Digital Network Interface

On the right side of the OpenCable Device in Figure 16.2, a connection to "Consumer Devices" is shown. Clearly, to make any use of any outputs from the digital cable set-top box the consumer has to connect it to devices he or she might own, such as television sets, audio systems, and recording devices. Nearly all set-top boxes today have traditional analog audio/video outputs (either through an RF re-modulator, or via baseband audio and composite or component video, or both). With OCI-C1, however, CableLabs was looking a step towards the future to an all-digital network interface.

With the explosive world-wide growth of the Internet, the power of networking has become evident. CableLabs realized, as have so many others, that any source or destination of audio/video or data within the home should be interconnected in a peer-to-peer network with every other source or destination. That means that all of the television displays in a home could access and receive audio/video from any given cable set top box that might be in the home. Likewise, any television display could access and receive audio/video from any cable, satellite, or terrestrial broadcast receiver, or any personal video recorder resident in the home network. Introduction of digital networking offers the goal that all it takes is one wire per unit to add it to the network bus. The rat's nest of wiring often found behind today's A/V equipment stacks can largely disappear, and with it all the separate "input" and "output" jacks.

By 1998 IEEE 1394 had emerged as the technology most appropriate for delivery of compressed audio and video among devices in a home network, and work began in earnest to define standards suitable for DTV receivers and cable set-top boxes. CableLabs focused on the application of IEEE 1394 to OCI-C1, while the EIA R-4.8 DTV Interfaces subcommittee put their attention towards the DTV receiver. CableLabs generated a specification for the OCI-C1 interface called the Home Digital Network Interface (HDNI) that was submitted in 1998 to SCTE Digital Video Subcommittee as DVS 194. It is now known as ANSI/SCTE 26[10].

The cable industry has pushed to make IEEE 1394 mandatory for cable-compatible DTV receivers in order to ensure that they are capable of supporting a cable-operator supplied set-top box at a future time. In fact, the FCC responded in September 2000 with a Report and Order defining the terms to be used for cable-ready

DTV receivers (FCC-0342[11]), including one variety with an IEEE 1394 interface (see also the EE Times article, "FCC opens the door to 1394 for cable-ready equipment"[12]).

EIA-775-A—DTV 1394 Interface Specification

As mentioned, EIA-775-A[13] was developed by the EIA R-4.8 DTV Interfaces subcommittee and specifies requirements for DTV receivers with IEEE 1394 interfaces. EIA-775-A specifies requirements including:

- the DTV receiver's use of IEEE 1394 interface.

- how compressed digital audio/video are to be transported on the 1394 bus.

- the format and method of delivery of uncompressed graphics to the DTV receiver for overlay onto decoded video.

- how devices on the bus discover each other's capabilities.

- how digital bus connections are to be managed.

Digital set-top boxes, storage devices, and other A/V equipment are able to connect to the DTV display via IEEE 1394 cabling, and to use it to decode and display MPEG-2 video and to overlay bit-mapped graphics supplied by the source device. A cable or satellite set-top may not include an MPEG-2 video decoder capable of handling high-definition formats, but when connected via IEEE 1394 to a DTV receiver, it can send compressed HD MPEG-2 video to the DTV for decoding and display there. This scheme is sometimes called "HD pass-through." Using the bit-mapped graphics capability of EIA-775-A[13], the source device can include channel names and logos, program names, or other graphics.

Even though many source devices will be unable to output content on a digital interface such as IEEE 1394 without some form of copy protection, the EIA-775-A standard does not specify a requirement or a standard method.

SCTE 26 Home Digital Network Interface (HDNI)

The Home Digital Network Interface (HDNI) specified in SCTE 26[10] describes a digital interface between a digital cable set-top box and a digital television based on the IEEE 1394 high-speed serial bus interface protocol. SCTE 26 builds on EIA-775-A[13], adding the requirement for copy protection based on the Digital Transmission Copy Protection (DTCP) standard. DTCP is commonly known as "5C" for the five companies who created it: Sony, Matsushita, Intel, Toshiba, and Hitachi.

Navigating the networked transport stream

Figure 16.3 illustrates a cable set-top box at the top, connected via an IEEE 1394 bus to a DTV receiver below it, where the DTV receiver complies with the EIA-775-A[13] standard and the set-top box complies with SCTE 26[10]. The source device

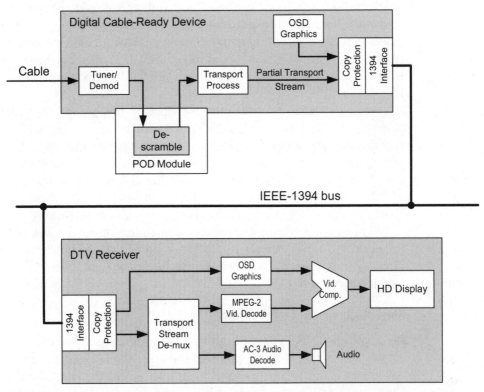

Figure 16.3 Cable Set-top Box Connected via IEEE 1394 to a DTV Receiver

can apply copy protection to the link between itself and the DTV receiver with DTCP, in accordance with SCTE 26. A user interacts with the set-top box to choose a virtual channel for viewing. The Transport Stream carrying that channel is tuned, demodulated, and (if necessary) descrambled. The Transport Stream coming from a source device such as that shown in the Figure is typically processed to delete every MPEG program but one. If it has just one, we call it a single-program Transport Stream (SPTS) and if it carries more than one, a multiple-program Transport Stream.

In the block labeled Transport Process, the full TS is converted to a "partial" Transport Stream. In a partial TS, only those TS packets pertinent to the chosen audio/video service are preserved; all other packets are discarded (deleted) from the

TS. For a single-program Transport Stream, the average bitrate of the partial TS is usually significantly reduced below the bitrate of the full (multiple-program) TS. In a typical example, the full TS may be a 19 Mbps stream carrying five A/V programs. The partial TS can reduce that to below 4 Mbps.

A partial TS preserves packet timing, so an MPEG audio or video decoder is unaffected by the missing packets. The advantage of reducing the average bitrate is so that the isochronous bandwidth needed to transport the partial TS over the 1394 bus is reduced down towards that lower rate. Using this technique, the most efficient use of bus bandwidth is ensured. The IEC standard that defines how an MPEG-2 Transport Stream is carried in IEEE 1394 isochronous stream packets is IEC 61883-4[14].

Figure 16.3 also illustrates the fact that bit-mapped graphics data can travel across the network bus along with compressed audio and video. EIA-775-A specifies that bit-mapped graphics (formatted according to another EIA standard, EIA-799[15]) travel across the bus in accordance with a 1394 Trade Association standard called AV/C Compatible Asynchronous Serial Bus Connections[16]. Whereas the MPEG-2 Transport Stream is delivered in an isochronous (guaranteed bandwidth) channel on the 1394 bus, graphics travel in asynchronous packets.

As the Figure illustrates, packets containing on-screen display graphics are routed in the DTV receiver to the block called "OSD Graphics." They are combined (composited) with video in the block labeled Video Compositor ("Vid. Comp." in the Figure).

Single-program Transport Streams

For applications like HD pass-through, the user navigates among the service offerings by interacting with the cable set-top box. The DTV receiver is used only as a decoder/monitor. When one service has been selected, the set-top box constructs a single-program Transport Stream by:

1. deleting TS packets carrying audio/video or data components corresponding to other MPEG-2 programs (services) in the multiplex.

2. editing the Program Association Table (PAT) carried in TS packets with PID value 0, to delete references to the other programs. This step involves extracting the PAT, modifying it, and re-inserting it. Note that the CRC must be recomputed.

Certain data services or data program components may be relevant to more than one service, so care must be taken if any data services or data stream components are deleted from the multiplex. For the single-program TS case, PSIP data may or may not be included in the TS after processing in the set-top box. Including it, however, is highly recommended because even though it will not be used when watching the program real-time, it can be very helpful if the Transport Stream is time-shifted or recorded for later playback.

Requirements in the EIA-775-A standard state that when a DTV receiver accepts an SPTS from a source device, it must decode and output audio and video without any need for a user to interact with the receiver. It is the responsibility of the receiver to find and parse the PAT, and to follow the link to the PMT where audio and video PID values are found. It may be that the MPEG-2 program includes one video stream but multiple audio tracks. In that case, the DTV receiver may choose a track based on the user's language preference, or may create an on-screen user dialogue to request that a choice be made.

Multiple-program Transport Streams

EIA-775-A states that if the DTV receiver encounters a multiple-program Transport Stream (MPTS), it is expected to allow the user to interact with the receiver to select which program he or she wishes for it to decode and present. It wouldn't make sense to navigate via the cable set-top box to a Transport Stream carrying several video services, but then have to put down that remote control unit and pick up the DTV remote to finish the selection of one program. This is the reason why cable set-top boxes normally supply single-program Transport Streams on the 1394 interface. Another reason is that they typically can only descramble one service at a time.

It is expected that an MPTS could be transported across the network to support applications like timed recording. An application running in the DTV receiver, for example, could control a slaved cable tuner/demodulator/descrambler, and use it to access any TS carried in the cable system. Such a scheme can run into practical limitations, however, since it may not be possible to descramble more than one or two programs in the Transport Stream simultaneously. In fact, using today's POD-Host architecture it is only possible to descramble one at a time.

EIA-849-A

We can envision various audio/video products coming into the retail market and offering IEEE 1394 interfaces (some are available as of this writing). Connectors and cabling are standardized, so it would be possible for a consumer to cable any one of these products to any other, or to wire all of them up to one another daisy-chain fashion. Types of products could include cable, satellite, or terrestrial broadcast set-top receivers, disk or VCR recording devices, camcorders, or even personal computers. But does a given device "play" as expected with the others? Would the DTV receiver be able to make usable pictures and sound from any source device on the bus?

The EIA committee that created the EIA-775-A Standard recognized that it brought together many standards above the IEEE 1394 foundation but it did not state requirements regarding content decoding or copy protection. For example, it stated that transport of MPEG Transport Streams on 1394 required the use of IEC 61883-4[14], but it did not state that a DTV must decode video in a 1080 by 1920 resolution in interlaced format at 24 frames per second. Likewise, no mention was made that the DTV must handle AC-3 audio streams.

EIA R-4.8 addressed this problem by creating EIA/CEA-849-A "Application Profiles for EIA-775-A compliant DTVs"[17]. This document cited specifications relevant to transport, service structure, service information, video and audio compression formats, copy protection and closed captioning for four application areas: terrestrial broadcast, digital cable, digital satellite, and digital camcorders.

One of the goals of the EIA/CEA-849-A[17] Standard was to emphasize the fact that applications developed in recent years must adhere to MPEG standards for transport, service structure, and video compression. Of course Dolby AC-3 audio compression is used in the US in preference to any of the MPEG audio compression methods. To further the Standard's goal, the satellite profile was split into two parts: one part was MPEG-compliant (corresponding to fielded systems from EchoStar Communications and Star Choice in Canada) and the other was not fully MPEG-compliant (DirecTV).

By splitting the satellite definition in this way, EIA/CEA-849-A[17] was able to define a set of requirements called the "MPEG profile." This set comprises the combined requirements needed to support terrestrial broadcast, cable, and MPEG-based satellite. If a DTV receiver is built to support the MPEG profile, it is compatible with terrestrial broadcast and cable set-top boxes, and EchoStar and Star Choice satellite-integrated receiver-decoder boxes that interface to the DTV via IEEE 1394. The Consumer Electronics Association has defined a logo called "DTV Link" that signifies compliance to the MPEG profile of EIA/CEA-849-A[17]. Consumers will be able to recognize whether a certain audio/video source device plays with a given destination device (DTV receiver) by looking for the DTV Link logo on both devices. Program details are available at http://www.ce.org/dtvlink.

OCI-C2 and the POD-Host Interface Specifications

As mentioned, the OCI-C2 interface is commonly known as the POD-Host interface. CableLabs brought the OpenCable specifications for OCI-C2 to SCTE DVS and standardized them as SCTE 28[18] and SCTE 41[19]. SCTE 28 is the primary document defining the interface, and SCTE 41 specifies copy protection for data traveling on the interface.

In Chapter 17, we explore those details of the POD-Host interface relevant to our PSIP discussion: specifically, how the out-of-band channel works to deliver SI data to the cable-ready device.

SCTE 54 Cable Transport Specification

SCTE 54[7] defines transport layer characteristics and requirements for digital cable; it is an important standard for our PSIP discussion as it cites ATSC A/65 as a normative reference and states requirements for system and service information carried in the in-band multiplex.

In-band System/Service Information

One of the most fundamental PSIP-related requirements takes into account the fact that some cable digital signals include free (in the clear) services. Such services do not require conditional access support in the digital cable-ready receiver and hence should be receivable without the need for a POD module to be in place. SCTE 54 states that PSIP data describing any unscrambled programming must be present in the Transport Stream in which that programming is carried. By the same token, if all of the virtual channels carried on a given TS are scrambled, no PSIP data is required by the standards to be carried—only MPEG Program Specific Information is sufficient (PSI is needed for basic MPEG compliance and is never optional).

Even though SCTE 54 states no requirements for in-band SI or EPG data to be present to describe scrambled services, such data may be present anyway. In fact, there has been much uncertainty and controversy surrounding the topic of how and when standardized Electronic Program Guide data might be carried on cable, and one solution is to include it in-band just as it is done for terrestrial broadcasting. In-band delivery of cable EPG data is the subject of the "PSIP Agreement" of February 2000 discussed in Chapter 19.

SCTE 54 states that the following SI tables shall be sent if PSIP is present in the TS:

- the Master Guide Table, describing all PSIP tables in the Transport Stream

- the System Time Table

- the Cable Virtual Channel table, or (if the Transport Stream originated from a terrestrial broadcasting source) the Terrestrial Virtual Channel Table

- the Rating Region Table corresponding to a non-US rating_region if a Content Advisory Descriptor quoting that region is present in a Program Map Table or an Event Information Table. For the US and possessions, rating_region 0x01 is fully defined in EIA/CEA-766-A[20], and because it is defined to be unchangeable, it need not be transmitted.

Maximum cycle times for PSIP tables and packet streams are specified in SCTE 54 as follows:

- the System Time Table must be repeated at a rate not less than once every 10 seconds.

- the Master Guide Table must be repeated at a rate not less than once every 150 milliseconds.

- the Cable Virtual Channel Table must be repeated at a rate not less than once every 400 milliseconds.

- if an RRT is present, it must be repeated no less often than once per minute.

Maximum bitrates are also specified to ensure that no cable operator sends PSIP tables at such a high rate that receivers cannot keep up. The combined total bitrate for all tables in the SI base PID, 0x1FFB, must not exceed 250,000 bps. Likewise all tables carried in any PID associated with EIT or ETT data must not exceed 250,000 bps. Note that because they are transported in TS packets with different PID values, EIT-m tables can use 250,000 bps while EIT-n uses a separate 250,000 bps.

Program Specific Information

SCTE 54 requires compliance with MPEG-2 *Systems* with regard to Program Specific Information (Program Association Table and Program Map Tables) and places constraints against MPEG in several areas:

1. **PMT repetition rate**: Each Program Map Table section must be repeated at a rate of not less than 2.5 repetitions per second (the maximum time between starts is 400 milliseconds).

2. **PAT repetition rate**: The Program Association Table must be repeated at a rate of not less than 10 per second. These repetition rates are set to minimize service acquisition time when PID values are unknown before the Transport Stream is processed.

3. **Adaptation headers**: No TS packet corresponding to a PID referenced by the Program Association Table can include an adaptation header. As we have seen, adaptation headers are not allowed in any packets carrying PSIP tables either. SCTE 54 makes note of one exception: it is allowable for the adaptation header to appear for the purposes of signaling with the discontinuity_indicator that the PMT's version_number may be discontinuous. In practice, this is rarely done, and receiving devices routinely detect any version change by processing the

header of every PMT section that arrives. The same restriction on adaptation headers (and the exception for signaling version number changes) applies to the PAT in PID zero.

4. **Private table sections with the PMT PID**: SCTE 54 states that private table sections may be present in TS packets with the same PID value used to transport sections of the Program Map Table. Any receiving device must therefore be sure to process the table_id fields when collecting table sections from TS packets known to carry PMT sections. This requirement is derived from existing cable practice wherein proprietary data recognized by cable set-top boxes owned by the cable MSO is sent along with the PMT. Such data cannot be interpreted by a retail cable-ready device unless the manufacturer obtains a technology license from that particular MSO and can thereby access the specifications for those private tables.

5. **Maximum data rates**: SCTE 54 reiterates the MPEG-2 *Systems* requirement that the combined bitrate of packets of PID zero, PID 0x0001, and any given PMT PID must not exceed 80,000 bits per second.

Private data

SCTE 54 states that private data may appear either in the adaptation header of TS packets or in TS packets with a unique PID value. With the latter approach, it is possible to define an MPEG program that includes standard program elements like MPEG-2 video and AC-3 audio as well as private elements. The private nature of an ES component is recognized by use of a stream_type code in the User Private range (0xC4 through 0xFF). A cable operator must identify the entity that has defined any private program element by use of an MPEG-2 Registration Descriptor (MRD) in the ES information descriptors loop in the PMT.

Audio type and language

As with the terrestrial broadcast transport layer specification, an AC-3 Audio Descriptor must be present in the PMT section to describe characteristics of Dolby AC-3 audio tracks. As mentioned in the discussion of the PMT in Chapter 3, the AC-3 Audio Descriptor specifies the type of the audio stream (including Complete Main audio, music and effects, visually or hearing impaired audio).

Audio language is identified with the presence of an appropriate ISO 639 Language Descriptor. An ISO 639 Language Descriptor is required to appear in the ES info loop of the PMT whenever there is more than one audio Elementary Stream component of the same type.

Component names

SCTE 54 also requires that the Component Name Descriptor be placed into the ES info loop of the PMT to give a textual name to audio components when the service has two or more audio components of the same type, labeled with the same language code. It is recommended to use the Component Name Descriptor for services that have an audio component without an associated language. This type of audio could be ambient sound from a sporting event, crowd noise, etc.

Other PMT descriptors

Cable requirements specified in SCTE 54 also include rules for the use of the Caption Service and Content Advisory Descriptors defined in ATSC A/65. When caption services are present for current programs, the Caption Service Descriptor must be present in the PMT in the ES info loop corresponding to the video program element that carries the caption data. The Caption Service Descriptor must also be present in an EIT-0 that might be included in the cable Transport Stream.

Content Advisory Descriptors are treated similarly. When a given program has content advisory data available, the Content Advisory Descriptor must appear in the PMT in the program info descriptor loop. Content Advisory Descriptors must be present in EIT-0 whenever content advisory data is available.

SCTE 54 states that any of the other descriptors defined in ATSC A/65 pertinent to the cable application may appear in a cable Transport Stream, subject to the usage rules defined in A/65.

Labeling "Digital Cable-Ready" Devices

In May of 2000, the National Cable Telecommunications Association (NCTA) and the Consumer Electronics Association (CEA) issued a joint press announcement stating they had come to agreement on terms that would be used to help consumers identify the various types of "cable-ready" devices they might find for sale. Digital TV sets with full interactive capabilities were to be labeled "Digital TV-Cable Interactive" while digital sets that lacked these capabilities would be called "Digital TV-Cable Connect."

Sets labeled "Digital TV-Cable Connect" would not require an IEEE 1394 interface, but could receive analog services (as long as they are unscrambled), digital basic and premium services from any cable operator offering these services. These sets must support a POD-Host interface to access the premium services. The type of sets labeled "Digital TV-Cable Interactive" could receive the previously listed types of programming plus other services provided via a cable set-top box connected via the IEEE 1394 (HDNI) interface. Such services might include video-on-demand, enhanced program guides, and television services including enhanced data services.

In September of that same year, however, the FCC issued a Report and Order of on DTV labeling[11]. In the R&O the FCC defined three different types of consumer cable-ready devices:

1. **Digital Cable Ready 1**: this type of receiver is analogous to the "Digital TV-Cable Connect" device described by NCTA and CEA. It does not include an IEEE 1394 interface, but it can receive clear analog services and free and subscription digital services. It supports a POD-Host interface and works with a cable operator-supplied POD module to descramble premium channels.

2. **Digital Cable Ready 2**: this type of receiver has all of the features of the Digital Cable Ready 1 device and it also includes an IEEE 1394 digital interface. Via the 1394 interface, the Digital Cable Ready 2 device can benefit from services and features offered by an attached cable set-top box.

3. **Digital Cable Ready 3**: this third type of receiver also offers all of the features of the Digital Cable Ready 1 device, but it can also access advanced and interactive digital services by direct connection to the cable—no MSO-supplied cable set-top box is required. Of course, the major problem with Digital Cable Ready 3 is that no standards are in place to allow a CE manufacturer to build such a device. Methods for support of features like video-on-demand and advanced program guides are not currently standardized. The cable community hopes that these features can be supported on an OCAP-compatible device.

Considerations for Cable-ready Receiver Implementation

In the sections to follow, we discuss a variety of issues related to the implementation of a digital cable-ready receiver. Most of these considerations apply equally as well to an integrated cable-ready DTV receiver as to a cable set-top box or other device connected directly to cable.

Any digital cable-ready device, regardless of what other features and functions it supports, must be able to host a POD security module. A CE manufacturer probably wishes to conform to the FCC's labeling convention for Digital Cable Ready devices, and all three of the FCC's receiver types are POD-ready. There is actually a more basic reason: unlike analog cable service where there may be a fairly large number of channels not subject to conditional access scrambling, in the world of digital television almost all services are likely to be scrambled. The major cable Multiple System Operators have stated on several occasions their intention to scramble everything they can legally scramble. A likely exception to the rule involves "must-carry" obligations, where a cable system is required to carry certain local terrestrial broadcast signals. These, it appears, will commonly be placed on the cable system without conditional access protection. Of course to receive them one must still pay the cable company for the "basic" tier of service, so even then they're not totally free.

In the following sections we discuss issues related to supporting the POD-Host interface, issues involved with processing PSIP and out-of-band SI data, issues related to handling Transport Streams that originated from terrestrial broadcast, and performance-related considerations. We wrap up with a look at the differences a designer should be aware of between the terrestrial broadcast and cable transport standards.

Supporting the POD module

Supporting a cable-operator-supplied POD security module involves implementing the POD-Host interface specification defined in SCTE 28[18] and the copy protection mechanism defined in SCTE 41[19]. Much of SCTE 28 is based on an earlier standard called EIA-679, Part B[22], and designers are encouraged to acquire and study this standard as well.

POD-Host Interface License Agreement (PHILA)

Any company wishing to interoperate with actual POD modules must sign a POD-Host Interface License Agreement (PHILA). As of this writing, provisions of this technology license agreement include:

- the manufacturer must design the POD-Host interface in accordance with the latest versions of the specification documents, for both the interface itself and the copy protection portion.

- the manufacturer must agree to "robustness rules," wherein the design must make provisions to ensure that the secret keys needed to decrypt controlled content are kept secret (both hardware and software).

- the design must obey the Copy Control Information coming across the secure interface, and protect (as signaled) analog and digital outputs from the device.

- the manufacturer must agree to update the design to conform to specification changes, within a reasonable amount of time following their adoption.

- the design of the cable-ready integrated DTV receiver or set-top box device must be submitted for certification to CableLabs, and must pass all the requirements of the governing OpenCable specification documents.

This last point is an important (and controversial) one. It says that features, functions, and performance parameters for a consumer device built by any Consumer

Electronic manufacturer must conform to a set of requirements defined by the group of Multiple Systems Operators that guide CableLabs policy.

Why is any kind of license needed at all to support a POD module? The first answer is that proprietary technology is used in the copy protection portion of the interface (a scrambling algorithm called DFAST). It would of course be possible to design a copy protection scheme that did not rely on proprietary technology. Why wasn't that done? One of the deterrents to piracy (the building and selling of devices to receive services for which they are not entitled) is a legal one. If any pirate device must use proprietary technology (in this case, the DFAST algorithm), that device can be disallowed and taken off the market simply because it is using patented technology without permission (a license). So including proprietary technology means a license is needed.

Copy protection wouldn't be any good if it only protected data going between the POD module and the Host but it did not protect any outputs from the box. So another important function of the POD-Host interface license is that it includes requirements to honor copy control information, extending content protection to the device's signal output ports (if any).

If a manufacturer follows all of the provisions of the PHILA, and passes Cable-Labs certification for their cable-ready device, CableLabs provides that manufacturer with the key data (digital certificates) the device needs to function with actual POD modules. A POD module performs a secure initialization process with the Host into which it is plugged, and refuses to function if that Host is not able to prove that it has a valid CableLabs-issued certificate.

The out-of-band channel

Another aspect of supporting the POD module is that the Host must implement an out-of-band QPSK tuner/demodulator. The design requirements for this data path, also called the Forward Data Channel (FDC), may be found in SCTE 28[18]. Information may also be found in SCTE 55-1[24] and SCTE 55-2[25], the cable out-of-band systems specifications. Clever designers may also find application notes and other design information from chip vendors such as Broadcom Corporation. Their BCM3250 and BCM3125 chips include the QPSK out-of-band tuner.

In Chapter 17 we go into details regarding the Extended Channel, and Service Information received from the POD module on this out-of-band path.

Virtual Channel Tables and navigation on cable

The most basic form of navigation, that is finding one's way among the possibly hundreds of service offerings in an efficient and effective way, is via the virtual channel tables. In accordance with the cable network interface specification SCTE

40^4, service information is available either in-band, out-of-band, or both, for every service carried in the cable system.

EIA-818[8] and the cable transport specification SCTE 54[7] state that in-band PSIP data must be provided to describe any unscrambled services that are to be accessible to devices without conditional access support. PSIP data may be present to describe scrambled services as well. In accordance with the February 2000 PSIP Agreement, the cable community has committed to ensuring that any PSIP data provided by a cable service content provider won't be stripped out of the cable Transport Stream. We discuss the PSIP Agreement in detail in Chapter 19.

The rule requiring PSIP data for unscrambled services comes from the realization that these services could be handled in the cable-ready device even if a functioning POD module is not present. PSIP data found in-band gives the virtual channel names and numbers needed for effective (basic) navigation. To have access to any scrambled services, a POD module must be present to descramble them, and with the POD module comes access to SI provided on the out-of-band channel.

Virtual channel tables available on the out-of-band channel conform to one of the six profiles described in SCTE 65[5]. We discuss the out-of-band channel and SCTE 65 in Chapter 17.

Let's look at some special considerations related to virtual channels when building a cable-compatible device.

When both in-band and out-of-band SI data are available

A digital cable-compatible device, if equipped with a functioning POD module, has access to the cable out-of-band channel. System/service information per SCTE 65[5] is available via this channel, conforming to one of the six profiles described in that Standard (see "Profiles for out-of-band SI" on page 344). A cable-compatible device therefore typically has access both to in-band and to out-of-band service information. The following recommendations apply to such a situation:

- The virtual channel table received via the out-of-band channel should be used as the basis for navigation, specifically to establish the list of available services and to associate a channel number with each service.

- If channel names are available in the out-of-band channel, those names should be used. If the profile for out-of-band SI does not include channel names, names of channels provided in in-band PSIP data may be used, where available.

- The cable-ready device may use event information and description data received by whatever path (out-of-band or in-band) it may arrive. The choice to use data from one path or the other can be made globally or on a channel-by-channel basis. If the cable operator is using SCTE 65 profile 4 or higher, it is likely that program guide data is available for all channels. Whenever guide data is not available via the out-of-band path for one or more channels or for a certain channel/time slot on any given channel, any applicable EIT/ETT data found in-band may be used.

Note that it is necessary to use the TSID/program_number pair to link a service listed in the out-of-band virtual channel with EIT/ETT data found in-band. To find EIT/ETT data associated with a channel listed in the out-of-band VCT, the following sequence of lookup-associations may be followed. These sequences start with a given channel number and the out-of-band virtual channel table. If the short-form VCT is used, Table 16.1 gives the procedure one can use to find EPG data starting with the out-of-band virtual channel number.

TABLE 16.1 Associations Using the Short-Form VCT

Association	Procedure
1. virtual channel number → CDS_reference/program_number	Look up virtual channel number in short-form VCT.
2. CDS_reference/program_number → TSID/program_number	Look up the TSID value of the TS at the given carrier frequency.
3. TSID/program_number → in-band source_id	Find an entry in the in-band CVCT (or TVCT) with the given TSID/program_number, note Source ID.
4. in-band source_id → EIT/ETT data	Look up EIT/ETT data using Source ID as usual.

If the long-form VCT is used the procedure shown in Table 16.2 may be used.

TABLE 16.2 Associations Using the Long-Form VCT

Association	Procedure
1. virtual channel number → TSID/program_number	Look up virtual channel number in long-form VCT.
2. TSID/program_number → in-band source_id	Find an entry in the in-band CVCT (or TVCT) with the given TSID/program_number, note Source ID.
3. in-band source_id → EIT/ETT data	Look up EIT/ETT data using Source ID as usual.

Channel numbering

As mentioned, if out-of-band SI data is available, that data stream must be used as the basis for channel number assignments for the cable-supplied services. There are some subtleties here, so we need to look at some details.

One of the considerations arises from the fact that the first digital cable systems used only one-part channel numbers (numbers in the 1 to 999 range). As digital terrestrial broadcasting protocols were developed and the broadcasters decided to use the two-part numbering scheme, the cable protocols were upgraded to support two-part numbering. To support the existing equipment, however, some cable operators must continue to supply one-part numbers. They can choose to define two-part numbers for some or all services anyway, by giving both a one- and a two-part number for any service.

How is a digital cable-ready device expected to deal with this? The following recommendations should be followed with regard to application of the available channel number information.

Important note: the approach we describe here is consistent both with the stated wishes of the cable community and with the February 2000 agreements but it does not necessarily reflect the desires of the broadcaster community. Broadcasters wish to retain their brand identity and would like to see two-part channel numbers used even if a one-part number is supplied for their service in the out-of-band path. The FCC has not stated requirements with regard to this channel labeling issue, but the National Association of Broadcasters has requested a reconsideration of the digital must-carry ruling. Designers are encouraged to check to see if new requirements have been established that impact the algorithm we're about to describe.

Two cases are possible:

- If out-of-band SI data is not available, the channel number given in in-band PSIP data should be used. The channel number will be a two-part number if provided in a Terrestrial VCT, and it can be either a one- or a two-part number if provided in a Cable VCT.

- If out-of-band SI data is available, it should always be used. In the event that a given digital service is associated with both a one-part and a two-part channel number (this is possible in the SCTE 65 protocol through the use of a two-part channel number descriptor), the two-part number should be used.

This logic is diagrammed in Figure 16.4 below.

Descriptor placement

The transport specifications for cable and terrestrial broadcast differ in several subtle ways. One of these ways has to do with the location of certain descriptors. For terrestrial broadcast, a Content Advisory Descriptor is required to be placed in the Event Information Table when a given program is content-rated, and a copy of that descriptor may also be placed in the Program Map Table when that program is on-air. Likewise, a Caption Service Descriptor must be placed in the EIT when a given program is captioned, and a copy of that descriptor may also be placed in the Program Map Table when that program is on-air.

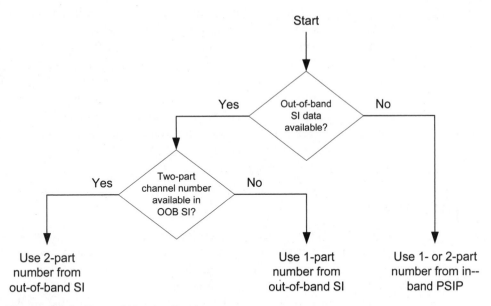

Figure 16.4 Channel Number Logic

For cable the situation is reversed. Those descriptors have to be present in the PMT, and may be optionally present in the EIT (the EIT itself is optional).

Therefore, to conform to cable standards a cable operator may need to either move or replicate EIT descriptors into the PMT. Alternatively, with or without an agreement with the broadcast station, these descriptors can be present in both the PMT and EIT locations in the terrestrial broadcast Transport Stream.

Why are these descriptors required to be in the PMT in cable, you ask? The reason has to do with MSO-owned digital cable boxes fielded in the mid- to late-1990s. These boxes did not process EITs because EPG data was received in a proprietary format via the out-of-band channel. They therefore could not access Content Advisory or Caption Service Descriptors unless they were present in the PMT. Designers of cable-ready equipment, therefore, can choose to process both the EIT and the PMT and to collect descriptors wherever they are found.

Acquisition performance—caching PID values

The amount of time that passes between a user's press of the "Channel Up" key on the remote control and the time audio and video from the new channel begin to be perceived should be as short as possible. Due to the way MPEG-2 video compression works, and the need to derive descrambling keys when a user arrives at a new channel, acquisition times for digital channels will always be greater than those for analog.

One way designers of cable-compatible equipment can reduce acquisition time for digital channels is by caching PID values. A table can be kept in memory to record the PID values last used for audio, video, PCR, and the PMT for each service (every virtual channel). Then, if the user chooses to change channels, these cached values can be used on the assumption that the PID values have probably not changed. The audio/video decoding process can start while the PMT is collected and the current PID values determined. If values have indeed changed (the cached values are incorrect), no harm has been done—the correct PID values can be set into the PID filters and the decoders can be reset. The caching step can only speed the process up; it cannot slow it down.

The cable channel map

There are a number of design issues applicable for the design of digital television receivers for terrestrial broadcast that don't apply to cable. For example, a DTV receiver that receives broadcast signals via an antenna must deal with shifting signal strengths and possibly even a movable receiving antenna. On cable, the situation is simplified. The only signals that are accessible are those that are delivered on the cable, and the strengths of these signals should be strong and stable.

Consequently, instead of worrying about keeping data in memory corresponding to signals that are not currently accessible (as one might do for the terrestrial broadcast case), on cable the out-of-band virtual channel table is the last and only word as far as the cable channel map is concerned.

If the out-of-band channel is not available (because a POD module is not present or is not functioning), the cable-ready device can operate in a similar way to the broadcast receiver and use in-band navigation data. A search for digital carriers can be done at setup time in response to a user command. The cable-ready device can compile a virtual channel table from PSIP data found on each carrier. The data for a given Transport Stream should be refreshed any time the receiver is tuned to that TS and PSIP data is present, even when the in-band data is not being used because out-of-band data is available.

Channel numbering issues

The ultimate goal of the systems designers with regard to channel numbering has been twofold. First, the desire is that every digital service on cable be known by one consistent channel number (either a two-part or a one-part number), regardless of whether or not a POD module is present. Second, the hope is that cable systems across the US will eventually migrate away from one-part cable channel numbers and adopt the two-part numbering system accepted by the broadcasters. Until this happens, the television-watching public will have to endure a bit of confusion.

A certain cable operator may use older digital set-top boxes in their system. These boxes navigate using a channel number in the range 1 to 999. Let's say this cable operator signs a retransmission consent agreement with the local Fox network affiliate station. If that Fox station broadcasts channels received over the air as 10.1, 10.2, 10.3 and 10.4, where will the cable subscribers find these channels on cable? It may be that the two-part channel number becomes part of the *name* of the channel rather than the number, for example by naming the channel "Fox 10.2." Or, perhaps they would be found on channels 101, 102, 103 and 104. A cable-ready device can possibly help a viewer with channel number issues such as these. An informational screen can be assembled from in- and out-of-band virtual channel table data to form a channel number cross-reference.

These difficulties with navigation by channel numbers point up the fact that navigation via Electronic Program Guide screens is often much more efficient and enjoyable. Channels on the EPG can be sorted alphabetically or filtered by preferences or other criteria.

Still pictures

SCTE 54[7] specifies that the MPEG-2 "still pictures" mode may be used on cable. The idea of still pictures is that instead of outputting black video when the video buffer goes empty, the decoder repeats (holds or freezes) the last reconstructed frame. It is possible by this method to create a "slide show" sort of video service that uses a very small amount of bandwidth.

SCTE 54[7] states that still pictures may be present but that any time the video stream contains still pictures, a Video Stream Descriptor in accordance with ISO/IEC 13818-1 Sec. 2.6.2 shall be present.

8-VSB modulation on cable

The "official" method for modulation of digital signals on cable is Quadrature-Amplitude Modulation (QAM), using either a 64- or a 256-point signal constellation. SCTE 40[4] specifies only QAM, and this document has been approved by SCTE—the only open standards-making organization that has consistent participation from the cable community.

Despite this apparent clarity, a bit of uncertainty currently remains. As of this writing, a few cable operators are sending digital signals down the cable using 8-VSB modulation. The FCC has acknowledged this practice as being a short-term situation, recognizing that it appears to be done for "demonstration" purposes. Everyone agrees that an 8-VSB signal present today may be changed to QAM tomorrow. In some cases, it appears that the frequencies used for 8-VSB transmission are off-air broadcast frequencies, rather than frequencies conforming to a cable frequency plan from EIA 542. Such a scheme can fool a terrestrial DTV receiver into acquiring a cable signal, and it can work for applications like HDTV demonstrations in store-front window or showroom displays.

The broadcaster community (specifically, the National Association of Broadcasters) has argued that 8-VSB modulation should be allowed on cable. It would, they say, allow a low-budget small-community cable operator to include terrestrial digital signals on the cable plant without the cost associated with converting the signal to QAM. The FCC has not yet made a ruling regarding the legality of 8-VSB modulation.

From the point of view of the ATSC and SCTE digital television standards, just one modulation method should be specified for cable (QAM). However, designers of cable-compatible devices are encouraged to survey current practice in the US and decide whether to support the ability to receive both QAM- and 8-VSB-modulated signals on cable. Also (separately), they must look at the need to support reception of 8-VSB carriers appearing on cable at off-air frequencies.

References

1. 47 U.S.C. § 549, Section 629, Telecommunications Act of 1996.

2. Adams, Michael, *OpenCable Architecture (Fundamentals)*, Cisco Press, 1999.

3. IEEE Standard 1394:2000, "Standard for a High Performance Serial Bus," Institute of Electrical and Electronics Engineers.

4. SCTE 40 2001 (formerly DVS 313), "SCTE Standard—Digital Cable Network Interface Standard," Society of Cable Telecommunications Engineers.

5. SCTE 65 2002 (formerly DVS 234), "Service Information Delivered Out-of-Band for Digital Cable Television," Society of Cable Telecommunications Engineers, 28 March 2000.

6. SCTE 18 2001 (formerly DVS 208), "Emergency Alert Message for Cable," Society of Cable Telecommunications Engineers.

7. SCTE 54 2002A (formerly DVS 241), "Digital Video Service Multiplex and Transport System Standard for Cable Television," Society of Cable Telecommunications Engineers.

8. EIA/CEA-818-C, "Cable Compatibility Requirements," Electronic Industries Alliance and Consumer Electronics Association, 2001.

9. EIA/CEA-819-A, "Cable Compatibility Requirements for Two-Way Digital Cable TV Systems," Electronic Industries Alliance and Consumer Electronics Association, 2000.

10. ANSI/SCTE 26 2001 (formerly DVS 194), "Home Digital Network Interface Specification with Copy Protection", Society of Cable Telecommunications Engineers.

11. FCC 00-342, "Compatibility Between Cable Systems and Consumer Electronics Equipment," PP Docket No. 00-67, September 14, 2000.

12. Yoshida, Junko and Leopold, George, "FCC opens the door to 1394 for cable-ready equipment," EE Times, 18 September 2000. http://www.eetimes.com/story/OEG20000918S0020.

13. EIA-775-A, "DTV 1394 Interface Specification," Electronic Industries Alliance, 1999.

14. IEC 61883-4, "Digital interface for consumer audio/video equipment-Part 4: MPEG-2 Transport Stream Data Transmission."

15. EIA-799, "On-Screen Display Specification," Electronic Industries Alliance, 1999.

16. "AV/C Compatible Asynchronous Serial Bus Connections 2.1," TA Document 2001009, 1394 Trade Association, July 2001.

17. EIA/CEA-849-A, "Application profiles for EIA-775-A compliant DTVs," Electronic Industries Alliance and Consumer Electronics Association, 2001.

18. SCTE 28 2001 (formerly DVS 295), "Host-POD Interface Standard," Society of Cable Telecommunications Engineers.

19. SCTE 41 2001 (formerly DVS 301), "POD Copy Protection System," Society of Cable Telecommunications Engineers.

20. EIA/CEA-766-A, "U.S. and Canadian Rating Region Tables (RRT) and Content Advisory Descriptors for Transport of Content Advisory Information Using ATSC A/65A Program and System Information Protocol (PSIP)," Electronic Industries Alliance and Consumer Electronics Association.

21. FCC 00-342, "Compatibility Between Cable Systems and Consumer Electronics Equipment," Federal Communications Commission, Washington D.C., PP Docket No. 00-67.

22. EIA-679-B Part B: "National Renewable Security Standard," Electronic Industries Alliance, March 2000.

23. Wasilewski, Anthony J., "Planning for PODs: Versatile Solutions for Portable, Secure Digital TV," Communication Technology, October 2001.

24. SCTE 55-1 2002 (formerly DVS 178), "Digital Broadband Delivery System: Out Of Band Transport – Mode A," Society of Cable Telecommunications Engineers.

25. SCTE 55-2 2002 (formerly DVS 167), "Digital Broadband Delivery System: Out Of Band Transport – Mode B," Society of Cable Telecommunications Engineers.

Service Information for Out-of-Band Cable

Digital television can be received through a variety of transmission media, including directly from satellite, through a terrestrial broadcast receiving antenna, in MMDS systems via a microwave dish, or by a cable coming into the home. Only on cable, however, can a receiver access a second downstream communication path separate from the 6-MHz channel supplying the digital audio/video service. This separate communication path is called the out-of-band channel because it is separate from the channel band used for audio/video services. In many systems, an upstream path from the cable-connected device back to a cable headend is also available. Such upstream communication is essential to support cable services such as impulse pay-per-view (IPPV) and video on demand (VOD).

Aside from the clear benefit of adding the upstream path, one might wonder why the out-of-band channel has been found to be so useful in cable systems. The answer is that cable operators, as they deployed more and more powerful set-top boxes to their customers, wished to keep a continuous and reliable communication path open to these boxes. Set-top boxes implement conditional access, so that for example customers who pay for premium movie services are able to access them while those who don't are barred from viewing the services. The out-of-band channel delivers the control signals and authorization keys needed for support of conditional access systems.

A cable system could be built that would deliver the control signals and authorization messages inside the audio/video channels themselves. For example, there are well-known techniques for including data packets in the Vertical Blanking Interval of the NTSC waveform. But the difficulty of such a system is clear: in order to achieve guaranteed access to all receiving devices, this data stream would have to be embedded within each signal in each 6-MHz band, so it would have to be replicated 60 to 100 times or more, depending upon the number of RF channels supported in the cable plant. The data would have to be present in each audio/video signal (in other words it would have to be present "in-band") because any given receiver might be tuned to any of the 6-MHz channels at any particular time. So to avoid this complexity, as well as inefficient use of cable bandwidth, the out-of-band channel concept was developed. When the out-of-band channel is used, the cable-compatible receiving device (set-top, TV, VCR or whatever) includes a separate tuner specifically designed to receive it. If the device supports the upstream out-of-band path, it will have a separate modulator/transmitter as well.

In the 1996 Telecommunications Act, the FCC was directed to adopt regulations and standards that would lead to the retail availability of "navigation devices." These include consumer-owned digital televisions (DTVs), digital VCRs and digital cable set-top-boxes. In its 1998 Report and Order on Competitive Availability of Navigation Devices, the FCC ruled that cable operators must supply, to any consumer who requests it, a security card that regulates access to digital services on cable. These two rulings spawned the development of a large number of open standards defining the technical aspects of compatibility with digital cable services, including standards for the cable out-of-band channel.

In this chapter we take a look at the standards for the cable out-of-band channel, focusing most of our attention on its relationship to PSIP. We first look at the way the cable spectrum is used to deliver audio/video services, data, and control signals. Then we review the architectural model that describes the out-of-band communication path. Out-of-band communications is intimately tied in to the access control security card, called the Point of Deployment or POD module, so we briefly discuss the POD architecture. Cable-ready devices that support POD modules are called "Host" devices, because they are able to act as hosts to POD modules.

We'll see that in the downstream direction, from the point of view of the Host, the out-of-band channel can be considered to be an MPEG-2 Transport Stream. The out-of-band Transport Stream carries system and service information pertaining to all of the channels in the cable system, providing navigation data in the form of virtual channel tables. In some cable systems, the out-of-band channel includes more extensive information about the channel line-up, including channel names and sometimes program guide data. We look in some detail at the system and service information standards for cable out-of-band use.

The Cable Spectrum

Figure 17.1 illustrates the way the cable spectrum is partitioned in US systems, and reflects US cable standards including SCTE 40[2] "Digital Cable Network Interface Standard." Frequencies in the 5- to 42-MHz range are used only for up-stream communications, those messages transmitted from individual cable-compatible devices back up to the cable headend. Frequencies above 54 MHz are used for all down-stream communications.

The 6-MHz in-band channels that carry audio/video and data services, also known as the Forward Application Transport or FAT channels, are found in the 54-

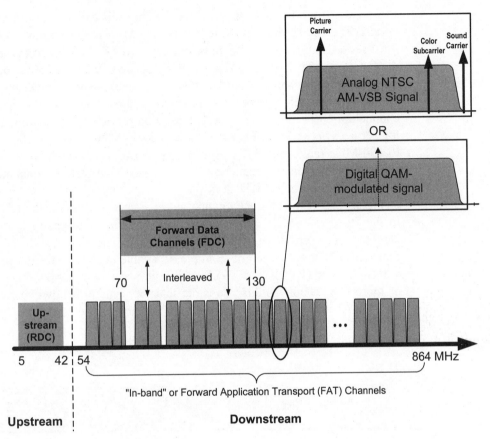

Figure 17.1 Cable Frequency Allocation

to 864-MHz range. Cable operators are not allowed to place these 6-MHz channels just anywhere they choose: Placement must conform to one of three frequency plans standardized in EIA/CEA-542-A[4]. The three are called the "Standard," Harmonic Related Carriers (HRC), and Incrementally Related Carrier (IRC) plans. A cable operator must pick one of the three and use it for all of the digital and analog carriers that can be received at any given cable drop. Note that EIA/CEA-542-A[4] states that HRC is discouraged for future use.

Some cable plants today may operate at frequencies in excess of the 864-MHz upper limit stated here, and in fact the frequency plans given in EIA/CEA-542-A[4] run up to 1.0 GHz. But agreement has been reached between consumer electronics manufacturers, the National Cable Television Association (NCTA) and CableLabs (representing cable Multiple System Operators) that services accessible to retail cable-compatible devices will stay within the 54- to 864-MHz frequency range.

Separate from the FAT channels are the channels called Forward Data Channels (FDC). The figure illustrates the 70- to 130-MHz range within which each FDC must reside, interleaved between or in place of FAT channels in that range. One of the functions of EIA/CEA-542-A[4] is to associate 6-MHz frequency bands on cable with the familiar channel numbers our analog cable-ready devices have used. Typically for the lower-numbered channels there are no gaps in the sequence of RF channel numbers offered by a cable television service (we're talking about the analog services here). From Figure 17.1, it looks like there isn't much room in the 70- to 130-MHz range for FDCs. But in fact, the RF channel numbering isn't monotonic with increasing frequency, and due to historical reasons there are gaps between some of the 6-MHz slots.

Figure 17.2 details the RF channel number assignments for the 6-MHz channels in the 70- to 130-MHz range for the Standard frequency plan. It should now be clear there is room for one or more Forward Data Channels if all these channel slots don't carry A/V services. For example, if a cable operator does not use the 114- to 120-MHz band (RF channel 99) for television services, that band is usable for one or more FDC channels. Also consider that the channel bandwidth for one FDC is just 1.0, 1.8 or 2.0 MHz, depending upon the method the cable operator chooses.

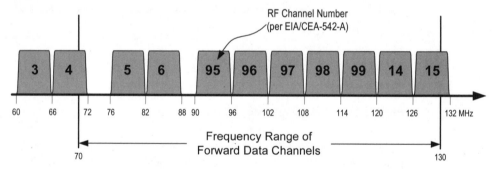

Figure 17.2 Forward Data Channel Spectrum and Standard RF Channel Assignments

A cable operator can operate several Forward Data Channels in the same cable plant. Sometimes, a portion of the population of cable-ready devices is assigned to one FDC channel frequency while other portions are assigned to different frequencies. Each digital cable-compatible receiver has just one out-of-band tuner/demodulator and can listen to just one FDC at a time. The cable operator often partitions the population of set-top boxes to take advantage of the efficiency of parallel communication paths, and to reduce the bitrate needed for any given FDC channel.

Each Forward Data Channel is capable of delivering to the cable-connected device a downstream data rate of either 1.544 Mbps, 2.048 Mbps, or 3.088 Mbps, depending upon the method chosen. Cable standards SCTE 55-1[5] and SCTE 55-2[6]

define the details of the QPSK modulation method used for the FDC. The first standard, SCTE 55-1, defines a method employed by General Instrument Corp. (now Motorola), while the second standard, SCTE 55-2, specifies a method used by Scientific-Atlanta.

We've now seen how a digital cable-ready device has access to a 1.5- to 3-Mbps out-of-band data channel which might carry some information of interest. But we haven't yet answered some important questions:

- If there are many out-of-band frequencies, how does a retail cable-ready device know which one to tune to?

- If there are several data rates and modulation methods, how does a retail cable-ready device know which one to use?

- Since cable systems use proprietary methods for conditional access, how is a cable-compatible device sold at retail able to descramble the pay services on a digital cable network?

The answers to these questions are covered in our next topic, the Point-of-Deployment architecture.

Point of Deployment (POD) Architecture

Figure 17.3 is a simplified block diagram of a Digital Cable-Ready Device (DCRD) connected directly to the cable on the upper left. A DCRD can be a cable-ready digital TV, VCR, or cable set-top box. This device also interfaces to another system owned by the cable operator, the Point of Deployment (POD) module diagrammed in the lower half of the figure.

As shown, digital audio and video services are processed in the DCRD. At the upper middle in the Figure, the RF cable signal is processed by the Forward Application Transport (FAT) tuner/demodulator to form a baseband signal. The result is an MPEG-2 Transport Stream (TS) multiplex. When a POD module is present in the DCRD, the TS is passed across the Host-POD interface to the POD, where descrambling is performed. That's assuming of course that the proper authorization has been granted (you've paid your cable bill, for example).

The POD module's primary function is to enforce conditional access to the digital services available to the Host's tuner. Once descrambled by the POD module, the TS is transferred back to the Host for decoding (MPEG-2 for video and Dolby Digital [AC-3] for audio). Not shown in Figure 17.3 is scrambling applied on the TS data path traveling from the POD module back to the DCRD. This is needed to enforce copy protection provisions. Without it, an unprotected Transport Stream would be available at the pins of the POD module connector.

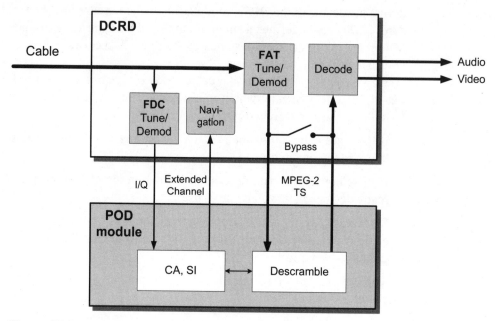

Figure 17.3 POD Architecture (Uni-directional)

The switch labeled "Bypass" in the Figure can route the Transport Stream from the Demodulator directly to the decoders if a POD module is not present. A Digital Cable-Ready Device is expected to operate without a POD module, but of course the only services that would be accessible in that condition are those that are provided without conditional access (in-the-clear programming).

In the lower left of the DCRD is a second tuner, the Forward Data Channel (FDC) tuner/demodulator. This QPSK receiver is there to demodulate whichever data channel the cable operator wishes to use. The frequency that is tuned by the DCRD's FDC tuner, in this architecture, is under the control of the POD module. It is the POD module that has the knowledge, for this cable system and for the particular subscriber to whom the module has been sent, of the frequency to use for the Forward Data Channel. Once the POD is in place and it is receiving commands from the cable headend, it may be instructed to ask the DCRD to re-tune its QPSK receiver to a new FDC frequency. The POD-Host architecture is flexible enough to support such adjustments.

In addition to the specific frequency to which the Host should tune its QPSK receiver the POD module also indicates, via messages across the interface (not shown), which standard for the out-of-band channel is being used: SCTE 55-1[5] or SCTE 55-2[6]. Based on this knowledge, the Host sets the operating mode of the QPSK receiver accordingly.

Note in the Figure that the output of the QPSK tuner/demodulator is raw In-phase (I) and Quadrature-phase (Q) bit streams. The Digital Cable-Ready Device, in general, has no way to process or understand this elemental data stream since the Standards do not require this bit-stream to conform to any particular standard format. It is up to the POD module to perform any error detection/correction, message framing, de-interlacing, re-assembly, or any other process needed to extract meaningful data from the output of the Host's FDC tuner/demodulator. Whatever methods might be employed are coordinated between the cable headend and the POD Module itself. Since the cable operator has supplied a POD module that is compatible with his system, the POD is able to make sense of the data stream.

At the lower left of the Figure, in the box labeled "CA, SI," the POD module processes the out-of-band channel and extracts messages related to the conditional access function. These messages are typically Entitlement Management Messages (EMMs) containing the authorization keys needed to descramble services carried on the cable system. It is the EMMs that grant or deny access to individual services, since they specify, in a cryptographically secure manner, the rights this particular user has to access scrambled content on the cable system.

Coming back across the interface from the POD module to the DCRD Host is the data path called the Extended Channel. SCTE 28[3] is the Standard that defines the POD-Host Interface, and the Extended Channel is one part of that interface.

Extended Channel Interface

Data flowing to the Host from the POD module that is associated with MPEG-2 private table sections comes across in exactly that format. The SCTE 28[3] standard also describes ways in which Internet Protocol (IP) packets can flow across the POD-Host interface. Note that data delivered into the POD from the Host's QPSK demodulator may or may not be organized in a Transport Stream compliant with MPEG-2 *Systems*. The best way to think of the data coming out of the Host's QPSK demodulator is that it is in some unknown proprietary format.

Figure 17.4 is a simplified block diagram, again showing a Host device and an attached POD module. Here we see the output of the Host's Forward Data Channel receiver/demodulator coming into the POD module in a format not described in the POD-Host interface standard. Inside the POD module, that transport is processed in whatever way is needed. Among the pieces of data relevant to the POD module are Entitlement Management Messages and other pieces of data relevant to the POD module's CA function.

As we have said, the Host considers that once the POD module has been installed and initialized, it has access to an out-of-band Transport Stream delivered via the FDC and accessed through its out-of-band tuner and QPSK demodulator. In addition to data kept proprietary to the cable operator, the out-of-band TS carries out-of-

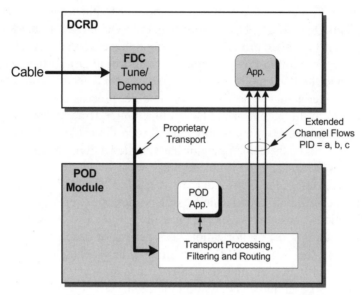

Figure 17.4 Extended Channel Data Flows

band System Information conforming to SCTE 65[1]. This SI standard has a great deal in common with PSIP, hence its inclusion as a topic in this book. We look in detail at the SCTE 65[1] protocol shortly.

The Figure shows the concept of Extended Channel "flows." A flow, as the term is used here, results from the Host asking the POD module to forward MPEG table sections that are collected from Transport Stream packets having a specified PID value. The POD-to-Host interface specification also defines a way for IP packets to flow from the POD to the Host, or (if the Host has a built-in cable modem) from the Host to the POD. For purposes of this discussion we focus only on the MPEG table section flows because that's the way out-of-band SI tables come into the Host.

For MPEG table sections, POD modules compliant with the POD-Host interface specification must support three or more simultaneous flows for MPEG table sections, each with a different PID value. Nearly without exception, one of the flows will be associated with the out-of-band SI base_PID, 0x1FFC (note that the out-of-band SI channel uses a different base PID than the in-band SI, which uses 0x1FFB). Once the SI base PID flow is running, the POD module forwards to the Host any table sections received as a result of filtering the out-of-band data on any of the up to three PID values it specified. The table sections can conform to either the long- or short-form private table section syntax defined in MPEG-2 *Systems* and can be as long as the section length maximum of 4096 bytes.

With rare exceptions, MPEG table sections are always error-checked in the POD module. For MPEG-2 long-form table sections, the 32-bit CRC is always present.

Some short-form tables correspond to standards, and these are known also to include a 32-bit CRC. The POD module understands which table sections have CRCs: where a CRC is known to exist the checksum is performed. In some cases a CRC check at another protocol layer is performed in the POD module and this check is assumed to provide equivalent protection. The rare exception noted above only occurs if the Host requests an Extended Channel flow using a PID value other than the SI base_PID, and a short-form section is found. The POD has no way to know whether or not this table section has a CRC, so in this case it is up to the Host to check it. Note that MPEG-2-compliant 32-bit CRCs are almost universally used, even for proprietary applications.

Out-of-Band vs. In-Band SI

Figure 17.5 brings these concepts together and clearly illustrates the two different data paths available to the Digital Cable-Ready Device when it is host to a POD module. The out-of-band data path is formed as the Forward Data Channel tuner/demodulator, whose frequency is controlled by commands from the POD module, goes to the POD module as raw output from the FDC demodulator and comes back as SI table sections. The in-band data path comes through the Forward Application Transport channel, and is available whenever the FAT tuner/demodulator is tuned to a digital carrier.

Figure 17.5 shows the Transport Stream from the FAT channel being demultiplexed to extract SI table sections used for navigation. Note that as shown, it is the version of the TS prior to descrambling that is used as the source of the SI table sections. That approach works because we know the Standard disallows SI tables to be transmitted in scrambled form.

For some applications, for example a data broadcasting application that works in conjunction with the conditional access system, it is necessary to demultiplex the Transport Stream *after* descrambling rather than before. Consider the current efforts towards defining a standard platform for the support of downloaded applications. Two examples of such work are the OpenCable™ Application Platform (OCAP) currently under development by CableLabs, and the DTV Application Software Environment (DASE) specification under development within ATSC. An application compliant with one of these Application Program Interface (API) specifications may well involve the processing of table sections delivered in scrambled Transport Stream packets. Whether or not a given DCRD can use the application depends upon whether the POD module in that device will descramble the data associated with that application. That, in turn, depends upon whether the user is willing to pay, via a transaction similar to impulse pay-per-view, or because that user has agreed to pay a monthly fee for the rights for this type of data.

SI data delivered on the in-band path typically only pertains to audio/video services delivered on the Transport Stream carrying the SI data itself. On the other hand, SI data coming via the out-of-band path can describe all of the services on the cable network.

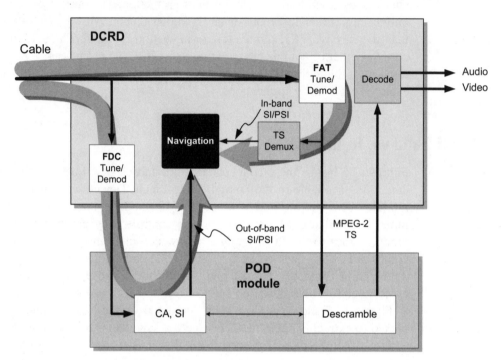

Figure 17.5 In-band vs. Out-of-Band SI

What happens if no POD module is installed? Figure 17.6 illustrates this case. Note that the DCRD is designed in the absence of a POD module to bypass it, thus sending the output of the FAT tuner/demodulator directly to the MPEG-2 audio/video decoder. Only in-band SI is available to the navigation function in this configuration.

To support the no-POD-present case, all of the standards for cable-ready devices stipulate that in-band PSIP data must be present in any Transport Stream that carries unscrambled services. If a Transport Stream contains a mixture of scrambled and non-scrambled services, for the DCRD to be able to successfully navigate to them the PSIP data must at least describe the services that are delivered in the clear (not scrambled).

To have access to digital cable signals, a cable-ready DTV must be able to demodulate 64- and 256-QAM signals, and must be able to tune to any of the three standard cable frequency plans in the 54- to 864-MHz range defined in EIA/CEA-

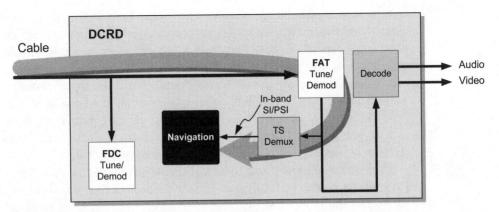

Figure 17.6 Operation Without a POD Module

542-A[4]. One might ask whether a DTV that supported QAM demodulation and the cable frequency plans, but that did not support conditional access (no POD support) could succeed in the market.

The answer to this question is rather complex, and it depends, in part, upon some FCC rulings that have not, as yet, been finalized. Many cable operators would like to scramble all of their digital services without exception, including the channels that originate from local off-air broadcast signals. Their desire to scramble comes from the point of view that scrambling thwarts theft of cable signals. Another reason cable operators like to scramble relates to the need to support the Emergency Alert System (EAS). For scrambled channels, the EAS messages are carried out-of-band. For a channel to be delivered in the clear, it may have emergency alert information appropriate to the local area that is served by the cable operator within the audio/video programming, because that channel originates from a local broadcaster who is obligated to send the EAS for local broadcast signal. But for any other cable service (one not originating from a locally-produced source), sending that service in the clear obligates the cable operator to insert EAS messages into that Transport Stream, a step that adds cost and complexity to the cable headend.

According to the FCC's rulings on retail availability of navigation devices, the cable operator can require any consumer device that wants to access certain digital cable services to utilize a POD module to do so. Since the cable operator knows just how many POD modules they have deployed, and which subscriber has each one, they have good control over everyone to whom they have granted access to cable services. This causes many to feel that cable operators will scramble all of the types of channels currently included in the programming tier known as "enhanced basic."

Broadcasters see things differently, and have called upon the FCC to require their terrestrial broadcast signals to be carried on cable in in-the-clear format. This battle isn't over, but it looks as if the only digital services that a POD-less DTV may be able to access on cable are the local terrestrial broadcast signals. Even so, it isn't clear what happens when a terrestrial broadcaster broadcasts a multi-channel multiplex. It appears likely, based on an FCC Notice of Proposed Rulemaking (NPRM) issued in January 2001, that only the "primary" channel must be carried in the clear over cable. But, during the analog-to-digital transition, the terrestrial broadcaster's primary signal may be supplied on the cable system in analog form. Cable operators have successfully opposed "dual carriage" requirements that would force them to provide the digital equivalent of an analog channel already carried as a result of retransmission consent agreement or by the must-carry rule.

So the general view is that to be truly cable-ready, a DTV must support the POD-Host interface. The FCC has stated a viewpoint with regard to this issue, saying that support of the POD-Host interface is required for a DTV set to carry the "Digital Cable-Ready" label. We discussed the details of the FCC's ruling in Chapter 16 (see "Labeling 'Cable-Ready' Devices" on page 318).

Now let's look at the out-of-band SI protocol, SCTE 65[1].

The SCTE 65 Standard

SCTE 65[1] standardizes a number of tables accessible to any cable-compatible device that has implemented a POD-Host interface and for which a POD module has been installed. These tables play an essential role in supporting navigation functions for digital cable-compatible devices. As we have seen with in-band System/Service Information used in terrestrial broadcasting, virtual channel tables in the out-of-band cable channel are fundamental to the user-friendly access to the digital services offered on a cable system. In addition to VCTs, SCTE 65[1] offers a method for delivery of network-wide Electronic Program Guide data.

Overview of SCTE 65 tables

SCTE 65[1] defines nine different table types for use on the out-of-band cable channel. Three are identical to ones defined in ATSC A/65, one is nearly identical, and two are functionally identical to their PSIP counterparts but structured slightly differently for efficient transport across the POD-Host interface. The other three reflect table types of PSIP's predecessor, ATSC A/56[7], and are present because SCTE 65[1] documents cable practice, and these older-format tables are in wide use in the US. Several of these A/56-based tables are structured such that each instance can deliver one of several types of sub-tables, as we'll see. Table and sub-table types defined for out-of-band cable in SCTE 65[1] include:

- **System Time Table** (STT)—the A/56[7] version of the PSIP STT; the two are almost the same.

- **Master Guide Table** (MGT)—the PSIP MGT, extended to allow it to refer to the table types defined in SCTE 65[1].

- **Network Information Table** (NIT)—This A/56-based table delivers two types of sub-tables which provide non-textual information relating to the cable network:

  - **Carrier Definition Subtable** (CDS)—specifies locations of the centers of the 6-MHz spectral bands in use in the cable system.

  - **Modulation Mode Subtable** (MMS)—specifies modulation parameters in use in the cable system; referenced by one form of the out-of-band virtual channel table.

- **Network Text Table** (NTT)—Another A/56[7] table; it delivers textual data relating to the cable system. One subtable type is defined:

  - **Source Name Subtable** (SNS)—supplies textual names for virtual channels, using Source ID as the reference. The SNS supplies the same data as the short channel name field in the PSIP VCT.

- **Short-form Virtual Channel Table** (S-VCT)—this A/56[7] table is similar in function to the PSIP VCT. Three subtable types are defined:

  - **Virtual Channel Map** (VCM)—lists services available on the cable system, much like the PSIP VCT lists channels available in the multiplex.

  - **Defined Channels Map** (DCM)—a data structure that gives a quick indication of which channels are defined and which are not.

  - **Inverse Channel Map** (ICM)—a table that maps Source IDs into virtual channel numbers.

- **Long-form Virtual Channel Table** (L-VCT)—this is a copy of the Cable VCT defined in A/65.

- **Rating Region Table** (RRT)—same as the PSIP RRT.

- **Aggregate Event Information Table** (AEIT)—this table delivers the same event information data (in the same format) that the PSIP EIT delivers, but packaged differently for efficient transport on the out-of-band channel. It was developed specifically for the SCTE 65[1] protocol.

- **Aggregate Extended Text Table** (AETT)—this one is analogous to the PSIP ETT, and again it is only different in that the structure is optimized for out-of-band delivery.

Overview of SCTE 65 descriptors

Four descriptors are defined in SCTE 65[1] for use in the A/56[7]-based table types (STT, NIT, NTT, and S-VCT). Each supplies a piece of data to the A/56[7]-based table that is present in its PSIP counterpart, so that when the descriptor accompanies the A/56[7] table, it can deliver the same information as its A/65 cousin. The four descriptors are:

- **Revision Detection Descriptor**—provides version and sectioning information for the short-form table types analogous to the version and section number fields in the long-form table section header.

- **Two-Part Channel Number Descriptor**—allows the S-VCT to associate a two-part channel number with any virtual channel, again to help align the functionality of the S-VCT with A/65.

- **Channel Properties Descriptor**—this descriptor is also for the S-VCT. It provides a number of attributes for a virtual channel that are in the PSIP Standard, again to bring it up to the same feature set as PSIP.

- **Daylight Saving Time Descriptor**—this one is for the A/56[7] System Time Table used in SCTE 65[1]. It's needed because the A/56[7] STT did not include the Daylight Saving Time data present in the PSIP STT.

Profiles for out-of-band SI

A given cable operator is not required to supply all of the table types defined in SCTE 65[1] in order to comply with the Standard. The out-of-band SI Standard describes six "profiles" of operation. Each profile is a set of tables a cable operator can choose to use. Cable-ready devices must handle whatever environment they find themselves in, so they must operate within any of the six profiles.

Table 17.1, taken from Annex A of the SCTE 65[1] Standard lists the optional and required tables associated with each profile.

Table 17.2 shows which descriptors are mandatory and which are optional for use with the six profiles.

TABLE 17.1 Usage of Table Sections in Various Profiles

Table Section	Table ID	Profile 1 Baseline	Profile 2 Revision Detection	Profile 3 Parental Advisory	Profile 4 Std. EPG Data	Profile 5 Combin-ation	Profile 6 PSIP Only (a)
Network Information Table	0xC2						
Carrier Definition Subtable		M	M	M	M	M	-
Modulation Mode Subtable		M	M	M	M	M	-
Network Text Table	0xC3						
Source Name Subtable		O	O	O	M	M	-
Short-form Virtual Channel Table	0xC4						
Virtual Channel Map		M	M	M	M	M	-
Defined Channels Map		M	M	M	M	M	-
Inverse Channel Map		O	O	O	O	O	-
System Time Table	0xC5	M	M	M	M	M	M
Master Guide Table	0xC7	-	-	(b)	M	M	M
Rating Region Table	0xCA	-	-	(c)	(c)	(c)	(c)
Long-form Virtual Channel Table	0xC9	-	-	-	-	M	M
Aggregate Event Info. Table	0xD6	-	-	-	M	M	M
Aggregate Extended Text Table	0xD7	-	-	-	O	O	O

Legend:

    M—Mandatory (shall be present)

    O—Optional (may or may not be present)

    "-"—Not applicable (shall not be present)

Notes:

a) Exception: System Time Table (table ID 0xC5 is used here instead of table ID 0xCD defined in PSIP) and other modifications.

b) Mandatory for outside of North America to describe any transmitted RRT. For region 0x01 (US and possessions), delivery of an RRT is optional, because this table is standardized in EIA/CEA-766-A[8].

c) Exception: delivery of the RRT corresponding to region 0x01 (US and possessions) is optional, because this table is standardized in EIA/CEA-766-A[8].

TABLE 17.2 Usage of Descriptors in Various Profiles

Descriptor (and associated table)	Desc. Tag	Base-line	Revision Detection	Parental Advisory	Std. EPG Data	Combin-ation	PSIP Only (a)
AC-3 audio (PMT, AEIT)	0x81	-	-	-	O	O	O
Caption service (PMT, AEIT)	0x86	-	-	-	O	O	O
Content advisory (PMT, AEIT)	0x87	-	-	(b)	(b)	(b)	(b)
Revision detection (NIT,NTT,S-VCT)	0x93	-	M	M	M	M	-
Two part channel number (S-VCT)	0x94	-	-	-	O	O	-
Channel properties (S-VCT)	0x95	-	-	-	O	O	-
Daylight Saving Time (STT)	0x96	-	-	O	M	M	M
Extended channel name (L-VCT)	0xA0	-	-	-	-	O	O
Time shifted service (L-VCT)	0xA2	-	-	-	-	O	O
Component name (PMT)	0xA3	-	-	-	O	O	O

Legend:
   M—Mandatory (shall be present)
   O—Optional (may or may not be present)
   "-"—Not applicable (shall not be present)
Notes:

   a) Exception: System Time Table (table ID 0xC5 is used here instead of table ID 0xCD defined in PSIP) and other modifications.

   b) The Content Advisory Descriptor shall be present in the AEIT and PMT for a given program when Content Advisory data is available for that program. It is not required for programs for which Content Advisory data is not available.

Profile 1

Profile 1, the so-called "Baseline" profile, represents the set of tables that were in common use by cable operators in the 1997-1999 timeframe. All of the tables are the short-form variety, reflecting their origin with PSIP's predecessor, Standard A/56[7]. Note that this profile supplies only the bare minimum of navigational data: a virtual channel map in which each channel is labeled only with a one-part number in the 1 to 999 range. Channel names are not even supplied. Cable operators pushed hard to have Profile 1 specified this way because they wanted to be able to say that the then-current existing practice for out-of-band cable System Information complied with the standard.

It is important to realize that the cable set-top boxes owned by the Multiple System Operators typically support a much richer feature set than can be realized if one has only the standard tables to work with. For example even when Profile 1 tables are used, the MSO-owned boxes typically support a fully-featured Electronic Program Guide. That's because these boxes use proprietary transmission formats and methods for EPG data and other features. It may be possible for DCRDs to access such proprietary data, but to do so requires a technology license. Furthermore, such a device could access EPG data delivered by one MSO, but those advanced features would probably not function if the consumer wished to use it in a cable system operated by a different MSO. In other words, without standard methods cable-ready devices that work equally well anywhere in the country cannot be built. We call DCRDs that can work with any cable system "portable."

Profile 2

With each step across the table, new functionality is added. At the second profile, the Revision Detection Descriptor is added to the NIT, NTT, and S-VCT. For a cable operator to move from Profile 1 to Profile 2 operation is just a matter of a headend software upgrade to begin inclusion of the Revision Detection Descriptor. Once revision detection data is present, set-top boxes or DCRDs can begin using it to increase the efficiency of table processing.

Note that when the revision detection information is not present (as in Profile 1), devices must either process every instance of every table section or rely on other methods to determine that a given instance represents new data. It is possible to record the CRC value associated with a certain table and process that as a "signature" value. If an instance of the same type of table is seen with a different CRC, something in the table has changed and that instance is processed. If an instance with a matching CRC is seen, that instance may be discarded (with a low probability that the discarded section actually carries different data).

Profile 3

Support for content advisory data is added at Profile 3, which builds on the set of tables and descriptors required at the prior level. To comply with Profile 3, the cable operator must include Content Advisory Descriptors in the Program Map Table of every rated program. These descriptors must also be present in the Aggregate EIT when AEITs are present. At Profile 3 it is required to transmit any Rating Region Table other than the one describing rating_region 0x01 (US and possessions) because the US RRT is fully described in EIA/CEA-766-A[8]. If an RRT is transmitted, a Master Guide Table must also be present with a reference to it.

Profile 4

Profile 4 represents the first level at which Electronic Program Guide data is made available. Consistent with the pattern, Profile 4 includes all of the types of tables and descriptors required for Profile 3 plus new ones. At Profile 4 an Aggregate Event Information Table must be transmitted. Optionally program descriptions can be delivered using the Aggregate Extended Text Table. An MGT is required to give the PID values, version, and size information for the AEITs and AETTs.

An important mandatory requirement that comes in at Profile 4 is that the Source Name Subtable must be present. So (finally) the names of the virtual channels are given.

Also at Profile 4 Caption Service Descriptors, Dolby AC-3 Audio Descriptors, and Component Name Descriptors may be present in the PMT. The audio and caption descriptors may also be present in the Aggregate Event Information Table.

Profile 5

At Profile 5 just one new table is added, the long-form virtual channel table, L-VCT. The L-VCT is identical to the Cable Virtual Channel Table defined in ATSC A/65. So Profile 5 includes two virtual channel tables, each describing exactly the same set of services.

Why would a cable operator choose to deliver the set of tables and descriptors corresponding to Profile 5? Consider that the cable operator specifies and purchases set-top boxes to lease to their customers. The specifications for these boxes can state that they shall support the SI tables based on the A/56[7] Standard, or they can require that PSIP table formats shall be supported. Cable equipment vendors may favor PSIP formats because of the consistent use of the long-form table section syntax, the consistent use of the Master Guide Table, and because of the commonality with terrestrial broadcast (in-band) SI tables.

A cable MSO may own an older group of set-top boxes that uses the A/56[7] tables, and decide to buy another group of boxes that use only the long-form variety. In that case, the choice of Profile 5 would be appropriate because it contains both types.

Note that the overhead in terms of out-of-band bandwidth to move from Profile 4 to Profile 5 is quite small, since it only involves adding the Long-form VCT.

Profile 6

At Profile 6 the pattern is broken. The Network Information Table (with its Carrier Definition and Modulation Mode Subtables), the Network Text Table (with the Source Name Subtable), and the Short-form Virtual Channel Table (with its three sub-tables) are no longer sent. As we've seen, the data provided by all of these short-form tables and sub-tables is present in the L-VCT. Its compact size and efficiency makes Profile 6 the most bandwidth-economical of the group.

As we have noted, there is a large number of existing digital cable boxes that rely on the short-form tables for their operation. A cable operator that has any of these boxes in use cannot move to Profile 6. As years go by and these boxes are replaced and equipment is upgraded, a time may come when all of the set-top boxes in a certain system use the PSIP-style long-form tables. At that time, that operator can safely switch off delivery of the NIT, NTT, and S-VCT and evolve to Profile 6 operation.

Profile 6 has been a contentious topic from the beginning among those of us forging the cable standards. The subject has been hotly debated in the working group that developed SCTE 65[1], and has even spilled over into other standards such as the SCTE 40[2] cable network interface standard, since requirements for out-of-band SI have been specified there too. Some equipment vendors apparently felt that Profile 6 threatened the established base of equipment they had fielded. An unspoken concern may have been that the existing practice appears not to be "state of the art" when compared with the efficiency of Profile 6. After all, the ATSC A/65 Standard is "next-generation" as compared with the older A/56[7], and it introduced a fairly large list of improvements.

The Aggregate Event Information and Extended Text Tables

As we've noted, SCTE 65[1] describes methods whereby Electronic Program Guide data can be supplied to a Host device through the cable out-of-band channel. Instead of the PSIP Event Information Table and Extended Text Tables, tables called the Aggregate Event Information Table and Aggregate Extended Text Tables are used. Why, one might ask, didn't SCTE 65[1] simply use the standard PSIP EIT and ETT tables instead?

To answer this question, one must look at how PID filtering is used to collect EPG information from a Transport Stream. With the PSIP EIT and ETT tables, all of the EIT instances for a given three-hour time slot are carried in TS packets with a certain PID value. That PID value cannot be used for any other type of table referenced in the MGT. Likewise, the same is true for ETT instances. Let's say a receiver wished to collect all of the EIT and ETT data for the next nine hours of programming. Table sections from TS packets from seven different PID values would want to be collected in-parallel (three each for EIT and ETT PIDs, plus one for the SI base_PID which is monitored continuously).

Now translate this problem to the out-of-band channel: The equivalent operation would involve opening seven concurrent Extended Channel data flows between the POD module and the Host, each with a different PID value specified. But the POD module is only required to support three concurrent flows. The Host would be forced to time-multiplex the three PID filters. Probably the first flow would stay tied to the SI base_PID value 0x1FFC, while the other two would be set, in turn, to other PID values as data came in. Such a process would be slow, inefficient, and a poor use of system resources.

During the development of SCTE 65[1], this difficulty was recognized. The author submitted a paper to SCTE in August, 1999 called DVS 246[9] outlining the problem and proposing the "aggregate" forms of the EIT and ETT as the best solution. As presented in DVS 246[9], the system requirements were stated as follows:

1. Cable-ready receivers must be able to *recover EPG data efficiently* from the out-of-band path, given hardware, transport protocol and data throughput limitations. Transport protocol constraints include the number of concurrent MPEG section flows associated with different PID values (this limit is three).

2. Cable-ready receivers must be able to *refresh EPG data within a reasonable amount of time* after powering on, and should be able to recover EPG data for the current time period more quickly than data for future programming. Note that these first two requirements are interdependent: If the data recovery is not efficient, the bandwidth usage on the multiplex is not as efficient.

3. Cable operators must have the *flexibility to structure the EPG data multiplex*, for example to allow the delivery of currently applicable data to be cycled faster than data applicable further in the future.

4. The receiver should be afforded a robust mechanism to *avoid re-processing* EPG data that has not changed from the last time it was sent.

5. The protocol should *not require the cable operator to regenerate* the entire EPG database at each time slot transition (e.g. every three hours).

6. The protocol should support the delivery of *at least 16 days of EPG data*—one month's worth is desirable, if possible.

DVS 246[9] discussed three different approaches to address these issues, but it was clear the AEIT/AETT solution was the best. The AEIT/AETT solution involved restructuring the outer part of the EIT and ETT for optimal delivery of data on the out-of-band path. The new "aggregate" table types were based on the current EIT and ETT. The differences between the aggregate forms and the regular EIT/ETT are highlighted in Table 17.3.

In EIT-n, the "n" indicates the time slot, with n=0 meaning "current." This association is removed in the MGT linkage to AEIT and AETT instances. Given the following considerations:

• due to the structure of the aggregate forms of the tables, each instance of a section of AEIT or AETT with data for a given time slot can appear in the same PID as another AEIT or AETT with data for a different time slot.

TABLE 17.3 Comparison of Aggregate vs. Standard EIT and ETT

Aspect	EIT/ETT	AEIT/AETT
Number of table instances per three-hour time period	One EIT for each Source ID. One ETT for each ETM ID.	One AEIT One AETT
Can EIT data for one time slot share a PID in common with data for another time slot?	No	Yes
The meaning of "n" in EIT-n vs. AEIT-n	n=0 means current three-hour time slot; 1 means next three-hour slot, etc.	n links MGT entries to AEIT/AETT instances. Each successive time slot is associated with the next sequential value of n.
How is the current time slot identified?	Through MGT: the "n" value identifies the time slot. n=0 indicates current.	In the MGT, the order of appearance of the AEIT and AETT references correspond to increasing time slots.
How are future time slots identified?	Through MGT: the "n" value identifies the time slot.	Each successive time slot is associated with the next sequential value of n (using modulo arithmetic). Each AEIT/AETT includes a tag field tying it to "n."
MPEG-compliant?	Yes	Yes
Maximum size of data, per time slot	1MB per Source ID (EIT). 4096 bytes per ETM ID (ETT).	1MB (EIT) 1MB (ETT)
Maximum number of time slots	128	256

- it must be possible to distinguish data for one time slot from that associated with a different time slot.

- to meet system goal #5, we must avoid the need to rebuild all of the tables every three hours (or to have to refresh them in the receiver if the data hasn't changed).

Therefore, the value of "n" is decoupled from time, and the time slot associated with each table is derived by its order of appearance in the MGT. The first AEIT or AETT reference in the MGT is to the current time slot, and the order of appearance is sorted by time slot.

Figure 17.7 compares the basic structure of the EIT and the AEIT. Note that the data in the hatched rectangles is the same between the two; only the wrapper is slightly different. Whereas one EIT instance supplies event information for a three-hour period for one virtual channel (as identified by its Source ID), one AEIT instance supplies event information for a three-hour period for many (maybe all) virtual channels.

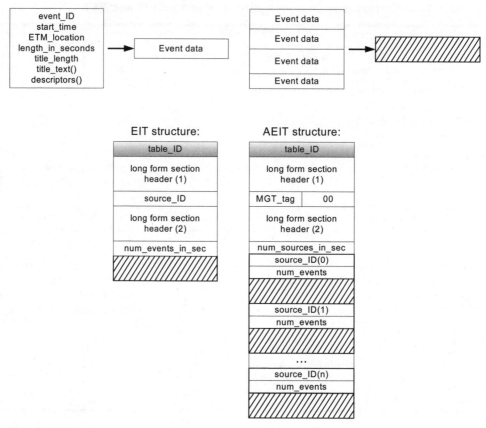

Figure 17.7 Structure of EIT vs. AEIT

Figure 17.8 compares the ETT with the AETT. Again, the extended_text_message() is common; the AETT simply includes a "for" loop so that multiple ETM_ids can appear in one table.

Figure 17.9 describes an example scenario using AEIT and AETT, and shows the aggregated tables and tables for different time slots sharing a common PID. In the example, all of the tables except those for the current time slot are in common PID Y.

Using AETT saves about 16% on transmission overhead as compared with ETT. Consider:

1. the overhead for each ETT instance (table section) is 13 bytes (table ID, MPEG section header, CRC).

2. each individual piece of text appears in a separate ETT.

3. the average total length of an ETT is somewhere around 68 bytes.

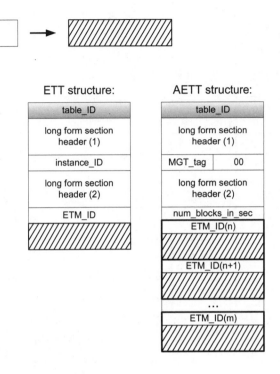

Figure 17.8 Structure of ETT vs. AETT

For a given time slot, text for approximately 71 average-sized ETTs can be sent in one 4096-byte AETT table section. The 13-byte section overhead appears once for the AETT where it appeared 71 times in the ETT case. The 71 ETT instances total 4,828 bytes. Sending the same data in one 4,060-byte table section yields a savings of 16%.

Similarly, using AEIT saves about 9% on transmission overhead as compared with EIT.

PID assignment for AEIT and AETT

For EIT and ETT tables, tables for each time slot must be transported in TS packets with different PID values. Furthermore, the same PID value cannot be used for both EITs and ETTs. The situation is quite different for the aggregate forms of the tables. The rules for PID values for Aggregate Event Information and Aggregate Extended Text Tables are as follows:

- AEITs and AETTs for the first two time slots (the current and following ones) must share a common PID value.

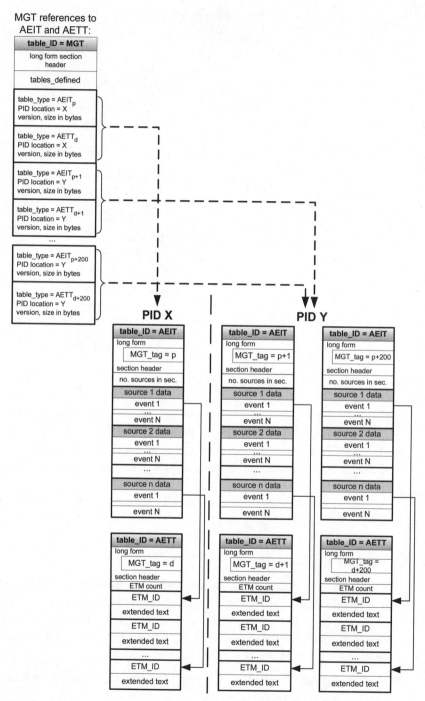

Figure 17.9 AEIT and AETT Example

- AEITs and AETTs for the second two time slots must be associated with a second PID value, separate from the first.

- AEITs and AETTs that provide data for time slots farther into the future may be associated with one or more PID values—the second PID value may be used for all or some of the future data.

References

1. SCTE 65 2002 (formerly DVS 234), "Service Information Delivered Out-of-band for Digital Cable Television," Society of Cable Telecommunications Engineers, 28 March 2000.

2. SCTE 40 2001 (formerly DVS 313), "Digital Cable Network Interface Standard," Society of Cable Telecommunications Engineers.

3. ANSI/SCTE 28 2001 (formerly DVS 295), "Host-POD Interface," Society of Cable Telecommunications Engineers.

4. EIA/CEA-542-A, "Cable Television Channel Identification Plan," 2000.

5. SCTE 55-1 2002 (formerly DVS 178), "Digital Broadband Delivery System: Out Of Band Transport – Mode A," Society of Cable Telecommunications Engineers.

6. SCTE 55-2 2002 (formerly DVS 167), "Digital Broadband Delivery System: Out Of Band Transport – Mode B," Society of Cable Telecommunications Engineers.

7. ATSC Standard A/56, "System Information for Digital Television," 3 January 1996 (obsolete).

8. EIA/CEA-766-A, "U.S. and Canadian Rating Region Tables (RRT) and Content Advisory Descriptors for Transport of Content Advisory Information Using ATSC A/65A Program and System Information Protocol (PSIP)."

9. SCTE DVS 246, "Out-of-band delivery of EPG data," Society of Cable Telecommunications Engineers, 4 August 1999.

Emergency Alert System

In 1994, the FCC issued the Report and Order[1] that replaced the old Emergency Broadcast System (EBS) with an updated system. The new system, called the Emergency Alert System (EAS), uses a digital signaling method. Its primary function is to provide the President with a way to immediately communicate with the general public in order to provide information during a national emergency. At the option of radio and television broadcasters and cable operators, the EAS system can be (and often is) used to disseminate weather-related information such as alerts originating from the National Weather Service. Radio and television stations are required to participate in EAS, and over the last several years, in accordance with FCC regulations, cable systems in the US have had to move into compliance as well. Standards for digital cable-ready devices require that the Cable Emergency Alert message standardized in SCTE 18[2] and the equivalent ANSI/EIA/CEA-814[3] (commonly called EIA-814) be supported.

The EAS system is relevant to our PSIP discussion for several reasons:

- The Cable Emergency Alert Message is transported along with in-band or out-of-band System Information in the SI base_PID.

- The Cable Emergency Alert Message fits the general format of the other PSIP tables; for example, it uses the text coding/compression scheme defined in A/65.

- The Cable Emergency Alert Message ties into the PSIP data structures: one of the functions of the message is to force a receiver to tune to a different channel, and channel references are made via either major/minor channel numbers or via Source ID.

In this chapter we give an overview of the Cable Emergency Alert message and briefly discuss the data that accompanies the EAS broadcast. Then we look at the message structure and its syntax and semantics. We discuss the usage requirements and what is expected or required of digital cable-compatible equipment with regard to EAS. We finish up with a discussion of some optional emergency alert-related features that a digital cable-ready device could choose to implement.

EAS Background

The FCC *First Report and Order*[1], issued in December 1994, amended Subpart G of Part 73 to replace the Emergency Broadcast System with the new Emergency Alert System and defined EAS in a new Part 11 of Title 47 of the Code of Federal Regulations[4]. All cable systems were required to participate, regardless of size.

The *Second Report and Order (Second R&O)*[5] adopted in September 1997, amended Part 11 and established further guidelines regarding cable system participation in EAS. FCC's rules require that cable systems must send some type of visual alert, as well as an audio alert on all channels when activating the EAS.

Specifically, the rules now require cable systems of greater than 10,000 subscribers to provide an audio and video emergency alert message on all channels carrying audio/video programming. By October 1, 2002, cable systems of between 5,000 and 10,000 subscribers must also meet this requirement. By that same date, cable systems with less than 5,000 subscribers must be able to interrupt video on all channels and provide emergency alert audio and video on at least one channel. The audio alert message must state which channel is carrying the audio and video EAS message.

In general, Emergency Alert information is transmitted to make television viewers or the listeners of radio broadcasts aware of weather, national-level, or other emergencies that may affect them. In some cases, the alert will be sufficient to suggest appropriate actions that are needed to ensure personal safety. Sometimes, further information will be needed, and for those cases the EAS includes directives indicating where additional information may be found.

FCC rules established two communication networks to be used to disseminate EAS information, the Emergency Action Notification (EAN) network and the Primary Entry Point (PEP) system*. EAN is dedicated to the purpose of communicating the EAS message used by the President for national-level emergencies; this EAS message is called the Emergency Action Notification message. The EAN network also communicates the President's news report in audio format, and a signal called the Emergency Action Termination (EAT) message when the report is over.

FCC rules govern the structure of the EAS messages traveling on the PEP or EAN network. Because the system must be periodically tested, even those who do not live in areas where weather-related alerts are common have had a chance to hear EAS messages on radio or TV. They are immediately recognizable by their distinctive three-squawk signature and the two-tone alert signal called the Attention Signal. The three squawks are actually a Frequency Shift Key (FSK) modulated signal that encodes a digital message at a data rate of 520.83 bits per second.

---

* Although the FCC rules do not reflect it, the PEP system has now been disbanded.

Each EAS broadcast message consists of four parts: the Preamble and EAS Header Codes, audio Attention Signal, message, and Preamble and EAS End Of Message Codes. For all but the Emergency Action Notification message, the message portion of the EAS broadcast is limited to a maximum length of two minutes. If the President were to take to the EAN airwaves, the message could continue as long as necessary, since no two-minute limit is imposed for the EAN message. A special message called the Emergency Action Termination (EAT) is used to indicate that the EAN is over. For all other types of EAS messages, the standard End Of Message Codes are used.

Within the EAS Header Codes can be found the following pieces of data relevant to the Emergency Alert:

- what entity has originated this alert; for example EAN, PEP, or the National Weather Service[*].

- the nature of the alert; for example a national-level or a weather-related alert and if weather-related, the weather condition involved.

- the geographic area affected by the alert (state, county, or portion of a county).

- the time the originator of the message transmitted it.

- the time the alert is to remain in effect.

- an identification of the broadcast station or office transmitting or retransmitting the message.

EAS for Cable

One could envision building a cable system in which EAS requirements are handled just as they are in a television or radio broadcast station. Whenever EA information comes in, it interrupts video and audio as necessary. Where the television broadcast involves one channel (or with digital compression, a few channels), such a cable system would need to arrange for the interruption of all channels. If it were necessary to use every channel in the cable system to provide the audio/video feed that gives the critical information about the alert, the cable system could use switching techniques to make that happen.

While such an approach is technically possible, in practice it is unworkable due to the cost and complexity of the switching and processing equipment involved. For example, the cable operator's desire would be to reduce the level of interrupted viewing for alerts of lesser importance, such as storm warnings. For alerts such as

---

[*] The originator can even be the local cable operator or broadcaster, especially in the case of a Required Weekly Test.

these, some text scrolling across the top or bottom of the video may suffice (see FCC 47 CFR 11.51[d]). Adding scrolling text to a video signal takes a dedicated piece of headend hardware for each video signal. For the analog signals in a cable plant, switching and/or text-insertion methods are often used to meet the EAS requirements. But for the digital signals, it's a different story.

For a signal source coming into the cable plant in digital form, to add text the equipment would have to decode MPEG-2 video, add graphics, and re-encode the video. Some of the incoming signals arrive with several programs sharing bit stream bandwidth in the Transport Stream in a statistically-multiplexed fashion. For these, the decode/re-encode step would have to result in exactly the same video bitrate after re-encoding. That is a very hard problem to have to solve, and it would be best to avoid it altogether. It's also necessary to interrupt audio, which again would involve some difficult manipulation of the Elementary Stream components in every MPEG program in every Transport Stream in the cable network.

Digital cable set-top boxes owned by cable MSOs and leased to subscribers handle EAS via control messages sent out-of-band. These messages can cause the box to create and display text, and can direct the box to replace the current audio with audio from a downloaded file or other source. These messages can also direct the set-top box to force-tune to a Details channel, where further information about the Emergency Alert may be found. It may be that the cable operator has made arrangements with one of the local broadcasters so that, in the event of an emergency of sufficient importance, that cable operator can cause all of the analog and digital channels to switch to that broadcaster's signal.

With the advent of the FCC's navigation devices rulings, we're moving towards the retail availability of cable-ready DTV and even (potentially) set-top boxes. This is where the Cable Emergency Alert Message standardized in SCTE 18[2] and EIA-814[3] comes in. These standards define the signaling method and requirements for retail cable-ready devices if they are to be compatible with the EAS.

Overview of the Cable Emergency Alert Message

The Cable Emergency Alert Message is specified in the format of a long-form MPEG table section, just like all of the PSIP tables defined in A/65. It delivers the following data:

- the Originator Code from the EAS broadcast, indicating the entity that originally initiated the alert.

- the nature of the activation, for example national-level, or if weather-related, the type of weather condition.

- a short textual description of the nature of the alert (for example "Tsunami Watch").

- how much time is left for display handling of this particular alert message, up to 2 minutes (or, for the EAN, indefinite).

- the date and time of day the EA event started.

- the duration, relative to the EA event start time, of this alert.

- a priority code for this event, to indicate for different situations whether or not video and audio should be interrupted, and allowing low-priority messages such as test messages not to interrupt viewing.

- identification of the Details channel, either in the form of Source ID or as a major/minor channel number.

- identification of an audio-only channel that can supply the audio portion for this alert.

- a textual description of the alert which can be displayed or scrolled across video by the cable-ready device.

- indication of the geographic area for which this alert applies.

- a list of "exception" channels, such that if the currently tuned channel is listed, this alert message should be dropped (because the current channel handles alerts within itself).

- an "event ID" to be associated with this event; it can be used by the DCRD to discard a subsequent occurrence of an event that has already been processed and displayed.

To help understand some of the concepts embodied in the Cable Emergency Alert Message, an illustrative example is given in Figure 18.1.

At the top, the Figure shows a time-line of approximately six hours duration. Two EAS events occur in this period: a Flood Watch (EAS Event Code FLA) occurs at noon and is in effect until 5:30 p.m., and a High Wind Warning (HWW) occurs at 2:00 p.m. and lasts until 4:45 p.m. The Flood Watch's "event duration" is thus 5-1/2 hours, while the event duration of the High Wind Warning event is 2 hours and 45 minutes. The "event end point" for the Flood Watch is 5:30 p.m., while for the High Wind Warning it is 4:45 p.m.

The time period from noon to 12:15 p.m. is expanded at the left to show that the Cable Emergency Alert Message announcing the Flood Watch is sent at noon and repeated ten minutes later. The period of time occupied by the box labeled "1" represents the time the DCRD is busy handling the Cable Emergency Alert Message that arrived at noon. In this example, this time period is two minutes, which is the maximum possible time for all types of events except the Emergency Action Notifi-

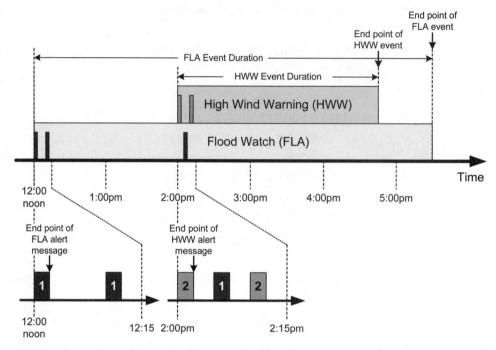

Figure 18.1 Illustrative Emergency Alert Time-line

cation (EAN). The EAN is special in that it does not specify a duration (it continues until terminated by the Emergency Action Termination).

One of the parameters delivered in the Cable Emergency Alert Message gives the number of seconds the alert interrupts video (or both audio and video). It gives the "message duration" and is the amount of time to spend tuned to the Details channel, to display alert message text, or to replace the regular program's audio with audio pertaining to the emergency.

The parameter in the actual message syntax that specifies message duration is called alert_message_time_remaining. It gives the number of seconds between the arrival time of the message and the time when processing it should be complete so a return can be made to the audio/video that has been interrupted. There is one situation where the interruption to video must continue past the time given in alert_message_time_remaining: if all of the alert text provided in the message has not yet been displayed, the text-scrolling operation must continue until it has.

Figure 18.1 labels the "message end points" of each of the alert messages, derived by moving the number of seconds given by alert_message_time_remaining past the moment the Cable Emergency Alert Message was received.

The Figure also illustrates the idea that more than one Emergency Alert can be "active" at any given time. By "active," we mean that the event end point has not yet been reached. At the option of a DCRD, a user interface screen can be designed to display a list of all active Emergency Alert events.

In addition, the Figure illustrates some of the concepts of duplicate detection. The alert messages labeled "1" deliver the information about the flood watch, while those labeled "2" provide information about the wind warning. The Cable Emergency Alert Message (CEAM) sent at 12:10 can be exactly identical to the one sent at noon, and because the sequence_number is the same as the prior message, the DCRD can discard it. A second type of duplicate detection involves the Event ID parameter. All of the messages about the high wind warning may carry the same information and can be labeled with a common value of Event ID. Likewise, those about the flood watch can be labeled with a different common Event ID value. The FLA event announced in the CEAM at 2:05 can be discarded in the DCRD because it carries the same Event ID as those sent at 12:00 and 12:10. Note that if anything related to the information pertaining to the flood watch has changed since noon, a new Event ID can be assigned to assure that all of the receiving devices process it for display.

The duration of the event must not be confused with the duration of the message. A flood watch might last all afternoon, but the announcement that interrupts viewing only lasts a couple of minutes at most.

CEAM Transport

Transport of the Cable Emergency Alert Message table section involves these considerations:

- CEAM sections are carried in Transport Stream packets with the SI_base PID value. This means, for in-band use PID 0x1FFB is used and for out-of-band, PID 0x1FFC is used.

- TS packets containing a CEAM section must not be scrambled (the transport_scrambling_control bits are set to zero).

- As with all PSIP tables, an adaptation field must not be present in TS packets carrying a CEAM section.

- The CEAM cannot be sectioned for delivery; any CEAM instance section can be only as long as the MPEG-2 limit for private table sections, 4096 bytes.

Structure of the Cable Emergency Alert Message

Figure 18.2 diagrams the structure of the Cable Emergency Alert Message table section. In the header portion of the section, the table_id_extension field is not used and is set to zero. Receiving equipment must recognize and process every instance of the cable_emergency_alert() as a separate and independent item of data.

In the body of the message following the standard MPEG and PSIP header is the EAS_event_id and the three-character EAS_originator_code from the EAS broadcast. We discuss EAS_event_id below. Next is a length field giving the number of ASCII characters in the EAS_event_code to follow, followed by that number of characters. The current FCC regulations show all EAS event codes as having three-character representations, but a variable-length data structure is used in the cable_emergency_alert() to accommodate the possible inclusion of longer or shorter representations in the future.

Next comes a field giving the length in bytes of the Multiple String Structure to follow. The string gives a textual name to the nature of the EAS event and is intended for on-screen display if the DCRD wishes to list current or recent EA events.

Three time-related fields follow: the number of seconds remaining for processing (display) of this alert message, the starting time and date of the event, and the duration of the event. Alert message time indicates how many more seconds the interruption of audio/video involved with this EA message should last. The starting time/date and duration of the event reflects the start and end of the time period over which the event is expected to occur. For example, with these fields the EA message could indicate a Severe Thunderstorm Warning was in effect beginning at 10:00 a.m. and lasting until 1:30 p.m. local time.

Next come an alert_priority field and pointers to the Details channel in the form of major/minor channel numbers or Source ID. A separate Source ID is given for an audio channel (we discuss how these are used below). Following these is another text length field followed by another Multiple String Structure. This MSS is the text that is displayed by scrolling it slowly across the top of the screen during the duration of the alert message.

Following the alert_text() field is a "for" loop giving location data, preceded by the loop count. Location data consists of state and county codes and a code that can indicate a portion of a county. Next is a list of "exceptions"—channels that should not be tuned away from if the DCRD is already tuned there. The structure of the message is such that these fields either hold the source_id of the exception channel or its major and minor channel number representations. The interpretation of those fields depends on the value of a flag called in_band_reference.

As with nearly all of the other PSIP tables, the cable_emergency_alert() ends with a descriptor length field and a number of additional descriptors fitting into that number of bytes.

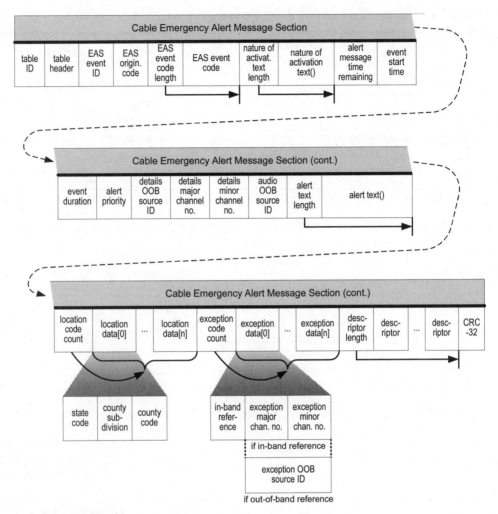

Figure 18.2 Structure of the Cable Emergency Alert Message

CEAM Syntax and Semantics

Table 18.1 diagrams the syntax of the cable_emergency_alert().

sequence_number

This 5-bit field is used by the DCRD to detect that a given cable_emergency_alert()
is a duplicate of one sent before. Generating equipment is required to increment
sequence_number by 1 (modulo 32) whenever anything in the
cable_emergency_alert() message changes. Receiving equipment is required to

TABLE 18.1 Cable Emergency Alert Message Syntax and Semantics

Field Name	Number of Bits	Field Value	Description
cable_emergency_alert() {			Start of the cable_emergency_alert().
table_id	8	0xD8	Identifies the table section type as being a cable_emergency_alert().
section_syntax_indicator	1	1b	The CEAM uses the MPEG "long-form" syntax.
zero	1	0b	This bit must be set to zero.
reserved	2	11b	Reserved bits are set to 1.
section_length	12		The CEAM can be 4096 total bytes in length, so section_length is limited to 4093.
table_id_extension	16	0x0000	Set to zero for the CEAM.
reserved	2	11b	Reserved bits are set to 1.
sequence_number	5		The sequence number reflects the version of this CEAM message. It is incremented when anything in the table changes. This field functions just like the version_number field in the standard MPEG long form table section syntax.
current_next_indicator	1	1b	Indicates that the CEAM is currently applicable.
section_number	8	0x00	The CEAM is not sectionable.
last_section_number	8	0x00	The CEAM is not sectionable.
protocol_version	8	0	Indicates the protocol version of this table section. The only type of cable_emergency_alert() currently defined is for protocol_version zero.
EAS_event_id	16		A 16-bit code that tags a particular EA event so the DCRD can recognize a repetition and discard the message.
EAS_originator_code	24		These three ASCII Characters reflect the Originator code delivered in the broadcast EAS message. See text.
EAS_event_code_length	8		Gives the length in bytes of the EAS_event_code field to follow.
EAS_event_code	var.		These ASCII Characters reflect the EAS Event code delivered in the broadcast EAS message. See text.
nature_of_activation_text_length	8		Gives the length in bytes of the nature_of_activation_text() field to follow.
nature_of_activation_text()	var.		A Multiple String Structure that reflects the Event Code field of the EAS broadcast but as a short text string rather than the abbreviation used in the broadcast.

TABLE 18.1 Cable Emergency Alert Message Syntax and Semantics

Field Name	Number of Bits	Field Value	Description
alert_message_time_remaining	8		Gives the number of seconds remaining for display processing of this message, up to a maximum of 120 (two minutes).
event_start_time	32		Indicates the start time of the EA event in GPS seconds since midnight January 6th 1980.
event_duration	16		Indicates the number of minutes, starting at event_start_time, that the EA event is expected to last.
reserved	12	0x3FF	Reserved bits are set to 1.
alert_priority	4		Indicates the priority of this event. See text.
details_OOB_source_id	16		Indicates the Source ID of the Details channel. Used with the DCRD, has access to an out-of-band channel.
reserved	6	0x3F	Reserved bits are set to 1.
details_major_channel_number	10		The major channel number of the Details channel. Used when no out-of-band channel is available.
reserved	6	0x3F	Reserved bits are set to 1.
details_minor_channel_number	10		The minor channel number of the Details channel. Used when no out-of-band channel is available.
audio_OOB_source_id	16		Indicates the Source ID of an audio service that can be switched to for the duration of the alert message.
alert_text_length	16		Gives the size in bytes of the Multiple String Structure to follow.
alert_text()	var.		This is a Multiple String Structure encoding the text that must be scrolled across the top of the screen for the duration of the alert message.
location_code_count	8		Indicates the number of blocks of location codes that is given in the "for" loop to follow.
for (i=0; i<location_code_count; i++) {			Start of location codes "for" loop.
state_code	8		Identifies the state affected by this alert Value zero means a national-level alert (all states). See text.
county_subdivision	4		The county area is sub-divided into nine sections on a three-by-three grid. See text for coding.
reserved	2	11b	Reserved bits are set to 1.
county_code	10		This is a three-decimal-digit code number representing a county within the state identified by state_code. See text.

TABLE 18.1 Cable Emergency Alert Message Syntax and Semantics

Field Name	Number of Bits	Field Value	Description
}			End of location codes "for" loop.
exception_count	8		Indicates the number blocks of exception channel data that is given in the "for" loop to follow.
for (i=0; i<exception_count; i++) {			Start of exception channels "for" loop.
in_band_reference	1		Indicates whether the channel reference to follow is in the form of a major/minor channel number (in-band reference) or a Source ID (for out-of-band reference).
reserved	7	0x7F	Reserved bits are set to 1.
if (in_band_reference) {			
reserved	6	0x3F	Reserved bits are set to 1.
exception_major_channel_number	10		The major channel number for in-band references to Details channels.
reserved	6	0x3F	Reserved bits are set to 1.
exception_minor_channel_number	10		The minor channel number for in-band references to Details channels.
} else {			
reserved	16	0xFFFF	Reserved bits are set to 1.
exception_OOB_source_id	16		The Source ID for out-of-band references to Details channels.
}			End of "if" statement.
}			End of exception channels "for" loop.
reserved	6	0x3F	Reserved bits are set to 1.
descriptors_length	10		This field indicates the length of any additional descriptors.
for (i=0; i<N; i++) {			Start of the additional descriptors "for" loop. The value of N is given indirectly by descriptors_length.
descriptor()	var.		Zero or more descriptors, each formatted as type-length-data.
}			End of the additional descriptors "for" loop.
CRC_32	32		A 32-bit checksum designed to produce a zero output from the decoder defined in the MPEG-2 *Systems* standard.
}			End of the cable_emergency_alert().

process the sequence_number in order to detect and discard duplicate transmissions. Cable operators often send the same message several times to help ensure delivery. Duplicate detection in the DCRD is discussed in more detail below.

In some instances generating equipment has no way to know the last transmitted value of sequence_number. Such a situation can occur, for example, if a new piece of equipment is brought on-line. Whenever the last-used sequence_number is not known, the Standard requires the generating equipment to set the sequence_number to zero and to send out an Emergency Alert with alert_priority zero. Such an alert is disregarded by all consumer equipment but sets the last-used sequence number to zero. If a real alert needs to be sent, the sequence_number can be set to 1 with assurance that the new alert will not be discarded as a duplicate.

EAS_event_id

A new EAS_event_id is assigned to every new EAS message distributed throughout the cable system. If anything in the cable_emergency_alert() message changes, a new EAS_event_id must be used. The purpose of EAS_Event_id is to give DCRDs another way to recognize an alert as being a duplicate of one already processed. Using EAS_event_id, it is possible to recognize a cable_emergency_alert() as giving information about an event that is no different than the information given the previous time, even when an alert of a different type has been received in the intervening period (and hence duplicate detection by sequence_number does not work). Duplicate detection using EAS_event_id is discussed in detail below.

EAS_originator_code

These three ASCII-coded characters indicate the entity that has originated this EAS event. It is coded in accordance with the table given in 47 CFR FCC Part 11 and reprinted here in Table 18.2. There are no stated requirements for the DCRD to process EAS_originator_code, but at the discretion of a manufacturer it could be used to support some interesting EAS-related features. We discuss some of these features later in this chapter.

TABLE 18.2 EAS Originator Codes

Originator Code	Meaning
EAN	Emergency Action Notification Network
PEP	Primary Entry Point System
WXR	National Weather Service
CIV	Civil Authorities
EAS	Broadcast station or cable system

EAS_event_code_length

This 8-bit unsigned integer indicates the number of ASCII characters to follow in the EAS_event_code field. While all of the codes in current use are three characters in length, the cable_emergency_alert() allows for different lengths for the EAS_event_code field in the future

EAS_event_code

The EAS_event_code field reflects the value of the EAS Event code defined in FCC Part 11 and indicates the nature of the emergency alert event. As with the originator code, DCRDs are not required to process the EAS_event_code field, but at the discretion of the manufacturer it can be used to trigger various alarms or attention-getting behavior. It can also be used in conjunction with other optional features (see the discussion later in this chapter). The EAS_event_code field is coded as N ASCII characters, where the value of N is given by the preceding EAS_event_code_length field. Table 18.3 reproduces the event codes defined in 47 CFR 11.31(e). The FCC presently has an open Further Notice of Proposed Rulemaking to determine whether or not this list of event codes should be expanded.

TABLE 18.3 EAS Event Codes

Event Code	Meaning
National Codes:	
EAN	Emergency Action Notification (National only)
EAT	Emergency Action Termination (National only)
NIC	National Information Center
RMT	Required Monthly Test
RWT	Required Weekly Test
NPT	National Periodic Test
Local Codes:	
ADR	Administrative Message
BZW	Blizzard Warning
CEM	Civil Emergency Message
DMO	Practice/Demo Warning
EVI	Evacuation Immediate
FFA	Flash Flood Watch
FFS	Flash Flood Statement
FFW	Flash Flood Warning
FLA	Flood Watch

TABLE 18.3 EAS Event Codes (continued)

Event Code	Meaning
FLS	Flood Statement
FLW	Flood Warning
HLS	Hurricane Statement
HUA	Hurricane Watch
HUW	Hurricane Warning
HWA	High Wind Watch
HWW	High Wind Warning
SPS	Special Weather Statement
SVA	Severe Thunderstorm Watch
SVR	Severe Thunderstorm Warning
SVS	Severe Weather Statement
TOA	Tornado Watch
TOR	Tornado Warning
TSA	Tsunami Watch
TSW	Tsunami Warning
WSA	Winter Storm Watch
WSW	Winter Storm Warning

nature_of_activation_text_length

This is an unsigned integer that gives the number of bytes in the Multiple String Structure directly following.

nature_of_activation_text()

This field gives the DCRD a textual representation of the EAS_event_code in the format of a PSIP Multiple String Structure. The expectation is that the nature_of_activation_text() reflects the strings in the "Meaning" column of Table 18.3, so for example the nature_of_activation_text() for EAS_event_code HWA would be "High Wind Watch." Because it is an MSS, the string can be represented multi-lingually if desired. Like the EAS_event_code, there are no requirements stating that a DCRD shall process nature_of_activation_text(). It is provided to support some EAS-related optional features in the cable-ready device.

alert_message_time_remaining

This 8-bit unsigned integer field indicates how much time remains for display of text or outputting of audio. The start time of an alert message is defined as the time the last bit of the CRC was received. If alert_message_time_remaining is zero, an alert message period of indefinite duration is indicated. The only kind of Emergency Alert that involves an indefinite duration is a national-level Emergency Action Notification (EAN) alert. As of this writing, such an alert has never been issued (and of course one hopes an occasion would never present itself).

EIA-814/SCTE 18[2] defines the "end point" of any given cable_emergency_alert() message as the point in time N seconds following the start time where N is given by alert_message_time_remaining. When the end point of the alert message is reached, the cable-ready device should return to the state it was in prior to the moment the alert was received. It should finish up displaying the text and/or change the channel back to the channel that was the current channel prior to the event.

Note that the DCRD is required to continue scrolling the text after the end point is reached if all of the text has not yet been displayed. If all of the text has been displayed when the end point is reached, it may choose to repeat the text until the end point. In that case, it can take down all text displays when the end point is reached.

event_start_time

This field is the date and time of day, given in GPS seconds since midnight January 6, 1980 UTC, that this EA event was first released from the originator. For example, the National Weather Service might create an EAS event for Dade County, Florida at 8:15 p.m. UTC the afternoon of May 20th to indicate a Hurricane Watch is in effect. The event_start_time field in the cable_emergency_alert() message would represent this date and time of day in standard PSIP time format.

Let's say this same EAS event is sent out again later (maybe it is repeated every 15 minutes) for the benefit of those who might have missed it. The next time it is sent, it would still indicate the event_start_time of 8:15 p.m. UTC because that represents the initial release time of this event from the National Weather Service. Note: because it is the same event, the EAS_event_id would remain the same too.

event_duration

Where the event_start_time indicates the date and time of day a particular emergency alert event started, the event_duration field indicates how long it is expected to last. The event_start_time field is coded as a number of minutes in the range from 15 to 6000, so an event as long as 100 hours can be specified. Continuing with the above example, perhaps the event_duration for the Hurricane Watch is four hours. In that case, event_duration would be specified as a value of 240. If a DCRD were to summarize an active event, it could create an on-screen display using data given in the cable_emergency_alert() message as follows: "A Hurricane Watch is in effect from 2:15 p.m. EST until 6:15 p.m. this evening." Note that we assume the DCRD has been configured so that the local time zone is known, because time zone is required for any conversion of GPS seconds (which are based on UTC, or time zone offset zero) into local time of day.

alert_priority

Any given instance of a cable_emergency_alert() message is tagged with a priority level ranging from priority level zero (used for test messages and discarded by all consumer equipment) up to priority level 15 (process unconditionally). EAS messages broadcast through the EAN or PEP systems do not have priority levels associated with them. Priority levels are assigned at the discretion of the cable operator. Their primary value is to allow certain types of EA messages not to interrupt audio/video. For example, a cable operator could make a choice to tag Winter Storm Warning events with a priority level that causes them not to interrupt viewers who are watching a Pay-Per-View event. Or, repetitions of certain events could be set to lower priorities.

The alert_priority field has another function as well: to distinguish between those events not requiring an audio component from those for which audio is necessary. Recall that in order to comply with EAS requirements for alerts reflecting a national emergency, audio (presumably, an announcement from the President) must accompany the alert. Usually, this audio track is a result of the cable-ready device tuning itself to the Details channel.

The alert_priority field is coded in accordance with Table 18.4.

Note that priority level 11 events always interrupt viewing, regardless of whether a viewer is tuned to a premium channel or even a pay-per-view event. But priority level 11 events do not require the DCRD to output alert-related audio. Whenever the nature of the alert must involve an audio output, priority level 15 must be used.

As can be seen in the table, some priority levels are undefined. The specifications state that the DCRD must treat any value in the reserved range the same as the next-highest defined value.

details_OOB_source_id

Whenever the DCRD has access to an out-of-band channel, a Virtual Channel Table is available. The details_OOB_source_id field is a pointer to the Details channel that is usable when out-of-band SI is available (when a functioning POD module is in place in the cable-ready device). For cases when no Details channel is available, the value of the details_OOB_source_id field is set to zero.

details_major_channel_number and details_minor_channel_number

These fields are used only when the DCRD does not have access to an out-of-band channel. Instead of Source ID, they indicate the major and minor channel numbers associated with the Details channel. If no Details channel is available, the

TABLE 18.4 Alert Priority

Alert Priority	Audio Required	Meaning
0	No	Test message: all receiving equipment except equipment designed to acknowledge and process test messages discard priority-zero alerts.
1-2		[Reserved]
3	No	Low priority: the alert may be disregarded if processing the alert would interrupt viewing of an access-controlled service.
4-6		[Reserved]
7	No	Medium priority: the alert may be disregarded if processing the alert would interrupt viewing of a pay-per-view event.
8-10		[Reserved]
11	No	High priority: the alert must be processed unconditionally, but can involve text-only display if no audio is available.
12-14		[Reserved]
15	Yes	Maximum priority: the alert must be processed unconditionally. If audio is available without tuning to the details channel, that audio must be substituted for program audio for the duration of the alert message. If audio is not available by means other than by tuning to the details channel, the details channel must be acquired for the duration of the alert message.

details_major_channel_number is set to zero (which is not a valid value for a major channel number).

audio_OOB_source_id

In some cases (probably rare), an audio track pertinent to the emergency alert event is available in such a way that it can be accessed without interfering with the currently-tuned video stream. A non-zero value for audio_OOB_source_id indicates that such an audio-only channel is available (zero indicates no audio-only channel is present). This field is only usable in conjunction with out-of-band SI and must be ignored if the out-of-band communication path is not accessible.

The SCTE 18/EIA-814 standards do not state requirements regarding DCRD handling of the audio_OOB_source_id. Some set-top boxes owned by the cable Multiple System Operators use private means to access EA audio, such as to download a file via the network. Note that according to FCC rules an audio-only activation is acceptable for a Required Weekly Test.

alert_text_length

This 16-bit field indicates the length, in bytes, of the Multiple String Structure to follow. The section length limit of 4096 bytes restricts alert_text_length to something less than 12 bits. In some cases no alert_text() is provided; for example when the function of the cable_emergency_alert() is to force the DCRD to re-tune unconditionally to a Details channel. In that case alert_text_length is set to zero and no alert_text() field is present in the message.

alert_text()

This Multiple String Structure represents the textual description of the Emergency Alert that is to be scrolled slowly across the top part of the on-screen display. Assuming one can comfortably read two words per second, a typical word is five characters long, and Huffman coding reduces the size of a text string to 50% of the uncompressed length, two minutes worth of English text translate to an MSS of somewhere around 900 bytes in length.

location_code_count

This field is a count of the number of blocks of location data present in the cable_emergency_alert(). Up to 31 blocks may be present; this matches the maximum number of location data blocks that an EAS broadcast message can hold. Each location data block includes a state code, a county code, and a county subdivision code.

Note that receiving devices are not required to process the location data. If they are able to process it, they can discard events that are found not to involve their physical location.

state_code

An 8-bit unsigned number in the range from 0 to 99 that represents the State or Territory affected by the emergency alert. The state_code field is coded according to State and Territory FIPS number codes. Table 18.5 reproduces the FIPS codes for the US states and territories. County FIPS numbers are contained in the State EAS Mapbook. The value of zero indicates all states and/or US territories, meaning a national level alert.

county_subdivision

A 4-bit value that indicates the portion of a county to which this alert applies. It is rarely used, and typically applies only for very large counties. It is coded in accordance with Table 18.6. Notice that a value of zero is used when the alert applies to the whole county. When an alert applies to portions of a county but not the whole thing, the location data "for" loop may be used to indicate each subdivision in turn.

county_code

This 10-bit number identifies a county within the state given by state_code. It is coded according to State and Territory Federal Information Processing Standard

TABLE 18.5 FIPS State and US Territory Codes

State	FIPS#	State	FIPS#	State	FIPS#	State	FIPS#	State	FIPS#
AL	01	HI	15	MA	25	NM	35	SD	46
AK	02	ID	16	MI	26	NY	36	TN	47
AZ	04	IL	17	MN	27	NC	37	TX	48
AR	05	IN	18	MS	28	ND	38	UT	49
CA	06	IA	19	MO	29	OH	39	VT	50
CO	08	KS	20	MT	30	OK	40	VA	51
CT	09	KY	21	NE	31	OR	41	WA	53
DE	10	LA	22	NV	32	PA	42	WV	54
FL	12	ME	23	NH	33	RI	44	WI	55
GA	13	MD	24	NJ	34	SC	45	WY	56

Terr.	FIPS#	Terr.	FIPS#	Terr.	FIPS#	Terr.	FIPS#	Terr.	FIPS#
AS	60	FM	64	GU	66	MH	68	MP	69
PR	72	PW	70	UM	74	VI	78		

District of Columbia - 11

TABLE 18.6 County Subdivision Code

County Subdivision	Meaning
0	All or an unspecified portion of a county
1	Northwest
2	North Central
3	Northeast
4	West Central
5	Central
6	East Central
7	Southwest
8	South Central
9	Southeast
10-15	[Reserved]

(FIPS) number codes which are maintained by the National Institute of Standards and Technology (NIST) in FIPS PUB 6-4[6]. A value of zero indicates the entire state or territory. FIPS publications are available online. As of this writing, the county codes may be found at: http://www.itl.nist.gov/fipspubs/co-codes/states.htm.

exception_count

If a digital cable-ready device is already tuned to the Details channel or is tuned to a local broadcast station, there is no reason to process the cable_emergency_alert() message. The exception_count field indicates the number of iterations of the exception data "for" loop to follow, and indicates the number(s) of the channels that should not be tuned away from to process this alert. Note that one physical channel may appear twice in the list, once identified by its Source ID and once by its major/minor channel number.

in_band_reference

This flag simply indicates whether the channel reference to follow is an in-band reference (specifying major/minor channel number) or an out-of-band reference (specifying Source ID). If an out-of-band channel is present, a DCRD is expected to disregard exception channels referenced in-band and process only those specified by Source ID. If no out-of-band channel is available, those channels associated with out-of-band references are disregarded and only the major/minor channel number reference is used.

exception_major_channel_number and exception_minor_channel_number

For in-band references, these fields indicate the major and minor channel number of an exception channel. A value of zero for exception_minor_channel_number indicates that the channel is an analog NTSC service.

exception_OOB_source_id

For out-of-band references, this field indicates the Source ID of an exception channel.

descriptors_length

This is a 10-bit field indicating the number of bytes of descriptors to follow.

descriptor()

Zero or more descriptors pertinent to this EA event. Currently, no standard descriptors are defined for use with the cable_emergency_alert().

Processing Requirements for the DCRD

Support for EAS in receiving devices is mandatory. For a receiving device to qualify as truly digital cable-compatible, it must respond to the cable_emergency_alert() message and process it in accordance with the SCTE 18/EIA-814 standards. Unlike PSIP, which is silent on receiver requirements and only focuses on the format and meaning of transmitted data, these cable standards state clear requirements compliant cable devices must meet.

A cable-ready device must support EAS because federal law requires the EAS system to be able to switch, at the command of the President, all radio listeners and television viewers to an audio feed supplying information of immediate national importance. If some cable-ready devices were designed to disregard the cable_emergency_alert(), viewers of those devices would not be aware of the President's urgent message.

It follows then, that the requirements for processing the cable_emergency_alert() only apply when the cable-ready device is "on" (not un-powered or in a standby-power mode). A cable-ready device that included a functioning POD module would get EAS directives through signaling from the OOB signal. A POD-less Host (or a POD-capable device without a functioning POD module present) would have to rely on in-band signaling. Here is a summary of the processing requirements:

- The protocol_version field must be processed, and the table section must be discarded if a non-zero value is seen. This requirement ensures that a future extension to the ATSC or SCTE standards could use the same table ID value that is used by the cable_emergency_alert() but with different syntax, and existing devices would discard it (as they should, since they would not have been designed to support the new function).

- Duplicate messages must be discarded (this is discussed further below).

- The exception list must be processed to see if the currently tuned channel is included. If it is included, the cable_emergency_alert() must be discarded.

- The alert_priority field must be processed, and if the priority level indicates that the alert is not of a high enough priority to interrupt viewing, the cable_emergency_alert() must be discarded.

- If the alert priority requires audio, either the audio-only service indicated by the audio_OOB_source_id field must be acquired or the Details channel must be tuned and acquired.

- If tuning to the Details channel is not required and alert text is provided, the text must be scrolled slowly across the top of the screen. If the alert message reaches its end point before all of the text has been displayed, scrolling must continue until all of the text is displayed.

- At the end point of the alert message, the audio/video that was interrupted must be restored. If the DCRD has tuned to the Details channel, at that point it must return to the channel that was interrupted.

- Indefinite duration alerts must be handled. These are discussed below in conjunction with a flow diagram.

- Overlapping alerts must be handled. These are discussed below. Note that an overlapping alert (one that starts prior to the end point of a previous one) can be used to terminate a prior alert of an indefinite duration.

Figure 18.3 is adapted from the SCTE 18/EIA-814 standards and illustrates all of the required and some of the optional processing steps. A given implementation is not required to perform exactly the steps in exactly the order shown in the flowchart; the diagram is intended to be an illustration of one approach that results in a compliant design.

At the top, the flow diagram starts with the reception of a cable_emergency_alert() in the cable-ready device. In the example implementation, protocol_version is checked first and if it is found to be non-zero, the table section is discarded. Several steps related to the detection of duplicate messages follow. Next, the test is made to see if the current channel is listed in the exceptions list. If it is included in the list, this message is discarded. Otherwise, processing continues with a test of the alert_priority level. Note that it is the alert_priority that determines whether or not processing this alert must include an audio feed pertaining to the emergency. Only the highest level alerts, priority level 15, require audio.

In some cases processing the priority level results in the message being discarded. For example, the required periodic testing of the EAS system means that broadcast radio and television stations occasionally interrupt programming to conduct a test. But a digital cable-ready device will not interrupt video for system tests because the cable_emergency_alert() can be sent at priority level zero.

If the priority is high enough to continue processing, the cable-ready device may check the state and county codes against the state and county codes representing the location of the cable-ready device. This is only an option, of course, if these codes are known. A cable operator can arrange for a given subscriber's POD module to be loaded with these FIPS codes. The POD module would then be capable of performing a location-filtering operation on any out-of-band cable_emergency_alert() message that might come down the FDC channel, discarding any that did not apply to its location. It is also possible that the user could be asked to supply the state and county of residence from on-screen menus, but this is unlikely as the number of counties in the US is quite large and many users would probably skip this step.

Next, audio output is provided if it is available. In some MSO-owned set-top boxes, an audio source may come via proprietary means. Sometimes, the audio_OOB_source_id can be used to acquire and output an audio track. If audio is

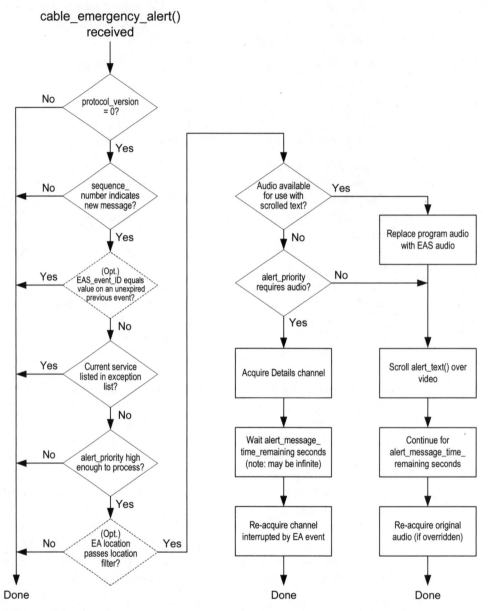

Figure 18.3 Processing the Cable Emergency Alert Message

not available but the priority level is 15 (meaning that audio is required), the Details channel must be acquired. The cable-ready device is required to stay on the Details channel until the end point of the alert message is reached. In the case of an indefinite duration message, it stays there until a new alert overrides the indefinite-duration one.

For those alerts with priority level less than or equal to 11, audio is not required. For these, display of text is sufficient. At this point in the process, the text is scrolled slowly across the top of the screen until all of it has been displayed. If the last word of the text has been displayed before the end point of the alert message is reached, the text display can be re-started from the beginning if desired. Finally, after all of the text is displayed and the end point of the message has been reached, the audio and video can be restored to what it was before the emergency alert event occurred.

Duplicate detection in the DCRD

Some further discussion of duplicate detection may be helpful. A cable operator may send several copies of the same cable_emergency_alert() to overcome noise on the communication channel. Duplicate alert messages are expected to be discarded. DCRDs are expected to recognize a cable_emergency_alert() as a duplicate of one previously sent by comparing the sequence_number field to the sequence_number of the last alert message seen on this same Transport Stream. Any time a DCRD acquires a new Transport Stream after having been tuned to another analog channel or a digital service on another TS, the DCRD must invalidate the sequence_number that would have been used for duplicate detection and accept any cable_emergency_alert() message seen, regardless of its sequence_number.

At the option of the DCRD manufacturer, a cable_emergency_alert() may be recognized as being exactly the same as one processed previously even though it has a different sequence_number. The EAS_event_id associated with emergency alerts that have been processed but that are still "in progress" can be saved and compared with the EAS_event_id of incoming cable_emergency_alert() messages. Here, "in progress" means that the end of the event has not been reached (recall that the time of the end of the event is computed by adding the event_duration to the event_start_time).

As an example of how EAS_event_id can be used, let's say an event with ID "A" comes in and is processed. For this example A has a one-hour lifetime (maybe it's a storm watch for the next hour). Twenty minutes after A, an event with ID "B" comes in and is processed. Now, if ten minutes later a cable_emergency_alert() identified with EAS_event_id A comes in, the cable-ready device can choose to disregard it. Note that if a cable operator wishes to ensure that no cable-ready device discards an event based on EAS_event_id, each new event can be assigned a unique value for that field.

If EAS_event_id is used for duplicate detection, the cable-ready device must discard stored events when they expire (that is, when the end point of the alert event has been reached).

Overlapping alert events

It is possible that a cable_emergency_alert() message is received (let's call it message "B") before the end point has been reached on a prior one (call the first one message "A"). If this happens, B must be processed to see if it is a duplicate. If it passes all of the initial tests (those that would result in the message being discarded if they don't pass), B overrides A and the cable-ready device is required to perform the following steps:

1. If text is being displayed, it must be erased. If audio has replaced the program audio, the original program audio must be restored.

2. If the processing of message A involved tuning to a Details channel, and processing B would result in either not tuning to a Details channel or tuning to a different Details channel, the channel that had been interrupted by A event must be re-acquired.

3. The new cable_emergency_alert() message must be processed.

The logic of the second step is designed to prevent tuning away from the Details channel (to reset and be ready for the next message) followed by coming right back to it (as a result of processing the overlapped event). Instead, if the overlapping event involves being on the same Details channel we're already on, there is no need to leave and come back. This logic means that a series of overlapping events can have the effect of extending the time period during which the cable-ready device should stay on the Details channel. If no further events are received, the last one reaches its end point and the alert is over.

Typical EAS events do not come in one on top of the other because, for example, a point of origination such as a National Weather Service station can provide some space between them. It is also possible for the cable headend to buffer them and provide some extra delay as needed. A national-level alert (EAN), however, can come in without warning and could collide with a weather-related alert. An EAN may interrupt a prior alert event that still may be in progress. Figure 18.4 shows an example of an EAN interrupting a Tornado Watch alert.

Referring to the Figure, at 3:00 p.m., a Tornado Watch (TOA) emergency alert is issued. It has a specified message duration of two minutes, but at 20 seconds after 3:01 p.m. an EAN alert is issued. Since the TOA had not reached its end point, the EAN overlaps it. The state of the receiver when the TOA was first received is restored, then the EAN is processed. This EAN has an indefinite duration, and it

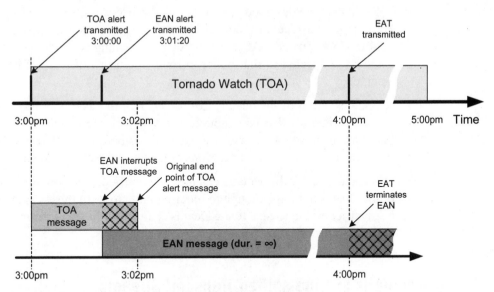

Figure 18.4 Example of Overlapping Emergency Alerts

goes on until terminated at 4:00 p.m. by an Emergency Action Termination (EAT) message.

Optional EAS-Related Features in the Cable-Ready Device

We've mentioned some of the processing options a digital cable-ready device has when processing the cable_emergency_alert(). These include:

- The EAS_event_id can be tested to see if the alert is a duplicate of an active alert.

- The state, county, and county subdivision codes can be processed and the message discarded if this DCRD is not within this alert's applicable region.

There is another class of optional processing as well relating to EAS-related user features. These are simply convenience features that could be offered in higher-end or specialized products. Some examples include:

- The DCRD can collect EA events when in "off" (standby power) mode, so that all of the information pertinent to any active events can be displayed to the user when the device is turned back on

- The DCRD can maintain a user interface screen to list currently active events. All of the information about each event can be stored along with the text; the user can re-display it if desired.

- A DCRD can be designed to respond in specific ways to certain types of alerts. For example, a receiving device can supply power to an AC outlet when any kind of flood-related EAS event is seen. Something like a bell or a bed-shaker could be connected to that outlet to rouse the user in the event of that type of emergency.

- At the discretion of the designer of the DCRD, the user may be given a way to cancel the audio/video interruption involved with the reception of a CEAM, at that point restoring audio/video to what it was before the alert message arrived. A "Cancel" key on the Remote Control Unit could be wired to such a function.

- In some cases, a valid Details channel is identified even when the priority level of the alert is not high enough to require tuning to it. In those cases, the user can be given an option to tune to the Details channel for further information pertinent to the alert event.

Emergency Alerts in Time-shifted Transport Streams

With the advent of digital audio/video disks and personal video recorders, it is possible to store a Transport Stream on tape or disk for later playback. With A/V hard disks, "later" may be just a second or two, since these disks can play and record (read and write) simultaneously. In any case, any stored TS may include Cable Emergency Alert Messages.

Depending upon how much time has elapsed between recording and playback, some alert events will have expired. When time-shifted material is played back, the current time of day should be compared with the event_start_time and event_duration given in the cable_emergency_alert() message. Any EA events that have expired should be discarded. Note that the System Time Table will be available on playback. It indicates the time of day the recording was made. By processing the stored STT, the exact amount of time shift can be determined. Knowledge of the time shift can allow proper processing of some types of EA events even when they are time-shifted.

References

1. "Report and Order and Further Notice of Proposed Rule Making (First R&O)," Amendment of Part 73, Subpart G, of the Commission's Rules Regarding the Emergency Broadcast System, FO Docket 91-171/91-301, 10 FCC Rcd 1786 (1994).

2. SCTE 18 2001 (formerly DVS 208), "Emergency Alert Message for Cable."

3. ANSI/CEA/EIA-814, "Emergency Alert Message for Cable," American National Standards Institute, Consumer Electronics Association, Electronic Industries Association, 2001."

4. FCC Title 47, Code of Federal Regulations, Part 11, "Emergency Alert System (EAS)," Federal Communications Commission, Washington D.C.

5. FCC Second Report and Order, "Amendment of Part 73, Subpart G, of the Commission's Rules Regarding the Emergency Broadcast System," Federal Communications Commission, Washington D.C., FO Docket 91-171 and FO Docket 91-301, adopted September 24, 1997.

6. "Counties and Equivalent Entities of the United States, Its Possessions, and Associated Areas," Federal Information Processing Standards Publication 6-4, 31 August 1990, http://www.itl.nist.gov/fipspubs/fip6-4.htm

NCTA/CEA PSIP Agreement

Two important industry agreements were announced on February 23, 2000 in a joint press release[1] from the National Cable Television Association (NCTA, now called the National Cable Telecommunications Association) and the Consumer Electronics Association (CEA). We have described the essence of the agreement commonly known as the "Technical Agreement" in our discussion of the cable network interface. That agreement documented RF, modulation, and signal strength and quality parameters that allow retail devices to tune and acquire the QAM-modulated digital carriers on a compliant cable system. SCTE 40[2] and SCTE 54[3] are the standards documents that have now captured the specific details of the Technical Agreement.

The second agreement is called the "PSIP Agreement" and it deals with the general problem of Electronic Program Guide data on cable (the Agreement is included in this book in Appendix C). In this chapter we discuss the philosophy of the agreement and explore some details of how it is intended to work.

In the months leading up to February, 2000, lively discussions between consumer electronics companies and representatives from NCTA, CableLabs, and the cable operators had taken place over the question of EPG data and how a retail cable-ready device might access the same type, amount, and quality of program guide data that an operator-owned set-top box could access. The CE community felt that they couldn't compete in the retail market if they couldn't offer attractive and fully-featured navigation applications such as program guides with full program schedules and descriptions.

Representatives from the cable side maintained that contractual obligations prevented them from sharing with the retail devices the same EPG data they supplied to their own set-top boxes. Furthermore, the cable operators supplied (via the out-of-band channel) EPG data formatted in proprietary ways to these boxes. Various cable operators throughout the country used different protocols and methods for delivery of the data, so even if the contractual problems could be worked out, a retail device would have to implement many different protocols. Most CE companies agreed that the number of different methods used throughout the country was so big that it was impractical to build a truly portable cable-ready device—one that could access EPG data from any system across the US it was plugged into.

Overview

In broad terms, the PSIP Agreement represents a kind of breakthrough in that it looks at the problem in two new ways. First, it says in essence that a retail cable-ready device can expect to get program guide data in-band, not out-of-band. Second, it places the responsibility on the service provider for making EPG data for a given cable service, not on the cable operator. Just as PSIP data for terrestrial broadcast is required to be supplied by each broadcaster, in this way of looking at the world a program provider who delivers audio and video programming to a cable headend or cable uplink distribution center is expected to provide PSIP data to describe his or her programming.

With the PSIP Agreement, cable multiple-system operators did not avoid any work or responsibility, however. For cable distribution systems in existence in 2000, even if PSIP data accompanied audio/video data coming into the cable system from a content provider, there was no guarantee that that data would make it through to the terminal devices in subscribers' homes. The path taken by the service provider's content is often circuitous. Starting from its origin in some location in North America, the service may be delivered via satellite to a cable distributor's main uplink facility, where it is processed and perhaps combined with other services and then uplinked again through a second satellite hop for reception at a cable headend. At the cable headend, the service may undergo further processing, again involving adding or deleting services that may have been carried with it in the multiplex. Finally, a Transport Stream carrying the service is modulated and transmitted down the cable.

Remultiplexing

One of the provisions of the PSIP Agreement was that cable MSOs would work to upgrade equipment in their headends and distribution and processing centers so that PSIP data would survive remultiplexing. Figure 19.1 is a simplified diagram of a remultiplexer, showing three Transport Streams coming in and one coming out. Each of the incoming streams includes PSIP describing the services on each Transport Stream. After remultiplexing, the output TS must include PSIP describing only those services actually in the output multiplex.

One common example of a remultiplexer at a cable headend is one that is capable of taking two 19-Mbps terrestrial broadcast multiplexes and combining them into a multiplex with a 38.8 Mbps information rate for cable transmission via 256-QAM

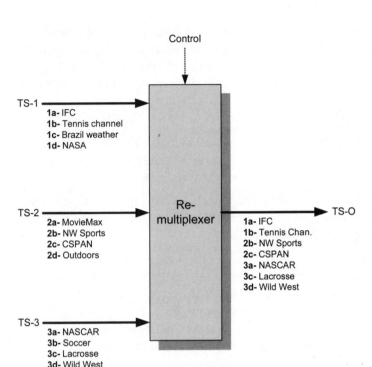

Figure 19.1 Remultiplexer at a Cable Headend

modulation. Again, PSIP in the two incoming Transport Stream must be taken apart and combined together.

Provisions of the agreement

As noted, the provisions of the agreement involved the carriage of PSIP data through the cable distribution chain, and did not involve any commitment on the part of cable operators or MSOs to create it. The agreement lists the following requirements regarding navigation data describing cable services:

1. **Virtual channel maps**: Virtual channel tables will be made available to describe all available audio/video services. The agreement states that channel maps are to be provided in the out-of-band channel in accordance with SCTE 65[5] for scrambled services, and that services carried in-the-clear must include PSIP in the same TS that carries the service.[*]

---

[*] Later, it was acknowledged that in-the-clear services might be present for which no PSIP data was available, but that there was no expectation that POD-less receivers would be able to access them.

2. **Channel names**: Every channel will have a one- or two-part channel number and a textual name (for example "HBO"). This commitment is important because the lower profiles of the out-of-band SI specified in SCTE 65 do not require channel names to be sent, only channel numbers. Therefore, this statement in the PSIP Agreement indicates a commitment on the part of cable MSOs to move beyond the bare-bones requirements for delivery of out-of-band SI data.

3. **Event information**: This provision calls for Event Information Tables to cover the same time period as is specified in ATSC A/65 for terrestrial broadcast: twelve hours, representing the first four EITs.

4. **PSIP data bandwidth**: The agreement says that cable operators may limit the total bandwidth of PSIP data in the TS to 80 Kbps for a 27-Mbps multiplex or 117 Kbps in a 38.8-Mbps multiplex.

5. **Program descriptions**: The PSIP Agreement states that carriage of PSIP Extended Text Tables is "desirable but optional." In a recent ruling on must-carry issues, the FCC stated that a cable operator would not be allowed to delete data relating to the primary program. ETT data describing the current program would most likely qualify, so it appears that the cable operator must ensure the survival of ETT-0 through the cable distribution chain.

6. **EPG data**: While the primary focus of the Agreement involves in-band EPG data in the form of EITs and ETTs, the Agreement says that, at the option of a cable operator, some portion of the EPG data describing the services on the cable plant can be provided out-of-band. When sent out-of-band, the Agreement states that the Aggregate EITs and ETTs defined in SCTE 65 are to be used. Receivers should be designed to collect EPG data from either in- or out-of-band sources.

7. **Analog TSID**: PSIP data can refer to analog services. The PSIP Agreement states that the cable-ready receiver is expected to use the Analog TSID (defined in Sec. 9.5.2.4 of EIA/CEA-608-B[4]) to unambiguously link the PSIP data with the analog service it references. Likewise, the Agreement states that cable operators must ensure that the analog NTSC signals include the Analog TSID within their line-21 VBI data whenever those services are referenced in an in-band VCT.

8. **Precedence of channel numbers**: The Agreement states that for scrambled services, channel numbers found on in- and out-of-band SI data may not agree with one another. It states that cable-ready devices must use the channel numbers given in out-of-band SI if an out-of-band channel is available. It also states that channel numbers found in in- and out-of-band SI for in-the-clear services "should" match each other. It says, however, that it may not be possible in all systems to ensure that they do match.

Cable distribution scenarios

Several different scenarios are described in the PSIP Agreement, each representing a different aspect of PSIP handling through a cable headend or the satellite distribution channels. These are discussed in the following sections.

Intact multiplex

In this simplest scenario, a TS multiplex arrives at the cable headend via either a satellite downlink or by reception at the headend of a digital terrestrial broadcast signal. The cable operator wishes to provide all of the services in the Transport Stream to their subscribers. In accordance with the PSIP Agreement, no PSIP data in the incoming TS may be stripped out prior to delivery at the cable plant. Figure 19.2 diagrams a satellite case on the left and a terrestrial broadcast case on the right.

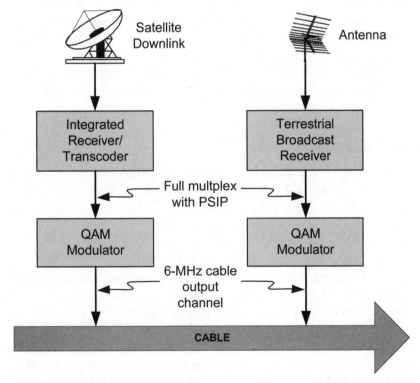

Figure 19.2 Intact Multiplex Case

Reception of a satellite multiplex typically requires a downlink facility where the dish feeds an Integrated Receiver/Transcoder (IRT). Aside from tuning and demodulating the signal from the satellite transponder, one of the primary functions of the

IRT is to descramble any conditional access scrambling that might have been applied over the satellite link by the content provider. In the Figure, the Transport Stream coming out of the IRT contains all of the services in the original satellite signal, including any PSIP data that might have been present. The up-converter QAM-modulates the TS onto one of the 6-MHz cable RF channels.

Note that in this scenario (as diagrammed in the Figure anyway) no cable-side conditional access is applied to the services being output. Therefore, cable-ready receivers without POD modules would be able to access them. It also follows then that these cable-ready receivers would use the Virtual Channel Table that has been supplied by the content provider, not one supplied by the cable operator. For this to work, some careful coordination of channel numbers is needed between the content supplier and the cable operator. If such coordination is not possible, the cable operator must replace the VCT supplied in the satellite feed with one generated locally to the cable operator's specifications. Note that for cable-ready receivers with functioning POD modules installed, the out-of-band path supplies the channel maps, the contents of which are under full control of the local cable operator.

On the right side of the Figure, a local digital terrestrial broadcast multiplex is received at the cable headend and down-converted to baseband by the terrestrial broadcast receiver. That resulting Transport Stream is then up-converted and QAM-modulated to a 6-MHz cable channel. PSIP in the off-air signal remains intact, and cable-ready devices should have no trouble using it.

Re-encoding of satellite feeds

Although the world is moving to all-digital formats throughout content distribution chains, we are not there yet. Cable headends today often receive content via satellite in scrambled analog NTSC format. For example, the de-facto standard for C-band satellite feeds has been General Instrument's VideoCipher^{TM} 2 system (GI is now part of Motorola), which uses analog video scrambling with digital audio scrambling. Another scenario is a satellite feed arriving at a cable headend downlink in digital form, but that feed is received, descrambled, and converted to an analog baseband feed before being re-encoded and re-scrambled at the cable headend. Figure 19.3 is a simplified diagram of these scenarios.

In the Figure, Integrated Receiver-Decoders (IRDs) are used rather than IRTs. Each IRD creates baseband audio and video (analog composite NTSC and left/right audio channels) from either digital or analog satellite feeds. For digital feeds, the IRD tunes and demodulates a Transport Stream and descrambles and outputs one service from the multiplex. The Figure illustrates the fact that the cable operator may use an MPEG-2 video encoder to combine a selection of analog feeds from a bank of IRDs into a digital multiplex according the desired specifications.

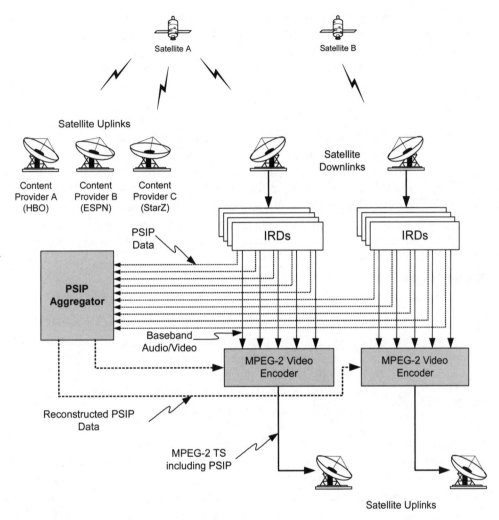

Figure 19.3 Re-encoding Satellite Feeds

We discuss the analog re-encoding scenario here in relation to the PSIP Agreement, wherein the cable community pledged to retain any PSIP data arriving with audio/video content, as that content is processed through the cable distribution chain. To meet this goal, as shown in the Figure, the IRDs must not only output baseband audio/video, but they must also provide any PSIP data that may have accompanied the content.

PSIP data can travel in the satellite signal in one of two ways. For digital feeds, which use MPEG-2 transport, PSIP is carried in the standard way. For analog feeds, PSIP data can be carried in the Vertical Blanking Interval of the NTSC waveform, in Extended Data Service (XDS) packets in Field 2 of line 21. ANSI/EIA/CEA-

608-B *Line 21 Data Services*[4] defines the syntax and structure of PSIP data carried in XDS packets. PSIP data including minor channel number, event number, event start time and duration, program title and description, and content advisory and caption services information can be sent in this manner.

The PSIP data must be collected and processed in some new piece of equipment, called here a "PSIP Aggregator." The PSIP Aggregator must be configured to create a valid PSIP data stream suitable for each encoder, and deliver that data to each encoder so it can be multiplexed into its output Transport Stream.

Summarizing the re-encoding case of Figure 19.3, the cable community committed to include new functionality in IRDs in order to allow them to process and output PSIP data arriving in the satellite signal via a suitable standard interface. Also, they pledged to specify and build an entirely new piece of equipment, the PSIP Aggregator. Finally, arrangements had to be made for MPEG video encoders to be able to accept PSIP data from the PSIP Aggregator via a suitable interface.

Remultiplexer

Of course, re-encoding is undesirable because it can introduce impairments in video quality, and multi-channel MPEG-2 video encoders are typically complex and expensive. All-digital solutions are preferable. Figure 19.4 illustrates a cable headend scenario wherein three digital multiplexes are received via satellite downlink. The RF signals from the dishes are fed to Integrated Receiver-Transcoders, which demodulate and descramble them down to baseband Transport Streams.

Each of the three satellite transport multiplexes is fed to the remultiplexer unit in the center of the Figure. It is the job of the remultiplexer to accept several Transport Streams, and under the control of the cable operator, produce one output TS that contains only those services of interest. To produce an MPEG-compliant output Transport Stream, it must look for PID and program_number collisions among the incoming streams and correct them by re-mapping, as needed. It must create a Program Association Table appropriate for the structure of the output multiplex, and it must edit the Program Map Tables as necessary to account for the re-mapping. Several companies offer products that are able to perform these functions.

In order to meet the objectives of the PSIP Agreement, however, this remultiplexer must also accept PSIP data within incoming Transport Streams, take it apart, and reassemble it for output. As we discussed earlier, the Virtual Channel Tables have to be re-built, any collisions in Source ID have to be dealt with, and the MGT must correctly reflect the transmitted tables.

Figure 19.4 illustrates a remultiplexing scenario in which satellite dishes at a cable headend facility bring signals to Integrated Receiver-Transcoders. The IRTs tune, demodulate, and descramble the digital signals, producing a full MPEG Transport Stream output. Each TS represents the signal received from one transponder on a given satellite. In the Figure, the three Transport Streams feed into the Remultiplexer in the center. Configured by and under control of the cable operator, the Remultiplexer extracts services of interest from the three incoming multiplexes to create an MPEG-2-compliant output Transport Stream in which all of the PSIP data pertinent to services included in the output multiplex is present.

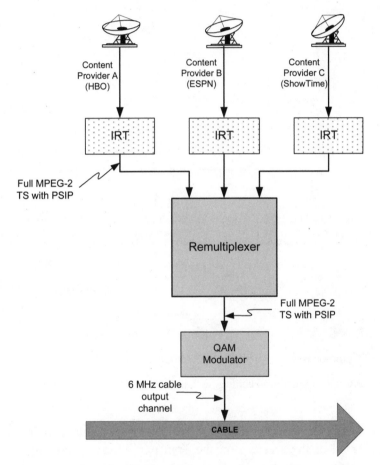

Figure 19.4 Remultiplexing Satellite Feeds

Finally, the Remultiplexer output is placed on a QAM-modulated 6-MHz cable RF channel.

Master downlink feeding multiple systems

An architecture used by several cable systems in the US, including some owned by AT&T Broadband and Shaw Communications, involves distribution of one Transport Stream received through an IRT at a satellite downlink to several separate cable plants. Figure 19.5 illustrates this scenario.

In the Figure, a satellite dish supplies an RF signal from a selected satellite to an Integrated Receiver-Transcoder (IRT). The IRT tunes, demodulates, and descram-

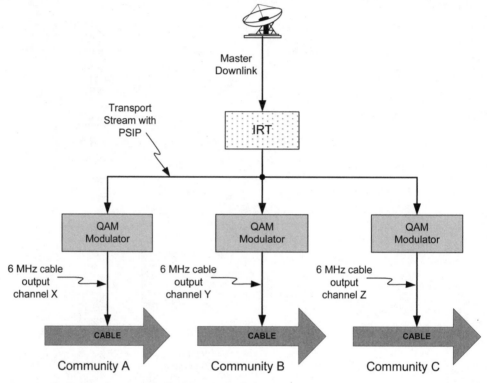

Figure 19.5 One Downlink Feeding Several Cable Systems

bles the services taken from one transponder of the satellite. It is capable of re-scrambling all of the services on the Transport Stream according to conditional access parameters appropriate to the system or systems it feeds.

In the example illustrated in the Figure, the IRT feeds three cable systems, each located in a different community. The PSIP Agreement document mentions, as one example, that a downlink site at the Headend In The Sky (HITS) facility in Denver feeds the nearby communities of Littleton, Boulder, and Castlerock, Colorado. At each community, the Transport Stream is QAM-modulated, placed on some 6-MHz cable channel, and transmitted down the cable. In the Figure, the signal fed to Community A is placed on cable RF channel X, while Community B uses channel Y and C uses channel Z. Don't forget that these "channels" reflect the 6-MHz bands used to transmit the signals, not the users' notion of channel numbers. Since the signals are digital multiplexes, it is a Virtual Channel Table that indicates the channel numbers subscribers use to access the services.

If the downlinked signal includes PSIP data describing services in the multiplex, in this scenario, clearly that PSIP data is preserved in the QAM-modulated signals transmitted at each community site. In this scenario, then, the overall goal of PSIP preservation is achieved.

The scenario in Figure 19.5 illustrates another system issue—channel numbering and the consistency of in-band and out-of-band channel numbering. Recall that provision #8 of the Agreement stated that in- and out-of-band channel numbers "should" match for services transmitted in the clear, but that it would not be possible to ensure such a match in all cable system architectures. Cable MSOs have stated their intention on several occasions to scramble all digital services except for those derived from terrestrial DTV content, and the PSIP Agreement re-iterates that statement of intent.

Looking one more time at Figure 19.5, it is clear that the same digital multiplex is delivered to the three cable systems. It is often true that each of the local communities wishes to assign virtual channel numbers to these services independently from those assigned in the other communities. Since no service in the multiplex is sent in the clear, a POD module is needed to take care of CA descrambling, and so any cable-ready device that is able to access the services also has access to the out-of-band channel. It follows, then, that each community can assign the channel numbers to services in this multiplex according to a Virtual Channel Table delivered by the local operator via the out-of-band channel.

It also follows that the out-of-band channel numbers will very likely not match up with the in-band channel numbers, because the PSIP VCT delivered with the in-band multiplex is the same one received at the satellite downlink dish and it cannot be correct for everybody. Digital cable-ready devices are expected to take channel numbers from an out-of-band channel, if and when it is present, and use these in preference to any channel numbers found in-band.

As we reiterate elsewhere in this book, a receiver can (and should) collect and use information from both in- and out-of-band sources (see "Channel numbering" on page 323). For example, if the out-of-band SI does not include channel name data, channel names found in in-band PSIP Virtual Channel Tables may be used. Perhaps more importantly, EPG data (EITs and ETTs) found in-band are entirely usable in the absence of out-of-band EPG data for a given service.

Distribution of a terrestrial broadcast multiplex

Before we leave the discussion of channel numbering issues, let's look at a scenario where one DTV terrestrial broadcast signal is distributed among several local cable operators. Figure 19.6 shows an example where several community cable systems in a certain geographic region each carry the DTV signal from a local terrestrial broadcaster. As with the previous example, the community cable systems are labeled Community A, B and C.

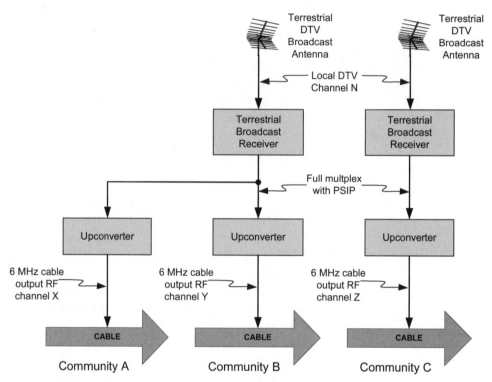

Figure 19.6 Cable Distribution of a DTV Terrestrial Broadcast Signal

In the example shown, a broadcast antenna feeds a terrestrial broadcast receiver to produce a Transport Stream that is distributed to Communities A and B. A second terrestrial broadcast receiver tuned to the same RF channel outputs the same Transport Stream for distribution at Community C. As with the satellite example, the terrestrial broadcast TS is up-converted at each community site with each potentially using a different 6-MHz frequency band.

Now, since the terrestrial broadcast multiplexes are typically sent in-the-clear (these days, anyway), it is possible for a DTV receiver in any of the communities to receive and decode these services even if a POD module is not present. Without a POD module in the cable-ready device, an out-of-band channel will not be available. Therefore, in-band PSIP data must be present and correct to support the navigation functions of POD-less cable-ready devices.

The terrestrial broadcaster may include only a Terrestrial Virtual Channel Table in the broadcast multiplex. At the broadcaster's option, a Cable Virtual Channel Table may also be transmitted. Here we see clearly the need for cable operators to work with broadcasters to coordinate the contents of transmitted PSIP data. As long as all the communities and all the cable operators in a region who carry a given terrestrial broadcaster's signal can agree to consistent channel naming and numbering, that broadcaster can include a Cable VCT in the transmitted signal, and it will not be necessary for any of the cable service providers to re-process the broadcast signal. It should be in everyone's best interest, the public included, to have regionally-consistent channel identification. The effectiveness of printed advertising and the usability of printed program guides is greatly improved by such agreement and coordination.

Current Status

At the present time, the National Cable Telecommunications Association continues to work with CableLabs and cable MSOs to implement the provisions of the PSIP Agreement. These parties are also working with equipment vendors including Terayon Communication Systems Inc., Scientific Atlanta, the Broadband Communications Sector of Motorola, Inc., and Triveni Digital, to complete system components such as PSIP aggregators and remultiplexers that are capable of preserving all relevant PSIP data in the output Transport Stream. At the time of this writing, no such product announcements have yet been made.

References:

1. Press release: "CEA and NCTA reach agreement enabling compatibility between cable television systems and digital televisions," National Cable Television Association and Consumer Electronics Association, February 23, 2000.

2. SCTE 40 2001 (formerly DVS 313), "Digital Cable Network Standard," Society of Cable Telecommunications Engineers.

3. SCTE 54 2002A (formerly DVS 241), "Digital Video Service Multiplex and Transport System Standard for Cable Television," Society of Cable Telecommunications Engineers.

4. ANSI/EIA/CEA-608-B, "Line 21 Data Services," American National Standards Institute, Electronic Industries Association, Consumer Electronics Association.

5.   SCTE 65 2002 (formerly DVS 234), "Service Information Delivered Out-of-band for Digital Cable Television," Society of Cable Telecommunications Engineers, 2000.

PSIP Implementation

In conclusion we discuss some implementation issues that may be of interest to engineers involved with PSIP in one form or another. We review considerations applicable to those building terrestrial broadcast receivers and cable-compatible devices, and issues of interest to broadcast station and cable system operators.

General Considerations for Receiver Designers

This book has covered a wide variety of PSIP-related design considerations applicable to those building terrestrial broadcast DTV receivers and cable-ready devices; we summarize some of these below. The reader is also encouraged to study the CEB-12 *PSIP Recommended Practice*[1] published by the Consumer Electronics Association.

A number of the considerations are specific to a particular type of receiver (broadcast or cable) but many apply equally well to any transmission medium. When building any device that is to receive and process PSIP data, design aspects described in the following sections should be considered.

Compatibility with protocol revisions

In Chapter 14 we discussed issues relevant to PSIP expandability via protocol update. The reader is encouraged to review Chapter 14 and receiver designers are encouraged to create compliance tests to verify the implementation for these aspects. In summary, the following issues related to protocol revisions should be considered:

- **Reserved fields:** Designers should make sure their implementation really ignores reserved fields. Setting reserved values to ones is usually sufficient because this generally uncovers errors in bit masks.

- **Protocol version:** Table sections in which the value of the protocol_version field is non-zero must always be discarded, as no table types are currently defined for versions other than zero. Designers should test to make sure that for all supported table types, an instance of that type of table with a non-zero protocol version is ignored.

- **Unsupported values:** One should include types of tables and descriptors in the test bit stream that are not supported by the implementation just to ensure that they are properly ignored. The types of tables and descriptors should include ones not yet defined in the standards as well as any that are currently defined but not supported.

- **Optional fields:** Certain fields and data structures are optional (they may or may not be present in the transmitted PSIP bit stream). Designers should verify that their implementation can properly handle the various optional elements in PSIP tables. For example, descriptors may or may not be present in certain locations in the PSIP tables. A design should be tested to make sure it properly ignores descriptors that may be present even if the design does not support extraction of useful data from a particular descriptor location.

- **Length extensions:** As a result of a protocol revision, new bytes can appear in certain places. For example, descriptors can have bytes added just following the byte last to be defined in the prior version of the protocol. Designers should construct and test with bit-streams that include descriptors with extra bytes added at the end. Table sections themselves can be length-extended in many cases as well.

New stream types

From time to time, the ATSC or SCTE may define new stream types. Receiver designers should create test programs with unsupported stream type code values to make sure they are properly handled (in other words, ignored).

Text coding

Any device that must accept textual data delivered in PSIP tables must support parsing and processing of the Multiple String Structure. Receiver designers must make sure their design can properly handle the various formatting options possible with the MSS. These include:

- multi-lingual text.

- Huffman-encoded text strings using either of the two standard encode tables.

- strings that include both compressed and uncompressed segments.

- strings that are too long to display and thus must be truncated.

- unsupported values of language code, compression_type, or mode.

Not all implementations will be able to render all of the characters defined in the Unicode Basic Multilingual Plane because a font definition is not likely to be available for all of these characters (there are more than 40,000). In any case, a receiver design should be tested to make sure it can gracefully handle the case wherein a certain character or characters in a text string cannot be rendered. An acceptable result may be to display a rectangular character-sized box on-screen for unsupported codes. An unacceptable result would be a software crash.

High bit-rates

A PSIP implementation may work fine in a test environment but fail in a field environment if the lab test uses a transmitted bitrate that is low compared to the real-world situation. Designers should create a test setup in which the maximum allowed bitrate is used in the test to ensure the design can handle it.

Decreasing channel acquisition time

The interval of time between the moment a viewer requests a change to a new channel and the moment the first video frame appears on that channel should be as short as possible. Channel acquisition time for a digital service is always going to be longer than it was in the analog world because of the pipeline delay inherent in the MPEG-2 video decoding process. A decoder must wait for intra-frame coded video to begin the process.[*]

As we have noted, one technique that can be used to decrease acquisition time is to record the last-used PID values for audio, video and the PMT for each channel. When it returns to a virtual channel, the receiver can assume that these PID values have not changed since the last time the service was acquired. More often than not that assumption will be valid and the time it would have taken to wait for and parse the Service Location Descriptor in the Terrestrial Virtual Channel Table (or for cable, the Program Map Table) will have been saved.

Whenever cached PID values are used the receiver should always process the Service Location Descriptor or PMT to re-verify them. If the stored values turn out to be wrong, the situation can be corrected easily—usually before any visible output appears.

---

[*] Intra-frame encoded video does not depend upon prior or later frames for decoding.

Managing changes to the virtual channel table

Even though a certain programming service's identity (its name and number) may not change, the physical path that service traverses to reach the receiver may occasionally move. In the world of analog NTSC television, a receiver might not offer the viewer a certain channel if it had not detected a signal when the last "channel scan" procedure was done. Likewise, until the viewer initiates another virtual channel scan a new digital channel may not be accessible.

Situations in which the transmission path used to deliver a certain virtual channel can change include:

- in the terrestrial broadcast application, a terrestrial broadcaster's RF channel is changed in response to a request by the FCC.

- also in the terrestrial broadcast application, a service like a community access or educational channel that is not affiliated with the broadcaster might move from one multiplex to another as carriage agreements change.

- in the cable application, the cable operator may change the channel line-up, adding, moving, or deleting services.

Cable or terrestrial broadcast receivers must be designed to either automatically (or under user control) manually accommodate these changes. If changes are not properly managed, the receiver will not be able to offer the viewer access to programming services that are accessible and therefore should be offered.

Considerations for Terrestrial Broadcast Receiver Designers

Designers of terrestrial broadcast receivers might consider the issues outlined in the following sections.

Unusual environments

As we have described the concept of virtual channels, we've seen some of the typical and atypical environments in which a terrestrial broadcast receiver may find itself. These include cases where the receiver:

- is operating in a mobile environment such as a recreational vehicle.

- is equipped with a movable antenna.

- can receive the same Transport Stream on two or more carrier frequencies due to the presence of one or more translators.

Mobile environment

A DTV receiver might be installed in a motor home or taken on a camping trip. Users will quickly become aware that their receiver will not be able to pick up local stations unless they initiate a channel scan similar to the one they did when they originally installed the receiver at home.

In some instances, it will be helpful to re-initialize the receiver to the same state it was in when it was first taken out of the shipping carton. In that state it had no memory of any virtual channels or RF carriers. A re-initialization such as this can be useful if the DTV is moved to a new broadcast area and the user knows that none of the channels originally receivable will be accessible in the new location.

In a mobile environment the situation is somewhat different. In this case, an "additive scan" may be appropriate. In an additive scan, Transport Streams found that have never been seen before are saved in memory along with the PSIP data collected from them. The record of a Transport Stream that had been found before but is inaccessible at the current moment is retained. We can be assured of the identity of any Transport Stream by its TSID value because TSID values in use in North America are unique.

After a number of additive scans have been done as the traveler moves around the country, the receiver may have knowledge of a large number of virtual channels. Most may not be accessible at any given time (from a given location). Similar to the "scan" button on a radio, however, the DTV can use the stored data to search quickly and efficiently for available services.

Movable antenna

A reception scenario involving a movable antenna was discussed in Chapter 5 (see "Directional antennas" on page 107). As we saw in Figure 5.6 on page 108, a DTV receiver may be fed with a signal from a directional movable antenna. A Transport Stream that had been available at some prior time may not be available now due to the position of the movable antenna. Just as in the mobile DTV receiver case, an additive scan can be a helpful way to handle the situation.

An initial (destructive) scan can be made with the antenna in one position, then the antenna can be moved to another position and an additive scan performed. When the user selects a certain virtual channel, the Transport Stream carrying that channel may or may not be available. If it isn't available, instead of indicating "invalid channel" the receiver can say "signal currently unavailable." Or, if the receiver supports movement of the antenna via control signals it could steer the antenna to the position needed to receive the signal and the requested virtual channel.

Broadcast translators

As we have noted, a Transport Stream with the same value of TSID may be found on more than one carrier frequency if signals from one or more broadcast translators are received. Considerations with regard to broadcast translators include:

- If the TSID value in one TS is the same as the TSID value in another TS (on a different carrier frequency), the two Transport Streams should be considered to carry identical services. The programming services on the two should not be offered to the viewer as if they are different services.

- A receiver could measure signal strength and tune to the strongest signal whenever the user wishes to access a virtual channel contained within one of the duplicated multiplexes.

- In the case the signal on one carrier is lost, a receiver could try another frequency known to transmit the same Transport Stream.

Normally, a viewer should be totally unaware that a translated signal is available or is being used. Typically the only way one can see what physical carrier frequency is being tuned is to bring up a lower-level diagnostic screen.

Miscellaneous considerations

Other considerations include:

- **CVCT:** A Cable VCT may be present in the terrestrial signal. If a Transport Stream is received via a terrestrial broadcast (8-VSB) demodulator, the Terrestrial Virtual Channel Table should be used for navigation even if a Cable VCT is present.

- **Carrier frequency:** The carrier frequency associated with a virtual channel should not be used even if it is supplied in the VCT (use of the carrier frequency field is deprecated but older equipment may still be sending non-zero values). Instead of using a carrier frequency that might appear in the VCT, the receiver should take note of the carrier frequency used to acquire a particular Transport Stream and associate its Transport Stream ID (TSID) with that frequency.

- **Inactive channels:** A viewer may attempt to access an inactive channel (one that is a defined channel but not currently transmitting). Instead of rejecting the request, the receiver should tune and display the channel name and number as usual. It could display a helpful text message such as "Channel WXYZ-DT2 is currently off the air." It is a good idea to stay on the requested channel anyway and wait for the status to return to active.

Considerations for Cable-ready Device Designers

We discussed in detail issues related to Digital Cable-Ready Receivers (DCRRs) in Chapter 16 (see "Considerations for Cable-ready Receiver Implementation" on page 319). One of the more challenging aspects of dealing with cable-ready devices lies in the fact that when the POD module is supported and a functioning module is present, an out-of-band data path carrying SI data is available. The out-of-band data can include virtual channel tables and may also include event information data.

The reader is encouraged to review Chapters 17 through 19 in this book and to study the applicable OpenCable specifications and SCTE standards. The CEB-12[1] *PSIP Recommended Practice* includes several sections that may be helpful to designers of DCRR devices.

Considerations for Broadcast Station Operators

A helpful collection of information and guidelines for terrestrial broadcasters has been compiled in an ATSC Recommended Practice called *PSIP Implementation Guidelines for Broadcasters*[2]. Those involved in broadcast station operation are encouraged to study it.

A few high points include:

- Complete and correct PSIP data is essential to the proper operation of DTV receivers.

- The Standard requires that the first four Event Information Tables be transmitted, providing from nine to twelve hours (depending upon current time of day) of future program schedule information.

- EIT-0 should always reflect the attributes of the currently broadcast digital service or services, including content advisory and caption services information.

- The transmitted System Time Table in the direct main broadcast signal must be accurate to within plus or minus one second of the atomic clocks maintained by the US Naval Observatory.

- Repetition rates for the various mandatory PSIP tables should be set to within recommended guidelines and monitored periodically.

- The value of the TSID in the transmitted Transport Stream must match the value assigned to the station in the TSID assignment list maintained by Maximum Service Television. A current link to this table can be found by navigating the ATSC website, http://www.atsc.org.

- If PSIP data includes data relevant to an analog NTSC service, that NTSC service must include an analog TSID transmitted in XDS packets of the VBI. The analog TSID value must be the one assigned in the MSTV list.

- Values for major and minor channel numbers must follow the rules established for the particular region (for a review of the rules for the US, see "Assignment of major/minor channel numbers in the US" on page 104).

Considerations for Cable System Operators

All of the discussions thus far involving SI and PSIP data on cable are of course relevant to the cable system operator because it is they who are responsible for creating the cable signals received and processed by consumer devices.

Carriage of PSIP and EPG data

Cable system operators should be aware that consumer-owned cable-ready digital TV receivers rely on the availability of Electronic Program Guide data (program schedules and descriptions) to help make the viewing experience match consumer's expectations for digital TV products. Although cable standards currently allow navigation data no more descriptive than a one-part channel number as a service identifier, consumers expect more. Digital satellite services, for example, all include the channel names as well as the numbers and extensive program guides. One can search for science-fiction movies or sitcoms in the database of upcoming programming.

As documented in the PSIP Agreement we discussed in Chapter 19, leading cable Multiple Systems Operators have agreed to supply PSIP data in the cable signal as long the content provider supplies it to them. Provisions for enabling the carriage of PSIP data through the cable distribution chain include proper specification and design of remultiplexing equipment that might be present at an uplink distribution center or at the cable headend.

Carriage of terrestrial broadcast signals

Cable systems often retransmit local terrestrial broadcast signals, either through must-carry provisions in the FCC rules or through retransmission consent agreements. Retransmission of digital terrestrial television services involves the following considerations:

- **Placement of descriptors:** in a terrestrial broadcast multiplex, certain descriptors must be present in the Event Information Table and may or may not be present in the Program Map Table (see "Summary of Usage Rules for Descriptors" on page 243). Cable signals, however, require the descriptors to be present in the Program Map Table. It may therefore be necessary for the cable operator to copy these descriptors from EIT-0 to the PMT on a real-time basis. Note that since descriptors change on a program-by-program basis an automated approach is needed. Alternatively, through cooperation with the broadcaster or via the retransmission consent agreement, the broadcaster can place the descriptors in both the PMT and in EIT-0.

- **Channel number issues:** the channel numbers in the terrestrial broadcast multiplex are two-part numbers described in a TVCT. A CVCT may also be present. Cable operators must be aware that if no POD module is present the cable-ready device uses the in-band PSIP CVCT for channel numbers or if no CVCT is present, the TVCT. If a POD module is present, scrambled digital services can be accessed and SI on an out-of-band channel will be available. In all of these scenarios, the cable operator must ensure that the set of channel numbers seen by any receiver represents a complete and consistent channel map with no duplicated channel numbers.

  Cable system operators should also be aware that the FCC may rule that the two-part number associated with digital terrestrial broadcast may have to be (in some way) retained if that signal is carried on cable in accordance with must-carry rules. As of this writing, the digital must-carry proceedings remain open.

- **Tracking changes:** in addition to changes in descriptors that happen at program boundaries, the types and numbers of digital services offered on a terrestrial broadcast multiplex can vary periodically. For example, a broadcaster may change, at some point in the broadcast day, from a mode in which several standard-definition services are sent (multi-casting) to a multiplex configuration in which just one high-definition service is transmitted. Later, the configuration may be changed back again. Remultiplexing equipment needs to be designed to handle such changes.

  Also note that such changes from multicast SD to HD involve changing the status of some of the SD channels from normal to "inactive" status and back again. If any of these SD channels are included in the cable channel line-up, the out-of-band SI data must reflect the normal/inactive status changes in real time.

References

1. CEB-12, PSIP Recommended Practice, Electronic Industries Alliance, 2002.

2. ATSC Recommended Practice, Program and System Information Protocol Implementation Guidelines for Broadcasters, Advanced System Television Committee, Washington D.C., 2002. Available at http://www.atsc.org/

AEIT	Aggregate Event Information Table
AETT	Aggregate Extended Text Table
APID	ATSC Private Information Descriptor
ARIB	Association of Radio Industries and Businesses (Japan)
ASCII	American Standard Code for Information Interchange
ATSC	Advanced Television Systems Committee
BMP	Basic Multilingual Plane
bslbf	bit serial, leftmost bit first
CA	conditional access
CAT	Conditional Access Table
CCI	Copy Control Information
CEA	Consumer Electronics Association
CRC	cyclic redundancy check
CVCT	Cable Virtual Channel Table
DCC	Directed Channel Change
DCCRR	DCC-capable DTV Reference Receiver
DCCSCT	Directed Channel Change Selection Code Table
DCRD	Digital Cable-Ready Device
DET	Data Event Table
DIT	Data Information Table
DTV	digital television
DVB	Digital Video Broadcasting
DVS	Digital Video Subcommittee
EAS	Emergency Alert System
EIA	Electronic Industries Alliance
EIT	Event Information Table
EMM	Entitlement Management Message
EPG	Electronic Program Guide
ES	Elementary Stream

ETM	Extended Text Message
ETSI	European Telecommunications Standards Institute
ETT	Extended Text Table
FAT	Forward Application Transport
FCC	Federal Communications Commission
FDC	Forward Data Channel
FIPS	Federal Information Processing Standards
GPS	Global Positioning System
HD	high definition
IEC	International Electrotechnical Commission
IP	Internet Protocol
IRT	Integrated Receiver-Transcoder
ISO	International Standards Organization
kbps	kilo (1000) bits per second
LSB	least-significant bit
MGT	Master Guide Table
MHz	megahertz
MMDS	Multichannel Multipoint Distribution Service
MPAA	Motion Picture Association of America
MPEG	Moving Picture Experts Group
MPTS	Multiple Program Transport Stream
MRD	MPEG-2 Registration Descriptor
MSB	most-significant bit
MSS	Multiple-String Structure
NPRM	Notice of Proposed Rulemaking
NTSC	National Television Systems Committee
NVOD	near video-on-demand
OCAP	OpenCable Application Platform
OOB	out-of-band
PAT	Program Association Table
PCR	Program Clock Reference
PES	Packetized Elementary Stream
PID	Packet Identifier
PMT	Program Map Table
POD	point of deployment
PSI	Program-Specific Information
PSIP	Program and System Information Protocol
PTC	physical transmission channel
QAM	Quadrature Amplitude Modulation
QPSK	Quadrature Phase-shift Keying
RCU	remote control unit

RF	radio frequency
rpchof	remainder polynomial coefficients, highest order first
RRT	Rating Region Table
SCSU	Standard Compression Scheme for Unicode
SCTE	Society of Cable Telecommunications Engineers
SD	standard definition
SI	System Information (or Service Information)
SMPTE	Society of Motion Picture and Television Engineers
SPTS	Single-program Transport Stream
STD	system target decoder
STT	System Time Table
TS	Transport Stream
TSDT	Transport Stream Description Table
TSID	Transport Stream ID or Transmission Signal ID
TVCT	Terrestrial Virtual Channel Table
uimsbf	unsigned integer, most significant bit first
UTC	Coordinated Universal Time[*]
UTF	Unicode transformation format
VBI	vertical blanking interval
VCT	Virtual Channel Table; used in reference to either TVCT or CVCT
VHF	Very High Frequency
VOD	video-on-demand
VSB	vestigial sideband
XDS	Extended Data Service (of EIA-608-B)

---

[*] Because unanimous agreement could not be achieved by the ITU on using either the English word order, CUT, or the French word order, TUC, a compromise to use neither was reached.

References on the Web

Advanced Television Systems Committee (ATSC)
http://www.atsc.org/

ATSC Standards
http://www.atsc.org/standards.html

Cable Television Laboratories, Inc. (CableLabs)
http://www.cablelabs.com

Consumer Electronics Association (CEA)
http://www.ce.org/

Federal Information Processing Standards (FIPS)
http://www.itl.nist.gov/fipspubs/.

Global Engineering Documents™
http://global.ihs.org/

International Organization for Standardization (ISO)
http:/www.iso.ch/

International Telecommunication Union (ITU)
http://www.itu.int/

Leap seconds
http://tycho.usno.navy.mil/leapsec.html and http://www.iers.org/

National Association of Broadcasters (NAB)
http://www.nab.org

National Cable & Telecommunications Association (NCTA)
http://www.ncta.com

OpenCable™
http://www.opencable.com

SCTE Standards
http://www.scte.org/standards/

SMPTE Registration Authority
http://www.smpte-ra.org/

Society of Cable Telecommunications Engineers (SCTE)
http://www.scte.org/

Society of Motion Picture and Television Engineers (SMPTE)
http://www.smpte.org/

TSID Assignment Table
http://www.mstv.org/downloads/TSIDASGN.doc

Unicode Inc.
http://www.unicode.org/

February 2000 PSIP Agreement

Carriage of PSIP over Cable Plants

1. Purpose and Scope

The purpose of this paper is to address issues related to the carriage of PSIP data over cable plants. This paper represents an agreement between the Consumer Electronics Association (CEA) and the National Cable Television Association (NCTA) on carriage of PSIP on cable in support of consumer digital receiving devices (digital receivers) connected directly to the cable TV system. It is also our view that the proposal described here represents an implementable solution that will add value to our collective customer base. In order to ensure that we have agreement on the implementation of PSIP, this paper details the requisite conditions necessary to carry PSIP on cable plants. Further work is needed on detailed aspects of the implementation.

Section 2 outlines a number of technical requirements regarding carriage of PSIP data on cable. Section 3 discusses implementation issues and outlines various scenarios involved in cable signal distribution at cable headends and at uplink centers such as HITS and Athena.

2. Requirements

The following requirements are based on the availability of PSIP data from the content provider. These requirements are aimed at the carriage of PSIP through the distribution chain and not its creation.

MSO's will require customers to obtain POD modules to receive scrambled digital services. For a consumer-owned digital receiver directly connected to the cable plant, we state the following requirements regarding PSIP data:

1. A map of all available audio/video services shall be made available to the digital receiver.

   a. Any given digital receiver may or may not include a functioning POD module at any given time. Therefore, if a digital Transport Stream (TS) includes one or more services carried in-the-clear, that TS shall include virtual channel data in-band in the form of ATSC A/65 (PSIP) and SCTE DVS-097 Rev 7 (once it is harmonized with ATSC A/65). The in-band data shall at minimum describe services carried within the Transport Stream carrying the PSIP data itself.

b. A virtual channel table shall be provided out-of-band via the Extended Channel interface from the POD module. Tables to be included shall conform to SCTE DVS 234r1[*].

2. Each channel shall be identified by a one- or two-part channel number and a textual channel name (for example: "ESPN").

3. PSIP data describing a twelve-hour time period shall be carried for each service in the transport stream. This twelve-hour period corresponds to delivery of the following Event Information Table (EIT) EIT-0, -1, -2 and –3 (or the equivalent data delivered out-of-band). This requirement matches those already in place for digital terrestrial broadcast. The total bandwidth for PSIP data may be limited by the MSO to 80 Kbps for a 27 Mbits multiplex and 115 Kbps for a 38.8 Mbits multiplex. 4.

4. Carriage of descriptive text in the form of PSIP Extended Text Tables (ETTs) is desirable but optional.

5. Event information data may be transported either in-band or out-of band. When sent in-band, Event information data format shall conform to ATSC A/65 PSIP and SCTE DVS-097 Rev 7 (once it is harmonized with ATSC A/65). When sent out-of- band, event information data shall conform to SCTE DVS 234r1 (profiles 4 or higher). In-band data may be used by the digital receiver to augment event information data sent out-of-band. In other words, both in-band and out-of-band data may be present to describe certain services. The digital receiver may collect and use data from both sources (with rules for use of the channel numbers noted).

6. If a reference is made in in-band PSIP to an analog channel, the digital receiver shall use the Transmission Signal ID method to unambiguously link the PSIP data to the analog service (see EIA-752). An analog feed shall include the EIA-752 TSID when PSIP data for that feed is present on an available digital feed. The digital receiver shall not use PSIP data referencing an analog channel unless a matching TSID is found in the analog feed.

7. The channel number identified with out-of-band SI data may or may not match the channel number identified with in-band PSIP data, for all scrambled services. The digital receiver shall use the channel numbers found in the out-of-band SI if a POD module is present.

8. The channel number identified with out-of-band SI data should match the channel number identified with in-band PSIP data, for all unscrambled (in-the-clear) services. This is desirable so that a digital receiver with no POD module installed will label a service the same as one with a POD module present. This may not be possible for all system architectures.

---

[*] Note: DVS 234 has since been updated to revision 2 and was published by SCTE in 2002 as SCTE 65.

3. Implementation Scenarios

3.1. PSIP in Multiplex

The most fundamental requirement for the MSO is to ensure that if PSIP exists within a multiplex, that it is not stripped from the multiplex and is carried on the cable plant without modification. Figure 1 represents the scenario in which a cable headend downlinks a digital multiplex such as Viewer's Choice utilizing an IRT (integrated reciever transcoder). In this scenario, Viewer's Choice contains PSIP data that was created and inserted into the multiplex by Viewer's Choice. In this scenario, the PSIP is simply passed through to the cable plant without modification. Each cable headend has the freedom to up-convert the multiplex to any physical channel. Enough information exists in the digital reciever (from inband PSIP and the Virtual Channel Table) to reconstruct the virtual channel number for each program in the multiplex. To this end, we believe that no changes are necessary to support the passthrough of PSIP on to the cable plant.

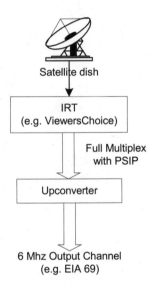

Figure 1. PSIP data on the incoming multiplex is passed through to the cable plant. The PSIP is not and does not have to be modified.

3.2. Content Re-Encoding

A number of content providers, such as HITS and Athena, create customized multiplexes by using content from multiple sources. Figure 2 depicts the scenario in which a number of IRD's are used to receive programs from multiple content providers. Presently, the baseband outputs of the IRD's are fed into the uplink encoder to create a customized multiplex. The Uplink Control System (UCS) is used to set the encoding parameters of each of the programs as well as to assign MPEG services numbers.

In order for PSIP to be correctly carried in the new multiplex, a number of issues need to be addressed. Presently, IRD's do not have a means of extracting PSIP. IRD's simply received and decrypt a given program. It should not prove to be difficult to build an IRD that would extract the PSIP data once the system requirements for this device have been developed. After the PSIP data is extracted from the IRD's, the data needs to be fed into a PSIP aggregator. The purpose of the PSIP aggregator is to coordinate all of the PSIP data and ensure that there are no collisions between the input PSIP streams.

Presently, a PSIP aggregator does not exist, but in principal this can be done and we do not expect there to be any fundamental technical hurdles. We do believe that an appropriate system design is needed before the IRD and aggregator can be built. In addition, we believe that modifications will be required of the UCS and/or Encoder to support the insertion of the aggregated PSIP stream. The cable industry has begun to discuss with potential vendors the requirements for such devices.

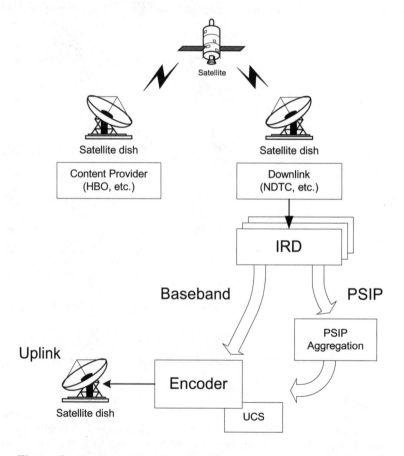

Figure 2. Content re-encoding is used to create custom multiplexes. In order to insert PSIP from each of the programs into the new multiplex, PSIP aggregation will have to be performed.

PSIP data may be present within the downlinked NTSC analog signal. The EIA-806 standard may be used to transmit PSIP data in XDS data packets in the VBI. If so, the PSIP aggregator function in Figure 2 will be designed to accept PSIP in either A/65 or EIA-806 formats, to accommodate digital or analog incoming feeds.

3.3. Content Provider PSIP Creation

In general, uplink providers uplink multiple services for multiple content providers. As an example, AT&T's National Digital Television Center (NDTC) houses playback and editing facilities for the Discovery Channel and Encore, just to name two. Once these content providers source program data for inclusion into PSIP, a means is needed to inject the PSIP into the uplink encoders. Figure 3 schematically depicts a scenario in which an interface is available to the content providers in which program data can be delivered to a PSIP generator. The PSIP generator would in turn create the PSIP stream that would be inserted into the transport multiplex.

Implementation in this scenario requires that an interface specification be developed that provides for a convenient method for content provides to supply program data. A PSIP generator needs to be developed to take program data and create the PSIP stream. The PSIP generator could be the same device used in the previous example to aggregate PSIP. Once created, the PSIP would be inserted into the transport multiplex. We believe that modifications will be needed to the UCS and/or Encoder to support the insertion of the PSIP stream. As in the previous scenarios, we do not see any fundamental technical hurdles, rather the need for a coordinated end-to-end system design.

3.4. Remultiplexing

Remultiplexing devices are becoming increasingly popular in order to optimize the use of plant bandwidth. A typical case is where an MSO would like to use one or more programs from one multiplex and combine these programs with one or more programs from another multiplex. Two companies (Terayon and VBITS) presently offer remultiplexing solutions. These products "fix" system information so that service numbers and PIDs are unique within the new multiplex. In order to support the carriage of PSIP, the remultiplexing unit would have to aggregate and coordinate PSIP from multiple sources. Figure 4 depicts this scenario. Remultiplexing units will require modifications to support coordination of PSIP, but we believe that there are no technical issues that would prohibit this feature from being included into future remultiplexers. Discussions with remultiplexing equipment vendors have begun in order to ensure that they have taken PSIP into consideration for future equipment designs.

3.5. Master Downlink, Multiple Channel Maps

A number of cable systems (including AT&T and Shaw) utilize a plant architecture in which there is a Master Downlink IRT feeding multiple channel maps. Figure 5 depicts such a scenario. As an example, the Denver Mile High headend provides cable service to Boulder, Littleton and Castle Rock, CO. Each of these local entities employ different channel maps. Thus the in-band PSIP virtual channel number may be irrelevant. Similarly, terrestrial DTV PSIP may not reflect the virtual channel that the broadcast is carried on in the cable plant.

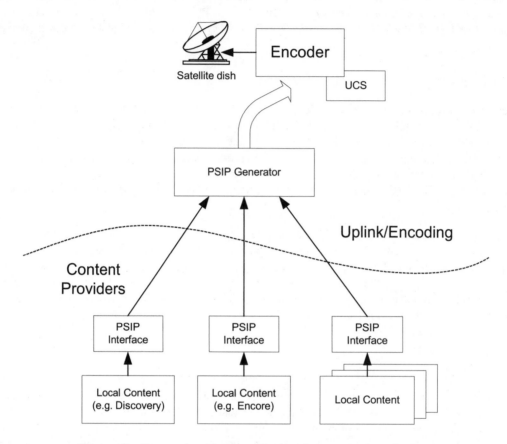

Figure 3. Content providers would transmit program data via the PSIP interface. This data would be used to create the PSIP for the multiplex.

According to requirement #8, "The channel number identified with the out-of-band SI data should match the in-band channel number identified with the in-band PSIP data, for all unscrambled services." Since it is our position that digital cable programs will be scrambled, there should not be a problem satisfying this requirement. The only possible exception to this is carriage of terrestrial DTV content. We believe that the best approach to satisfying this requirement is to have local coordination with terrestrial broadcasters. We have not worked through all of the scenarios relative to terrestrial content, such as two-part channel numbers, but believe that we can develop operational guidelines to ensure that the consumer is provided consistent information across multiple platforms.

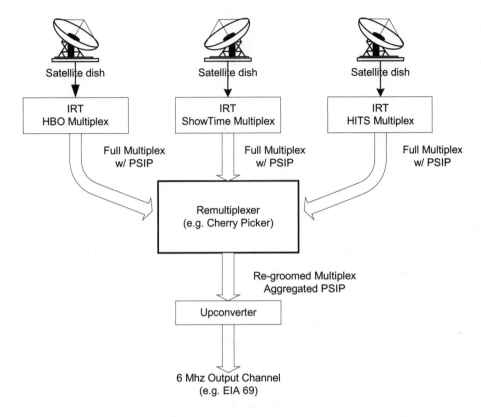

Figure 4. Remultiplexing units will need to aggregate and coordinate PSIP from multiple input sources.

4. Implementation Plan

The steps necessary to achieve the requirements set forth above include:

- Systems Engineering
- Product Development
- Product Qualification
- Procurement
- Systems Integration
- Infrastructure buildout
- System Acceptance Testing.

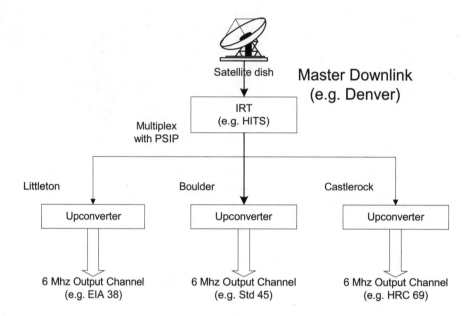

Figure 5. In many instances, a Master Downlink is used to feed multiple headends, thus the in-band PSIP virtual channel number may be irrelevant. Similarly, terrestrial DTV PSIP may not reflect the virtual channel that the broadcast is carried on the cable plant.

The NCTA believes that this process can be completed in a timely fashion, but will require the active participation of PSIP equipment vendors, content providers (e.g. HBO), cable operators, and consumer electronics manufacturers.

5. Conclusions

The NCTA and the CEA have reached an agreement on the carriage of PSIP for cable. We believe that this implementation of PSIP will add value to the cable offering. In addition, we believe that carriage of PSIP will speed the acceptance of DTV and the associated digital receivers. A number of issues need to be resolved and system components need to be designed in order to fully implement the system described here. The NCTA is committed to working with the CEA to add further detail to the component specifications. In addition, equipment vendors will be engaged as soon as possible to solicit them for hardware solutions that satisfy the requirements for carriage of PSIP.

History of the PSIP Standard

This book describes the ATSC A/65B PSIP Standard released in 2002. What follows is a short tour through the history of A/65, a journey that begins circa 1994. Then we look at the changes to the specification that have occurred with various amendments and corrigenda.

Predecessor standards

The A/65 Program and System Information Protocol was first issued by ATSC on December 23, 1997. Its genesis began more than three years earlier in 1994, a year after the Grand Alliance of AT&T, the David Sarnoff Research Center, General Instrument, Massachusetts Institute of Technology, North American Philips, Thomson Consumer Electronics, and Zenith Electronics. Two of these Grand Alliance members, General Instrument (GI) and Thomson Consumer Electronics (TCE) made many contributions relating to system/service information to the ATSC Transport Specialists group, T3-S8, throughout the years.

At about the same timeframe as the ATSC Digital Television Standard began to come together in 1994, engineers in Europe were busy documenting the Digital Video Broadcast (DVB) standard. The first edition of the DVB Service Information Standard, ETS 400 368, was issued in October 1995. Companies participating in the ATSC process generally felt that the DVB approach didn't address their needs. Clearly it had a European focus. For example, DVB defined a field called service_type that encoded the type of service (either analog or digital) being described, but until an amendment was made in 1996 the only analog television waveform options were PAL and SECAM—NTSC was not listed.

Back in the US, Thomson's initial proposal was known as T3-S8 Document 050, and the first versions were submitted in 1995. T3-S8 050 eventually became ATSC Standard A/55[1], *Program Guide for Digital Television*. Here, concepts such as the Master Guide Table made their first appearance, as did a predecessor to PSIP's Virtual Channel Table, the Channel Information Table (CIT). In A/55[1], text strings were formatted in a similar fashion to PSIP's Multiple String Structure, although no text compression method was defined. Jack Chaney (then working for Thomson) was A/55's chief architect.

General Instrument contributed a proposal for System Information that was based on protocols used in the company's DigiCipher II© product. DigiCipher II (or DC-II as it was known) was built on the MPEG-2 *Systems* and MPEG-2 video compression standards. It employed proprietary SI and conditional access methodologies. In May 1995, GI decided to contribute the system/service information protocol used by DC-II to ATSC on a royalty-free basis. At the time, the author was responsible for creating and maintaining GI's "control channel" protocols, including the SI proposal. GI's contribution was given the document number T3-S8 079. Initially the GI proposal did not include program guide information per se, but it did include information pertaining to individual programs. This information included the name of the program, content advisory information, and data pertinent to impulse-pay-per view functions.

GI's approach was geared towards delivery of virtual channels over satellite, cable, MMDS, or terrestrial broadcast. To support satellite delivery, it included an exhaustive collection of network information tables describing satellites, transponders, modulation modes, transmission systems, and carrier frequency plans for C- and K-band satellites. Cable was similarly supported with a network information table that also described frequency plans, modulation modes and transmission formats. Virtual channel tables were the cornerstone of GI's approach, and T3-S8 079 introduced the Source ID concept, discussed in this book in Chapter 6. GI's document made extensive use of the descriptor concept employed in DVB SI. It also endeavored to be international in scope, using Unicode for text representation, and supporting international currency systems for pay-per-view.

It was clear that Thomson and GI's approaches were different yet similar. Both had virtual channel tables and both employed methods for text delivery. Each had strengths and weaknesses. Using each unchanged would have produced a standard that lacked cohesion, yet political forces dictated that ATSC had to document their SI/EPG solution soon. In the end, Thomson agreed to make some changes that would reduce some of the redundancy and conflict with the GI approach. Likewise, GI agreed to prune back their proposal.

The ATSC A/53 *Digital Television Standard* was first issued in April 1995. At that time, the System Information and Program Guide standards had not yet been completed in T3-S8. The first release of A/53 indicated that the SI shall be carried in accordance with the ATSC Standard resulting from T3-S8 079 and that the program guide data shall be carried in accordance with the ATSC Standard resulting from T3-S8 050. Near the end of 1995, T3-S8 050 was submitted to T3 and eventually was adopted as ATSC Standard A/55[1], *Program Guide For Digital Television*, on January 3, 1996. T3-S8 079 traveled the same path and was issued that same day as ATSC Standard A/56[2], *System Information for Digital Television*.

Genesis of PSIP

It was clear nearly from the start that the broadcasters had really embraced neither A/55[1] nor A/56[2]. It was also clear that these two standards needed to be merged into one, taking the best parts of each. ATSC's T3-S8 Transport Specialists group, chaired by consultant Bernard Lechner, took up the task. Their approach this time was to start with a carefully considered set of requirements statements. First, a set of general requirements was outlined, followed by a set that captured the specific needs of terrestrial broadcasting, and finally a set that attempted to encapsulate the needs of cable Multiple System Operators. From the beginning, the members of T3-S8 formed a close working relationship with participants in SCTE's Digital Video Subcommittee (DVS), the group that was creating and documenting standards for digital cable. We discuss issues involving PSIP on cable in Chapters 17 to 19.

About this time Thomson proposed a set of data tables designed to be delivered within the MPEG-2 Transport Stream that would provide program guide data. Engineers on the Thomson team included Mehmet Ozkan, Edwin Heredia, and Andy Teng, all working out of TCE Corporate Innovation and Research in Indianapolis, Indiana. Representing General Instrument, this author collaborated with the Thomson group to re-formulate a program guide and service information standard that brings together the best aspects of the prior work.

Thomson's approach offered a complete method for delivery of program guide data. An especially helpful feature of the approach was the precursor to the concept of the Master Guide Table found in A/65 today. In the early drafts there was not a clear separation between the definitions of the services and the program guide database. One of the most powerful concepts described by GI's earlier approach was that of the virtual channel tables and their linkage via Source ID to the program guide data. As we mentioned in our discussion of Source ID, this separation means that one program guide database can be usable by many service providers, each of which delivers a particular service on a separate physical channel (or as a different MPEG-2 program in the Transport Stream).

During the last half of 1996 and continuing into 1997, GI and Thomson worked together to merge the two proposals into a seamless design. GI defined the terrestrial broadcast and cable Virtual Channel Tables, and their use of GPS seconds for all representations of time of day was adopted for the new standard. Thomson added the Daylight Saving Time information to the System Time Table, and defined the EIT and ETT data structures and their method of linkage to each other. GI's use of the descriptor mechanism used in MPEG-2 and DVB SI was adopted throughout.

Everyone agreed that all tables should be in the MPEG-2 long-form table section format.

Thomson's concept of the multiple string structure was incorporated into the new standard. In this approach, within the data block defining each piece of textual data in the SI, text can be represented in one or more languages. This contrasted to GI's approach in which separate table instances had to be delivered to support multilingual text. GI contributed the Huffman encode and decode tables that provide the text compression capability in PSIP, the use of the ISO 639 language code to identify languages, and the use of Unicode in text string encoding.

Early in 1997, we began to use the phrase "naming, numbering, and navigation" to describe the problem we were working on. It was about this time that it became clear that some form of "channel grouping" was needed. The SI information needed to indicate to receivers that certain channels should be associated with each other in a group or collection. For example, a local broadcaster might want to offer six channels of standard-definition programming. How would users be able to associate these six channels with that broadcaster, or navigate among all of the available channels quickly and easily? How could channels be listed in a program guide (the Sunday paper, for example) so as to group them by broadcaster? DVB had defined a grouping mechanism in their Bouquet Association Table. GI first proposed a grouping mechanism based on extending the virtual channel table to add a flag providing that function. About this time it was proposed that the virtual channels should consist of two numbers, one to reflect the broadcaster's RF channel number and the second to differentiate each service among those offered by that broadcaster. The terms "major" and "minor" channel numbers were adopted. Dan O'Callaghan was the first to describe two-dimensional navigation in which up/down navigation would be among the channel groups and left/right navigation would be among the sub-channels.

In parallel with the development of the definition of channel and program guide tables, T3-S8 was tasked with creating the digital equivalent of the "V-chip" content advisory system that was then being finalized for use with analog NTSC television. From this work, the Rating Region Table and the Content Advisory Descriptor were born. These were integrated into the PSIP proposal during 1997.

The document that was to become ATSC A/65 was called T3-S8 Doc. 193; the first draft was dated July 9, 1997. At that time it was called "An Integrated Program and System Information (IPSI)."

One of the last major changes to the PSIP table structures was to re-structure the Master Guide Table to make it the flexible, powerful, and extensible mechanism it is today. Prior to the change, the MGT listed just a fixed set of table types. The update involved defining the table_type field, and specifying exactly the same information about each type of table (size, version, PID).

T3-S8 Doc. 193, now called "Program and System Information Protocol," was approved by T3 and then by the full ATSC membership. It was issued as an ATSC standard on December 23, 1997.

Changes since first release

In the sections to follow we discuss some of the changes that have been made to A/65 since its first release. None has broken backward compatibility.

"A" Version of A/65

ATSC A/65A was introduced in February, 2000. Changes included in this update included:

- Some of the bits and bit fields in the header portions of some of the tables had been reserved for future use with the value zero, or had been specified as value zero. This had been done at the request of General Instrument, who wished the tables to be consistent with their $A/56^2$-based applications. For consistency with other reserved fields these are now specified as being value one.

- Initially the 16-VSB modulation mode had been inadvertently left out of PSIP, but it was added back in at revision A. Note that 16-VSB is not authorized by the FCC for use in the US, but there is no reason this modulation method could not be used in another country.

- The concept of "inactive" channels was introduced at revision A by addition of the hide_guide bit.

The "A" release of ATSC A/65 included a new informative section, Annex G: An Overview of PSIP for Cable, and it clarified a number of normative aspects of the Standard.

Amendment No. 1 to A/65A

Fox Broadcasting and Tribune Media proposed the Directed Channel Change extension to PSIP in 2000. ATSC originally published the DCC specification in the form of Amendment No. 1 to A/65A on May 31, 2000.

Note that Amendment No. 1 was made available to the public even though ballot comments had not been addressed (contrary to the usual practice). It was not until the latter part of 2001 that comments received during the ATSC ballot on that amendment were fully processed. The process of addressing ballot comments brought up a number of other substantive issues and resulted in an extensive revision of the Directed Channel Change mechanism. In 2002, a new Amendment No.

1 (called No. "1A") was prepared and balloted. The discussion of Directed Channel Change in this book reflects this revised approach as it appears in A/65B.

Amendment No. 2 to A/65A

During the latter half of 2001, a second Amendment to A/65A was prepared by T3-S8 and balloted to S8's parent body, T3. This amendment proposed changes to A/65A mainly for the purposes of adding clarity, but it did add some new requirements and make adjustments to some existing ones. Nevertheless, none of the changes was expected to create any problems for fielded receiving devices. In terms of substantive changes these may be noted:

- **Delivery of RRT**: Beginning with Amendment No. 2 it is no longer a requirement to transmit the US Rating Region Table, rating_region value 0x01. This change was made in recognition of the fact that the RRT for the US region is unchangeable by design. It is fully documented in EIA/CEA-766-A, and consumer electronics manufacturers must refer to all of the EIA standards relevant to the design of V-chip compliant devices in order to comply with FCC rules.

  PSIP, as now amended, states that the RRT for any region for which the RRT can be updated (by the regular table update and versioning method of MPEG) must be sent if any transmitted Content Advisory Descriptor refers to that region.

- **Removal of some "user private" ranges**: A decision was made that certain fields would no longer include user-private ranges. Any values in the range called "user private" will not be standardized by ATSC in a future version of the Standard. A user of the Standard would be safe in making assignments using values in the user-private range knowing that these private extensions would be compatible with future versions of the Standard. In some cases, SI including private data can be transmitted and can appear in consumer devices. However, there are strict rules dictating the conditions under which private data can appear in a cable or terrestrial broadcast Transport Stream, and they involve the requirement that the entity that has provided the user-private data must be identified. The identification is typically in the form of an MPEG-2 Registration Descriptor, which carries a registered format_identifier field uniquely identifying the party that has provided the private data. See "format_identifier" on page 82.

  Beginning with Amendment 2, there is no longer a user-private range for the modulation_mode field in the Virtual Channel Table. Those values are now reserved for future ATSC use. Likewise, the user-private range for the stream_type field in the Service Location Descriptor has been converted to a reserved range, along with the compression_type and mode fields in the Multiple String Structure. User-private ranges remain in the assignment of table ID values and in the MGT table_type definition.

- **Allow the MGT to refer to user-private table types**: Starting with Amendment No. 2, the MGT can reference tables whose table_type value is in the user-private range, 0x0400 through 0x0FFF. The new rules state that an MPEG-2 Registration Descriptor must be placed in the table type descriptor loop associated with the user-private table_type.

- **Deprecate the use of the carrier frequency field**: As most PSIP implementers are now aware (as they should be after reading this book), receiver implementations should not use the carrier_frequency field in the Terrestrial or Cable VCT. With the adoption of Amendment No. 2 these fields may be set to zero in the transmitted data stream. An informative note has been added describing the use of the TSID field for making positive identifications of received digital multiplex data streams.

- **Use of analog TSID**: Broadcasters in the US are now required to transmit a valid analog TSID in the analog NTSC broadcast signal if their digital broadcast includes VCT or EIT/ETT data describing that analog service.

- **One-part channel numbers for cable**: The cable community requested that PSIP should be able to describe cable virtual channels with single-part channel numbers. Service information in the out-of-band channel, as described in SCTE 65, already allowed a mixture of channels with one-part and two-part numbers to be described. Amendment No. 2 adopted the same approach used in SCTE 65, describing a method whereby the major and minor channel number fields can be combined to represent a one-part channel number in the range from 0 to 16,383.

- **CVCT in the terrestrial broadcast multiplex**: With the adoption of Amendment No. 2, at the option of a broadcaster, a Cable VCT may be included along with the Terrestrial VCT in the broadcast Transport Stream. Such a scheme can save trouble and expense when a broadcaster has coordinated consistent channel numbering among the cable operators for those digital services to be carried on cable. Receivers are expected to use the TVCT for signals demodulated via 8-VSB and the CVCT for those found on cable, and those demodulated via 64- or 256-QAM.

- **Overlapping events**: A statement was added requiring that the start of an event cannot occur prior to the end of the previous event. This isn't anything new—it's simply a clarification of the intent of the PSIP EIT approach.

- **ETT table_id_extension field**: Prior to Amendment No. 2, the ETT was out of compliance with the approach described in MPEG-2 *Systems* with regard to the use of its table_id_extension field. Starting with Amendment No. 2, table_id_extension for the ETT must be set so the value is unique among all ETT instances appearing in Transport Stream packets with a common PID value. While receivers can easily be built to differentiate among these instances even without looking at the table_id_extension, compliance with MPEG-2 *Systems* was felt to be important and this change was adopted to correct the oversight.

- **Component Name Descriptor usage**: New rules are in place with Amendment No. 2 that make it mandatory to include a Component Name Descriptor whenever a service includes two or more audio components of the same type labeled with the same language code. The type of an audio program element is indicated by the bit-stream mode (bsmod) field in the AC-3 Audio Descriptor. Different types include Complete Main audio services, services for the visually or hearing impaired, and music and effects services.

  A Component Name Descriptor may also be used when a service includes an audio component without an associated language code (an example of such a component is the audio feed of ambient sound or crowd noise from a sporting event).

- **New rules for assignment of major channel number values in the US**: A number of additional stipulations are listed that dictate how major channel number values are to be assigned, especially during the digital transition. These rules are discussed in Chapter 5.

- **Delivery of "next" Virtual Channel Table**: Syntactically, the VCT that is the next to apply can be sent. The current_next flag in the table section header, when set to zero, indicates that the table will be the next to apply. Amendment No. 2 however, suggests that "next" VCTs should not be sent. It explains that the recommendation arises from the fact that no mechanism exists to update the "next" tables prior to their becoming "current." Furthermore, there is no benefit in sending the next table unless it is too big to fit into a single table section.

A variety of editorial changes also appeared in Amendment No. 2 including:

- Updates to some of the reference documents were available, so these references were brought in line.

- The term "table instance" was clarified, adapting a definition from the ATSC A/90 Data Broadcast Standard.

- A review was made of the use of the term "table" and in several places was replaced by the more accurate "table section."

- The introductory section in A/65 that describes the relationship between PSIP tables and the MPEG-2 private section syntax was reworked and improved.

- All of the language relating to the Unicode standard was reworked for accuracy and precision. For example, references to the Unicode Basic Multilingual Plane (BMP) were replaced with references to Unicode Transfer Format-16 (UTF-16), and the relationship between Version 3.0 of the Unicode Standard and ISO/IEC 10646-1:2000 was clarified.

- The example Rating Region Table was changed to one that is country-neutral.

Finally, a number of new sections and examples was added to address questions raised by those implementing the Standard and to clarify certain aspects of it. Topics in this category include:

- **Interpretation of MGT version number**: many people had misinterpreted the meaning of the table_type_version_number given in the MGT for the various EITs and ETTs. They had thought that the table type version number was the version number of the tables associated with the indicated time slot, not the version of the particular tables associated with the indicated PID value. This is a subtle distinction, and we discussed it in detail in Chapter 12.

- **Near-Video-on-Demand**: a number of clarifications was added to sections describing NVOD and the use of the Time-Shifted Service Descriptor. A new subsection was added to Annex D describing two NVOD scenarios.

- **Use of the Program Map Table in service acquisition**: In prior versions of A/65, one was left with the understanding that there would be no reason to process the PMT when accessing a terrestrial broadcast service. Amendment No. 2 explains that in some situations it is necessary for the receiving device to acquire and parse the PMT.

- **GPS time discussion**: A new informative annex is added to discuss some of the finer points of leap seconds, and their relationship to system time.

- **Use of the analog Transmission Signal ID**: a brief discussion of the analog TSID was added including expected receiver behavior.

Amendment No. 3 to A/65A

Amendment No. 3, balloted at the end of 2001, introduced the Retransmission Control Descriptor. This descriptor is discussed in Chapter 11 (see "Summary of Usage Rules for Descriptors" on page 243).

"B" Version of A/65

The "B" version of A/65 was produced in 2002 from the application of the finalized Amendments 1, 2, and 3 to the A/65A standard.

References

1. ATSC Standard A/55, "Program Guide for Digital Television," 3 January 1996 (obsolete).

2. ATSC Standard A/56, "System Information for Digital Television," 3 January 1996 (obsolete).

Index

ABOUT THE AUTHOR

Mark K. Eyer is currently Director of Systems at the Technology Standards Office of Sony Electronics. He graduated Cum Laude with a B.S. degree from the University of Washington in 1973 and received an MSEE degree in 1978 from the same institution. For the past twenty years, Mr. Eyer has been involved with the development of technologies and products related to secure and digital television and he holds twelve US patents in these areas.

After joining General Instrument (now Motorola) in 1982, he was responsible for design of decoder firmware and system control software. Beginning in 1988, Mr. Eyer designed firmware for products employing digital video compression technology. In 1990, he was given responsibility for the development and maintenance of the protocols used to deliver data and control across the satellite link to individual decoders. This work formed a contribution to ATSC that led to the A/56 System Information for Digital Television standard in 1994.

Since 1994, Mr. Eyer has made contributions to various digital television standards including ATSC A/65 *Program and System Information Protocol (PSIP) for Terrestrial Broadcast and Cable*, part of which was derived from the earlier A/56 work. He became involved in digital interconnection standards in 1997, and co-chaired the committees in EIA/CEA that created the EIA-775-A *DTV 1394 Interface Specification*, EIA-775.2 *Service Selection Information for Storage Media Interoperability* and EIA-849 *Application Profiles for EIA-775-A Compliant DTVs*. Mr. Eyer was a primary contributor to various SCTE Digital Video Subcommittee (DVS) standards including ANSI/SCTE 26 *Home Digital Network Interface*, DVS 216 *POD Extended Channel Specification*, and SCTE 65 *Service Information Delivered Out-of-Band for Digital Cable Television* and he led the team that developed EIA-814/SCTE 18 *Emergency Alert Message for Cable*. Currently, Mr. Eyer chairs the ATSC T3/S8 Transport Specialists group, works with various SCTE, ATSC, and EIA/CEA standards committees, and contributes systems engineering expertise to the development of Sony's digital television and cable set-top box products.